T0336520

Additional Praise for *Smart Manufacturing: The Lean Six Sigma Way*

"China is no longer the low-labor cost manufacturer of the past and must adopt smart manufacturing to remain viable. In the design process and implementation of smart manufacturing, our company fully draws on the knowledge of this book, especially on how to combine lean six sigma tools with smart technologies. It is a rare book that fully and effectively combines production management concepts and practices. This book can be very effective in helping to realize smart manufacturing in the factory to lower cost, improve customer satisfaction, and improve employee morale."

—*Jianfeng Du, Founder and CEO, Millennium Power, China's Leading Hybrid Energy and Battery Storage Solution Innovator*

"This book covers topics that are at the heart of our firm's investment thesis. Modern supply chains will have to become fully digitized and required to be resilient and efficient. Just like software has changed the world, smart technologies will change how goods and services are manufactured and delivered swiftly in a fully automated way. The author covers all of the building blocks that will be at the core of the smart technologies wave that unfolds in the next few years. The book is a great reference to have and I strongly encourage you to read it!"

—*Najib Khouri-Haddad, General Partner, Sway Ventures*

"Dr. Tarantino's newest tome, *Smart Manufacturing: The Lean Six Sigma Way*, is a tour de force and comprehensive work that will appeal to both readers who are new to the field as well as accomplished experts. In addition to providing fresh perspectives on the latest smart manufacturing approaches, he and his chapter co-authors also expand on several of the most pressing challenges and important issues facing the United States and the global manufacturing economy, including supply chain resiliency, cybersecurity, big data, as well as the rapid adoption of game-changing technologies including artificial intelligence, machine learning, and edge computing. This encompassing volume is highly recommended reading for anyone interested in understanding the state-of-the-art in the rapidly evolving advanced and smart manufacturing landscape."

—*Daniel Dirk, PhD, Interim Dean of Engineering, Florida Institute of Technology*

"*Smart Manufacturing: The Lean Six Sigma Way* is a comprehensive and accessible overview of the technologies that are transforming industry. Relevant to both students and practitioners, the book places smart manufacturing in its historical context while clearly bringing across the powerful disruptive potential of Industry 4.0. This is already being felt in the aerospace sector, where a combination of the approaches and technologies outlined in *Smart Manufacturing* are bringing down development costs and time to market, while reducing entry barriers and enabling a new generation of start-ups with innovative business models. Anthony Tarantino's book provides insight into this emerging paradigm that will be of huge benefit to the reader."

—*Harry Malins, Chief Innovation Officer, Aerospace Technology Institute*

"Industry 4.0 is underway. Data analytics, augmented reality, artificial intelligence, collaborative robots, additive manufacturing, and other technologies are already helping manufacturers increase efficiency, reduce downtime, lower prices, and improve service, delivery, and quality. And there's more to come. These technologies are not science fiction. They are being applied right now by manufacturers, large and small, in a variety of industries. However, Industry 4.0 is not merely a matter of connecting machines to the Internet. Industry 4.0 will inevitably lead to new types of work and new ways of working. It will require changes to company structures and relationships between companies. Businesses must understand what they want to achieve and then develop an implementation strategy. This book will help you get there."

—*John Sprovieri, chief editor, ASSEMBLY magazine*

"A wonderful book to introduce undergraduate students to a career in operations or manufacturing, a long-overlooked field. The book is easy to read and will allow students to understand the challenges facing those implementing Industry 4.0. Particularly enlightening in describing how smart manufacturing will open up opportunities for women who choose a STEM field for a career."

—*Deborah Cernauskas, PhD, Professor of Business Analytics and Finance,*
Chair Undergraduate Business (retired), Benedictine University

SMART MANUFACTURING

The Lean Six Sigma Way

Anthony Tarantino

WILEY

For general information on our other products and services or for technical support, please contact our Customer Care Department within the United States at (800) 762-2974, outside the United States at (317) 572-3993 or fax (317) 572-4002.

Wiley also publishes its books in a variety of electronic formats. Some content that appears in print may not be available in electronic formats. For more information about Wiley products, visit our web site at www.wiley.com.

Library of Congress Cataloging- in- Publication Data is Available:

ISBN 9781119846611 (hardback)
ISBN 9781119846628 (ePub)
ISBN 9781119846635 (ePDF)

Cover Design: Wiley
Cover Image: © ImageFlow/Shutterstock

SKY10033880_032422

To my beloved wife, Shirley, whose continued encouragement and support have guided my writing and teaching efforts over the past 15 years in creating five tomes for John Wiley & Sons and in teaching at Santa Clara University.

Contents

CHAPTER 2
Lean Six Sigma in the Age of Smart Manufacturing   **21**
Anthony Tarantino, PhD

CHAPTER 3
Continuous Improvement Tools for Smart Manufacturing   **43**
Anthony Tarantino, PhD

CHAPTER 7

Big Data for Small, Midsize, and Large Operations 167
Omar Abdon and Randy Shi

CHAPTER 8

Industrial Internet of Things (IIoT) Sensors 185
Deb Walkup and Jeff Little

CHAPTER 9

Artificial Intelligence, Machine Learning, and Computer Vision 205
Steven Herman

CHAPTER 10
Networking for Mobile Edge Computing   219
Jeff Little

CHAPTER 11
Edge Computing   253
Vatsal Shah and Allison Yrungaray

CHAPTER 12
3D Printing and Additive Manufacturing 267
Bahareh Tavousi Tabatabaei, Rui Huang, and Jae-Won Choi

CHAPTER 13
Robotics 311
Thomas Paral, PhD

CHAPTER 14

Improving Life on the Factory Floor with Smart Technology   331
Miles Schofield and Aaron Pompey, PhD

CHAPTER 15

Growing the Roles for Women in Smart Manufacturing   345
Maria Villamil and Deborah Walkup

CASE STUDIES

Foreword

Benjamin and Mae Swig Professor of Supply Chain Analytics
Leavey School of Business
Santa Clara University

It is with great pleasure that I write this foreword for *Smart Manufacturing: The Lean Six Sigma Way*. I want to congratulate the editor, Dr. Anthony Tarantino, for compiling this impressive volume. To the best of my knowledge, this is the first book that discusses applications of the well-known Lean and Six Sigma (LSS) concepts in the new and emerging world of Smart Manufacturing, or Industry 4.0. I have no doubt that this book will turn out to be a great resource for practitioners, and hope that it will inspire academics to embark on new research opportunities in this sector.

What is distinct about the manufacturing environment is the potential for vast amounts of data that can be generated, stored, and analyzed. This data can relate to production processes as well as to the broader ecosystem of which the manufacturing process is a part. When I first started conducting research on issues related to the design of production systems and supply chains nearly three decades ago, my colleagues and I would often find the timely availability of sufficient data at the right level of granularity to be a major constraint. Consequently, we would have to rely on limited data sets, and extrapolate implications based on these results. However, the fantastic developments in our ability to generate, store, and access vast amounts of (big) data at unprecedented levels of granularity, optimize large-scale mathematical models of such manufacturing and supply chain systems at incredible speeds, and leverage cloud computing infrastructure have fueled the convergence of physical and digital systems. The various technologies underlying such developments form the core of Smart Manufacturing/Industry 4.0. Therefore, deployment of these technologies can lead to improvements in process flexibility, speed, cost, quality, scale, customizability, and responsiveness in unimaginable ways. Since such improvements are fundamental goals of the LSS methodologies, it is imperative for academics and practitioners alike to explore its applications in this emerging world of Smart Manufacturing/Industry 4.0.

In this book, Anthony and a group of amazing academics and practitioners with deep domain expertise provide insightful illustrations of how LSS principles can leverage a variety of Smart Manufacturing/Industry 4.0 technologies in a wide range of contexts. I had the pleasure of working closely with Anthony when we jointly advised a major cloud infrastructure provider on several LSS projects, which led to demonstrable and compelling cost savings and process improvements. It is gratifying to see him bring his unique perspective and deep knowledge of LSS honed over a nearly 40-year career in the high-technology industry to this volume. The applications, insights, and lessons contained in this volume are relevant to manufacturing and service industries alike. I am sure that readers will share my great enthusiasm for this book.

Acknowledgments

I wish to acknowledge the exceptional efforts of my proofreaders and editors. Besides writing their own great chapters, Deborah Walkup and Jeff Little made valuable suggestions to chapters in related fields based on their subject matter expertise. Alexander Tarantino and Apollo Peng proofed and edited several chapters, making critical revisions to the final content. Angelina Feng is our 13-year-old middle school student with a remarkable mastery of the English language. She spent hours reviewing each chapter, making hundreds of suggested changes. Most remarkable is that her grammatical suggestions were spot on. I believe she has a great career ahead of her as a journalist or author if she chooses to pursue it.

I also wish to acknowledge the support and guidance from my Wiley editors: Sheck Cho, executive editor; Susan Cerra, managing editor; and Samantha Enders and Samantha Wu, assistant editors.

About the Author

Anthony Tarantino received his bachelor's degree from the University of California, Santa Cruz, and his PhD in organizational communications from the University of California, Irvine. He started his manufacturing and supply chain career working first in small and then in large domestic manufacturers, including running Masco's supply chain for the world's largest lockset manufacturing facility. He was certified in purchasing management (ISM), materials management (APICS), and Lean in the 1980s. During the same period he began implementing ERP systems and Lean programs for divisions of Masco Corporation at several facilities. After 25 years in industry, he moved into consulting, becoming a supply chain practice lead for KPMG Consulting (BearingPoint) and later IBM. In the 2010s he led 30-plus Lean Six Sigma projects as a Master Black Belt for Cisco Systems Supply Chain and trained over 1,000 employees in their lunch-and-learn programs. He leveraged his consulting experience to create and deliver executive-level seminars in supply chain and risk management in Europe, Asia, Australia, New Zealand, and the United States.

He began as an adjunct faculty member at Santa Clara University in 2010, teaching risk management in finance and supply chain. More recently, he created a Lean Six Sigma Yellow Belt training program that introduced students to continuous improvement tools and techniques. Working with Professor Narendra Agrawal, he created and delivered an accelerated Lean Six Sigma Green Belt program. The most recent program was for a leading corporate client of the university. The five live projects in that program generated an estimated annual savings of $3 million.

Over the past five years he has supported Smart Manufacturing startups focused on computer vision identifying the most attractive industry verticals and accounts to pursue. He has also acted as a client-facing advocate for the new technologies to improve operations, safety, and competitiveness. His work with these startups was the inspiration for *Smart Manufacturing: The Lean Six Sigma Way*, his fifth book for John Wiley & Sons over the past 15 years.

About the Contributors

Omar Abdon is a product-focused growth-hacker working with successful startups in Silicon Valley with 15-plus years of experience in building and growing B2B4C products. He founded, grew, and successfully exited three startups in mobile software and digital growth marketing spaces across a wide range of industries like manufacturing, banks, telecom, financial institutions, and more. Currently, Omar is the head of innovation and customer success at Atollogy Inc., a platform to connect, collect, and leverage valuable enterprise big data through machine vision (MV) and to utilize artificial intelligence (AI) to digitize business operations and achieve the highest possible efficiency and end-user experience.

Narendra Agrawal is the Benjamin and Mae Swig professor of supply chain management and analytics in the department of information systems and analytics of the Leavey School of Business at Santa Clara University. He has conducted extensive research on problems related to supply chain management in the retail and high-technology industries and conducted numerous management development seminars on these topics internationally. His research has been published in leading academic and practitioner-oriented journals. Previously, he served as the interim dean as well as the associate dean of faculty at the Leavey School. Naren holds an undergraduate degree in mechanical engineering from the Institute of Technology, BHU, India, where he received the Prince of Wales Gold Medal; an MS in management science from the University of Texas at Dallas; and an MA and PhD in operations and information management from The Wharton School of the University of Pennsylvania.

Jae-Won Choi received his BS, MS, and PhD in mechanical engineering from Pusan National University, Busan, Korea, in 1999, 2001, and 2007, respectively. He is an associate professor in the department of mechanical engineering at The University of Akron. He has authored more than 50 articles and secured five patents. His research interests include additive manufacturing, 3D-printed smart structures including sensors, actuators, and electronics; 3D-printed rubbers for insoles and tires; and bio fabrication and low-cost binder-coated metal/ceramic for 3D printing. He is currently serving as an associate editor of the journal *Additive Manufacturing* and editorial board member of the *International Journal of Precision Engineering and Manufacturing – Green Technology*.

Steven Herman builds useful artificial intelligence to solve real-world problems. He is currently a software engineer at Atollogy Inc., leading the development and deployment of novel computer vision models to solve problems in manufacturing and yard management. He holds a BS in computer engineering from Santa Clara University.

Rui Huang received her BS and MS in mechanical engineering from the North China University of Technology and Syracuse University in 2014 and 2016, respectively. She is currently a PhD candidate in the department of mechanical engineering at The University of Akron. Her research interests include additive manufacturing, 3D printing of ceramic materials, conformal printing, and 3D printing of proximity sensor packaging for harsh environments.

Jeff Little is an electrical engineer with 40-plus years' experience in design, engineering management, and technical program management. His areas of experience and expertise include CPU design, voice and network telecommunications, software, microcode, power engineering, compliance, systems engineering, and highly reliable systems design.

Companies and organizations he has been involved with over the years include major corporations such as Intersil, AMD, IBM, Siemens, ROLM, Cisco Systems, and Tandem Computers as well as startups such as Procket Networks, Maple Networks, S-Vision, and RGB Labs. He is currently enjoying retirement while occasionally consulting.

Craig Martin is a seasoned operations and supply chain leader with more than 30 years' experience in the technology sector as the senior executive (VP/SVP) driving global initiatives through all stages of corporate growth. He is currently a senior consultant for On Tap Consulting and an adjunct professor at the Leavey School of Business: at Santa Clara University.

Craig helped establish a new company as cofounder, ramped global operations for a private security firm from startup to a successful IPO, scaling to $800 million, and managed global operations for two multibillion-dollar industry leaders. He has extensive experience in supply chain design and operations, hardware development and manufacturing, managing multiple international factories, commodity management, global facilities, and real estate. Technologies he supports range from simple, high-volume electronics to full cabinets with infinite combinations of highly complex electrical and electromechanical assemblies.

Alex Owen-Hill works with business owners and technology companies that want to stand out in their industries, helping them to create a unique voice for their business that feels authentic to them and attracts the people they most want to work with. He earned his PhD in robotics from the Universidad Politécnica de Madrid with a project investigating the use of telerobotics for the maintenance of particle accelerators at CERN and other large scientific facilities. His regular blog articles on the use of robotics in industrial settings are often shared throughout the online robotics community. Details of his work can be found at CreateClarifyArticulate.com.

Thomas Paral received his doctorate in mechanical engineering and applied computer science from the University of Karlsruhe in 2003. His career began in 2003 as director of R&D engineering for electromechanics at Aichele GROUP GmbH & Co.KG. After various functions in Germany, China, and the United States, he developed as CTO the Aichele GROUP into a global market and technology leader in its rail and automotive markets. From 2014 to 2018, as director of technology of industry solutions at TE Connectivity, he was responsible for new markets and smart factory technologies with a focus on industrial robotics.

From 2018 to 2020, as executive vice president of strategy and business development and GM of cobots and new markets at Schunk he was responsible for the reorganization and realignment of structures including the robotic gripping components and gripping solutions business units. He successfully established and managed the new business unit cobots and new markets. Since 2020 he has been chief business development officer at OnRobot, the leading robotic end-of arm solutions provider for collaborative robotic applications.

Aaron Pompey received his PhD from the University of California at Los Angeles. With several years' experience in executive management across both the corporate and public sectors, he has leveraged smart technologies to achieve efficiency, satisfaction, and growth with major brands across multiple industries, including education, government, healthcare, manufacturing, quick-service restaurants, and transportation. Aaron is based in the Bay Area and currently leads the Pan America region of AOPEN Inc., a global technology company specializing in small form factor hardware solutions for commercial, industrial, and medical environments.

Frank Poon is an enterprising and intuitive business and product leader with over 20 years of experience in growing both multinational companies as well as startups with successful exits. His focus is on business strategy, growth hacking, general management, product strategy, business transformation, operations strategy, and supply chain management. He has an MBA from the University of Chicago and master's and bachelor's degrees in industrial and operations engineering from the University of Michigan.

Miles Schofield is a professional engineer, dancer, musician, speaker, teacher, designer, artist, entrepreneur, and IT specialist with 10 years of experience in application engineering for the semiconductor industry in metrology, where he wrote qualification and control procedures for a number of processes in addition to integrating unique optical and phase imaging tools into global production flow. He has 10 years of application engineering experience in global hardware and IoT computing solutions for leading brands in retail, healthcare, hospitality, and transportation.

Vatsal Shah leads the management and engineering team as co-founder and chief executive officer of Litmus. He has extensive experience with industrial engineering, electronics system design, enterprise platforms, and IT ecosystems. Vatsal earned his master's degree in global entrepreneurship from Em-Lyon (France), Zhejiang University (China), and Purdue University (United States) jointly and his bachelor's degree in electronics engineering from Nirma University in India.

Bowen Shi, aka Randy, from Santa Clara University received dual BS degrees in Mathematics and Sociology in 2016 and a MS degree in Business Analytics in 2019. In 2016, he spoke at the 43rd Annual Western Undergraduate Research Conference with his Witold Krassowski Sociology Award winning research *Success of Digital Activism: Roles of Structures and Media Strategies*. Published in *Silicon Valley Notebook* Volume 14, 2016, the data analytical research investigated how different forms and purposes of digital campaigns affected their success. His expertise is analytics in IT, finance, and manufacture world. He initiated a series of successful analytic projects as the Sr. Data Analyst at Atollogy, Inc. and he is currently a Business Intelligence Analyst at Intuitive Surgical, Inc as of 2021.

Bahareh Tavousi Tabatabaei received her BS in biomedical engineering from Azad University, Isfahan, Iran, in 2014. She is now a PhD student in the Department of Mechanical Engineering at The University of Akron. Her research interests include additive manufacturing, 3D-printed sensors, and biomedical application.

Maria Villamil has a bachelor of science degree in computer information systems from Woodbury University and is a Certified Scrum Master. As senior vice president of WET Design, she is responsible for the planning, construction, and maintenance of the multibuilding WET campus, which includes everything from science labs to state-of-the-art manufacturing facilities consisting of capabilities like sheet metal, welding (manual and robotics), CNC machining, vertical machining, precision machining, tube bending, metrology, vacuum forming, injection molding, surface mount technology manufacturing, additive manufacturing, and powder coating facilities to computer server farms. Maria is in charge of the acquisition, installation, and ongoing maintenance of WET's scientific and industrial manufacturing equipment.

Maria began her career at WET in IT (which she now leads), and which at WET includes high-performance computing, enterprise networking, software development, animation rendering farms, and support for computational engineering systems. She is WET's governmental liaison, in which role she deals with issues ranging from regulatory compliance to the hosting of community and state leaders for events at WET's campus. Maria has led the recent launch of WET's line of PPE products to help the world deal more safely with the COVID-19 pandemic.

Deborah Walkup holds a bachelor of science degree in mechanical engineering from Iowa State University. She began her career designing circuit boards and enclosures for military and space applications at Texas Instruments and Boeing. For the bulk of her career she has worked in solution engineering, teaming up with sales representatives for enterprise software companies in the supply chain space. Her sales career began with a reseller of HP Unix workstations and mechanical CAD software used to support design engineering. She works and lives in Silicon Valley and survived the internet bubble and bust of the early 2000s. Other companies she has worked for include i2, FreeMarkets, Ariba, E2Open, GTNexus, and Infor. Deborah is an avid traveler and scuba diver, having visited all continents except Antarctica, with over 400 hours in the water.

Allison Yrungaray has 20 years of experience in high-tech marketing and public relations. With a bachelor's degree in communications from Brigham Young University, she has written hundreds of articles and achieved media placements in the *Wall Street Journal*, the *New York Times*, *Forbes*, and many other leading publications. She currently leads marketing communications at Litmus, a company with an Industrial IoT Edge platform that unifies data collection and machine analytics with enterprise integration and application enablement.

Introduction

Naren Agrawal

Benjamin and Mae Swig Professor of Information Systems and Analytics, Santa Clara University

In *Smart Manufacturing: The Lean Six Sigma Way*, Dr. Anthony Tarantino and his collaborators deliver an insightful and eye-opening exploration of the ways the Fourth Industrial Revolution is dramatically changing the way we manufacture products across the world, and how it is revitalizing and reshoring American and European manufacturing for both large operations and small to midsize enterprises (SMEs).

Lean Six Sigma has been the mainstay driving continuous improvement efforts for over 20 years. Over time, some shortcomings have become apparent, one of which is that it requires labor-intensive data-gathering requirements. Because of the cost and time required to collect this data, only small sample sizes are created. Operators also behave differently while they are being monitored and tend to backslide into old habits once a project or initiative ends. By creating a digital twin of physical operations using unobtrusive, continuous monitoring devices, data gathering becomes relatively inexpensive, sample sizes grow to 100%, and all behavioral modes for all operators are captured.

This text profiles 23 popular Lean Six Sigma and continuous improvement tools and how Smart Manufacturing technologies supercharges each one of them. The author also explains why much of the criticism of Lean that arose during the COVID-19 pandemic is unfounded.

Dr. Tarantino explores technology's evolution from the start of the Industrial Revolution through today's Industry 4.0 and Smart Manufacturing. He next explores how Smart Manufacturing can improve supply chain's resilience to quickly adjust to sudden disruptive changes that negatively affect supply chain performance. Expert contributors highlight the role of Smart Technologies in making logistics and cybersecurity more effective, critical with the growing volatility of global supply chains and the sophistication of cyberattacks. Leading experts in individual chapters showcase the major tools of Industry 4.0 and Smart Manufacturing:

- Modern networking technologies
- Industrial Internet of Things (IIoT)
- Mobile computing
- Edge computing

- Computer vision
- Robotics
- Additive manufacturing (3D printing)
- Big data analytics

The text explores the contributions women can make in Smart Manufacturing, and how adding their perspective can enrich Smart Manufacturing initiatives. In this breakthrough analysis, the coauthors share their personal stories, providing practical advice on how they achieved success in the manufacturing world.

Finally, several case studies provide examples of Smart Manufacturing helping manufacturers and distributors address previously unsolvable issues. The focus is on SMEs highlighting tools that are affordable and easy to implement. Case studies explore the use of:

- Barcoding to enable rapid inventory transactions
- Computer vision to automate visual inspection and to improve safety
- Mobile computing to replace legacy manufacturing systems
- Robots to do dangerous and boring jobs
- Factory touchscreens to improve shop-floor communications
- Edge computing to collect data close to physical operations for immediate visualizations and business value
- 3D printing to provide vital medical equipment during the COVID-19 pandemic

This book is a must-read for anyone involved in manufacturing and distribution in the twenty-first century. *Smart Manufacturing: The Lean Six Sigma Way* belongs in the library of anyone interested in the intersection of smart technologies, physical manufacturing, and continuous improvement.

CHAPTER 1

Introduction to Industry 4.0 and Smart Manufacturing

Anthony Tarantino, PhD

Introduction

The terms *Industry 4.0* and *Smart Manufacturing* (SM) are widely used today in industry, academia, and the consulting world to describe a major industrial transition underway. This transition is truly revolutionary in that it is now possible to create a digital twin of physical operations to improve operational efficiency and safety while fostering the automation of repetitive, labor-intensive, and dangerous activities.

Exhibit 1.1 shows the digital twin of a car engine and wheels in an exploded image above the physical car.[1]

EXHIBIT 1.1 Digital twin of a car engine and wheels

Source: Digitaler Zwillig/Shutterstock.com.

The first question most people ask is "What is the difference between Industry 4.0 and Smart Manufacturing?" The answer is that they are actually different phrases for the same thing. Klaus Schwab, president of the World Economic Forum, coined the phrase "Industry 4.0" in 2015.[2] The argument for the name Industry 4.0 is that it captures the four phases of the Industrial Revolution dating back 400 years and highlighting the coming of cyber-physical systems. The advantage of the name Smart Manufacturing is that it is catchy and easy to remember. The first references to Smart Manufacturing date back to in 2014, so both names originated at about the same time.[3]

The two terms are now expanding and being applied to nonmanufacturing areas. For example, we now have Smart Quality, or Quality 4.0, and Smart Logistics, or Logistics 4.0. The important thing to remember is that they describe the same goal of creating a digital twin of physical operations. The digital twin is not restricted to equipment and includes people and how they interact with equipment, vehicles, and materials. Only by capturing the dynamic interaction of people, materials, and equipment is it possible to truly understand physical operations and the detailed processes that they use.

A more detailed definition of Smart Manufacturing is that it encompasses computer-integrated manufacturing, high levels of adaptability, rapid design changes, digital information technology, and more flexible technical workforce training.[4] More popular tools include inexpensive Industrial Internet of Things (IIoT) devices, additive manufacturing (also known as 3D printing), machine learning, deep learning computer vision, mobile computing devices, Edge computing, robotics, and Big Data analytics. We will cover each of these tools and technologies in subsequent chapters.

Smart Manufacturing creates large volumes of data describing a digital twin, which in the past was not practical to create. The term *Big Data* has been used since the 1990s but has become central to the growth of Smart Manufacturing and Industry 4.0 in the past few years. By some estimates, the global per-capita capacity to store information has roughly doubled every 40 months since the 1980s.[5] More recent estimates predict a doubling every two years. The good news is that Moore's Law applies to Big Data. (Intel's Gordon Moore predicted a doubling of technological capacity every two years while costs remain constant.) It can be argued that cheap and accessible data is the most critical pacing item to the use of Smart Technology.

The next question readers of this book may ask is "What is the connection between Smart Manufacturing or Industry 4.0 and Lean Six Sigma?" The answer is fairly straightforward. Six Sigma is a framework for complex, data-driven problem solving. Six Sigma practitioners excel at analyzing large volumes of data. Smart Manufacturing offers rich new sources of data. Traditionally Six Sigma practitioners would have to settle on taking small samples of data for their analysis. Now they can capture and analyze all data without the labor-intensive efforts of the past. I ran over 30 projects over a seven-year period for a global high-tech company and always feared that our sampling of data was merely a snapshot in time, regardless of how great the data gathering effort. Running those projects with Smart Technologies would yield a more accurate picture of the truth.

Lean also plays a critical role in Smart Manufacturing. Simply put, Lean is a philosophy for continuous improvement by eliminating all types of waste in operations. As envisioned by Taiichi Ohno, the founder of the Toyota Production System in the 1950s and 1960s, Lean also advocates empowering workers to make decisions on the production line. Smart Manufacturing will eliminate many low-skilled jobs in manufacturing. Smart factories and Smart distribution centers will require higher-skilled workers comfortable in utilizing the many new sources of data to drive continuous improvement efforts.

The First Industrial Revolution

Manufacturing before the Industrial Revolution was typically a cottage enterprise with small shops producing leather goods, clothing, harnesses, and so on. The labor was all manual, that is, people-powered. Beginning in the mid-1700s, the First Industrial Revolution introduced machines that used water or steam power. Factories using steam and water power were larger and more centralized than earlier cottage industries. Factory workers did not require the high skill levels of cottage industry craftsmen and artisans. Women and children were used as a cheap source of labor.

Exhibit 1.2 shows what a blacksmith shop may have looked like in the Middle Ages.[6]

EXHIBIT 1.2　A blacksmith shop in the Middle Ages

Source: O. Denker, Shutterstock.com.

The First Industrial Revolution began in England, Europe, and the American colonies. Textiles and iron industries were the first to adopt power. The major changes from cottage industries of the Middle Ages to the First Industrial Revolution can be summarized as follows:

- Steam- and water-powered production centralized in one factory
- Factories replace cottage industry (e.g., the village blacksmith or leather shop)
- Specialization with the division of labor – workers and machines arranged to increased efficiency
- Harsh and dangerous work environment – primarily using women and children as mechanical power eliminated the need for most heavy labor performed by men

Exhibit 1.3 is a painting of a textile mill powered with either steam or water and a labor force primarily made of children and women.[7]

EXHIBIT 1.3 A painting of an 1800s textile mill

Source: Everett Collection/Shutterstock.com.

The Second Industrial Revolution

The Second Industrial Revolution began in the United States, England, and Europe with the introduction of electrical power over a grid, real-time communication over telegraph, and people and freight transportation over a network of railroads. The railroad and telegraph also increased the spread of new ideas and the mobility of people. Travel times of days using horse-power were reduced to travel times of hours.

The introduction of electric power to factories made the modern mass-production assembly line a reality. The number of people migrating from farms to cities increased dramatically in the early twentieth century. Electric power made possible great economic growth and created a major divide between the industrial world and the poorer nonindustrial world. The rise of the middle class and the migration to cities may be the most visible manifestations of the Second Industrial Revolution. At the time of the American Civil War, only 20% of Americans lived in urban areas. By 1920 that number had risen to over 50% and to over 70% by 1970.[8]

Exhibit 1.4 shows workers on an auto assembly line in the 1930s.[9]

EXHIBIT 1.4   A 1930s auto assembly line

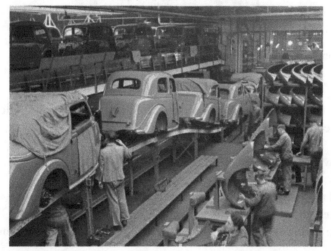

Source: Everett Collection/Shutterstock.com.

Frederick W. Taylor (1856–1915) is credited with creating the efficiency movement, which advocates systematic observation and scientific management for manufacturing. Taylor's approach included scientific study applied to all work tasks, systematically selecting and training each employee, and creating work instructions for each task. He is known as the father of scientific management.[10]

Frank Bunker Gilbreth (1868–1924) and his wife Lillian Gilbreth (1878–1972) were early efficiency experts and pioneered the use of time, motion, and fatigue studies. Lillian is widely accepted as the mother of industrial engineering. They were the inspiration for the *Cheaper by the Dozen* (1948) book and movies. Unlike Taylor, the Gilbreths worked to improve workplace safety and working conditions. Lillian was also a pioneer for women pursuing engineering educations and careers.

Exhibit 1.5 is a photo of Lillian Gilbreth, who continued to teach and lecture until 1964 at the age of 86.[11]

EXHIBIT 1.5   Lillian Gilbreth

Source: Purdue University Engineering.

The major changes from the First Industrial Revolution to the Second Industrial Revolution can be summarized by the following:

- Electrically powered mass production
- Assembly lines
- Telephone and telegraph providing real-time communication
- Efficiency movement of Fredrick Taylor, and Frank Gilbreth and Lillian Gilbreth
- Henry Ford perfecting the assembly line, converting molten steel into a car in 72 hours

The Third Industrial Revolution

The Third Industrial Revolution began in the 1970s and 1980s with the introduction of the first electronic computers. Even though they were very primitive by today's standards, they laid the foundation for a revolution in information management. Manufacturing efficiency dramatically improved with software applications, automated systems, Internet access, and a wide range of electronic devices. Programmable logic controllers (PLCs) began the conversion to Smart machines. Barcode scanning systems replaced error-prone, paper-based processes.

Exhibit 1.6 shows the use of a personal computer with wireless connectivity to manage the factory floor.[12]

EXHIBIT 1.6 Managing the factory floor with a personal computer

Source: Gorodenkoff/Shutterstock.com.

The major changes from the Second Industrial Revolution to the Third Industrial Revolution can be summarized as introducing the following:

- Semiconductors, mainframe and personal computing
- The World Wide Web and Internet
- Manufacturing software (MRP, ERP, and MES, eProcurement) replacing paper-based processes
- Additive manufacturing/3D printing
- Robots replacing people

The Fourth Industrial Revolution

The transition to Industry 4.0 and Smart Manufacturing began over the past 20 years and is based on the following core principles:

- **Secure connectivity** among devices, processes, people, and businesses
- **Flat and real time** digitally integrated, monitored, and continuously evaluated
- **Proactive and semi-autonomous** processes that act on near-real-time information
- **Open and interoperable** ecosystem of devices, systems, people, and services
- **Flexibility** to quickly adapt to schedule and product changes
- **Scalable** across all functions, facilities, and value chains
- **Sustainable** manufacturing: optimizing use of resources, minimizing waste[13]
- **Information transparency** offering comprehensive information to facilitate decision making

- **Inter-connectivity** allowing operators to collect immense amounts of data and information from all points in the manufacturing process, identifying key areas that can benefit from improvement to increase functionality
- **Decentralized decision making** in which smart machines make their own decisions. Humans will only be needed when exceptions or conflicting goals arise.[14]

The major changes from the Third Industrial Revolution to the Fourth Industrial Revolution can be summarized as the following:

- The Internet of Things (IOT) – increasing from 7 billion to 26 billion devices in one year
- Smart sensors, AI, and computer vision to create digital twins
- Affordable advanced robots and cobots (a robot that supports people)
- Mainstream additive manufacturing, also known as 3D printing
- Mobile computing
- Location detection technologies (electronic identification)
- Advanced human–machine interfaces
- Authentication and fraud detection
- Big Data analytics and advanced processes
- Multilevel customer interaction and customer profiling
- Augmented reality wearables
- On-demand availability of computer system resources

The Major Components of Smart Manufacturing

The chapters in this book will cover the major components in Smart Manufacturing that will impact manufacturers of all sizes and complexities. There are others, and the list will grow. The components we cover include:

- Lean in the age of Smart Manufacturing
- Six Sigma in the age of Smart Manufacturing
- Improving supply chain resiliency using Smart Technology
- Improving cybersecurity using Smart Technology
- Improving logistics using Smart Technology
- Big Data for small to midsize enterprises (SMEs)
- Industrial IOT sensors
- AI, machine learning, and computer vision
- Networking for mobile-edge computing
- Edge computing
- Additive manufacturing and 3D printing
- Robotics
- Improving life on the factory floor using Smart Technologies
- Growing the role of women in Smart Manufacturing

Lean and Six Sigma in the Age of Smart Manufacturing

Lean dates back to the 1960s when Toyota introduced Lean in its Japanese production plants. Because computer systems were primitive or nonexistent on the factory floor, Lean philosophy relied on simple visual controls. Lean also empowered all employees to make decisions that could shut down an entire assembly line, a far cry from US practices of the time.

Six Sigma dates back to the 1980s with Motorola followed by GE. It provided a common-sense framework for solving complex problems using the scientific method. Six Sigma data analysis drives the effort to reduce defects, improve quality, and optimize operational efficiencies.

Most organizations have combined the philosophy of Lean with the problem-solving framework of Six Sigma. Many organizations have rebranded their Lean Six Sigma programs as Continuous Improvement and more recently as Industry 4.0 programs.

With Smart Technologies, Lean and Six Sigma have a new lease on life, becoming more efficient and more effective. One of the best ways to envision the change is to picture the traditional process of an industrial engineer or continuous improvement team member watching a manufacturing process and noting cycle times with stopwatch and clipboard.

Imagine the ineffectiveness of trying to accurately capture the variations in a physical operation across various machines, across three shifts, and across each day of the week and each season of the year. Regardless of the skill and dedication of the analyst, they can only observe and document a small percentage of the entire population of operations. Now imagine smart cameras and IoT sensors watching all transactions on an Edge computer (a computer located near the action) and transmitting data to the Cloud for analysis.

Exhibit 1.7 shows a worker making notes on clipboard, the traditional method of data collection on the factory floor.[15]

EXHIBIT 1.7 Collecting data traditionally

Source: NDOELJINDOE/Shutterstock.com.

Smart Manufacturing offers quality and process improvement professionals robust digital tools to examine and evaluate operations without the labor-intensive and ineffective practices of the past. AI and computer vision provide the means to automate visual inspection with greater accuracy and consistency than using manual methods. The new technology also provides data sets for all transactions, not the small sample sizes used in the past.

Exhibit 1.8 demonstrates just how small a sample size is required to meet the Military Standards that have been in place since the 1950s. In this example a 500-part sample is less than 2% of the total population.[16] With smart cameras and IoT sensors watching the action on a 24/7 basis, the new sample size is the entire population of 35,163 parts. The combination of Edge computing and the Cloud provides an easy means to run statistical analysis leading to improved quality.

EXHIBIT 1.8   Military Standard 105e

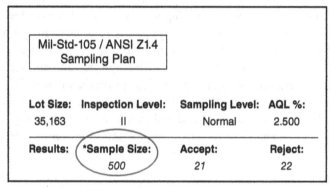

Source: Quality Assurance Solutions.

Improving Cybersecurity Using Smart Technology

Cybersecurity threats are coming at organizations from a variety of sources, including those sponsored by foreign governments hostile to Western democracies, and from criminal sources, both foreign and domestic. What they have in common is a very successful track record of overcoming firewall protections to steal and hold hostage critical company information and cripple operations.

It is a big mistake for manufacturers, especially smaller ones, to believe that they are not a cyberattack target. They are, for the simple reason that they are easy to breach. Here are 10 key cybersecurity takeaways to consider:

1. Globalization, specialization, and IoT trends have increased cyber risk.
2. Supply chains are much deeper and broader than you realize.
3. The supply chain is an attractive target for several reasons.
4. No supplier is too insignificant to be immune from risk.
5. Security controls are only as strong as the weakest link.
6. Threat actors have a wide range of motives and methods.
7. Cyber risk can be mitigated by making business tradeoffs.
8. Impact of a breach can be mitigated with proper controls.
9. Cyber risks need to be considered in sourcing decisions.
10. The costs of a breach can be far-reaching and catastrophic.

Smart Logistics

The modern science and practice of logistics had its origins in World War II. American logistics practices were a primary factor in the Allied victories over Germany and Japan. Logistics is the process of managing the end-to-end planning, acquisition, transportation, and storage of materials through supply chains.

Logistics 4.0 revolutionizes the practices that help win wars and power modern manufacturing and distribution. The digitization of logistics operations includes driverless trucks, delivery drones, automated warehouses, smart ports, smart containers using radio-frequency identification (RFID), blockchains, and AI-powered routing of parts. Smart Logistics come at a critical time to help mitigate supply chain shocks from pandemics, tsunamis, trade wars, shooting wars, and the instability inherent in less developed economies. Finally, Smart Logistics may be the only option to solve the chronic shortage of local and long-haul drivers.

Big Data for Small, Midsize, and Large Enterprises

Smart Manufacturing's ultimate goal is to digitize all physical operations, creating a constant stream of data in real time, typically captured on Edge computers and communicated to the Cloud. Big Data is not a goal of Industry 4.0 and Smart Manufacturing. Manufacturing organizations have generated large volumes of structured and unstructured data for several years. The problem is that much of the data ends up in silos, not extracted or normalized for analysis.

Smart Manufacturing is transforming traditional manufacturing by replacing isolated and siloed data with the ability to collect both structured and unstructured data from diverse sources. Therefore, the goal of Smart Manufacturing is to mine, merge, and transform data to provide a digital twin of operations in real time. Without Big Data analytics much of Industry 4.0's technology is wasted, just as large amounts of data were wasted with Industry 3.0 technology.

Big Data makes possible predictive modeling that exploits patterns found in historical and transactional data to identify risks and opportunities. While some of the most advanced Big Data technology solutions may beyond the reach of smaller organizations, there are many affordable and easy-to-use Big Data tools that smaller organizations can utilize. Today's global supply chains are dynamic, with multiple levels of dependencies. Without Big Data, it is not practical for an organization of any size to adequately identify risks and opportunities.

Industrial IoT Sensors

Smart sensors are devices that generate data transmitted to Edge computers and to the Cloud to monitor various processes. They are typically easy to install and require little configuration. The types of sensors are quite varied:

- Temperature sensors
- Pressure sensors
- Motion sensors
- Level sensors
- Image sensors

- Proximity sensors
- Water quality sensors
- Chemical sensors
- Gas sensors
- Smoke sensors
- Infrared (IR) sensors
- Acceleration sensors
- Gyroscopic sensors
- Humidity sensors
- Optical sensors[17]

With Smart sensors, data is automatically collected and analyzed to optimize operations, improve safety, and reduce production bottlenecks and defects. Sensors communicate data to Edge computers and/or the Cloud via IoT connectivity systems on the factory floor. IoT technology leverages wired and wireless connectivity, enabling the flow of data for analysis. It is now possible to monitor operations remotely and make rapid changes when warranted by conditions. The use of Smart sensors helps improve manufacturing processes and product quality while reducing waste and safety violations on the factory floor.

Exhibit 1.9 shows the flow of data from several types of IIoT sensors to Edge computers for analysis and to data monitoring applications and dashboards.[18]

EXHIBIT 1.9   Internet of Things (IoT) data analytic concept

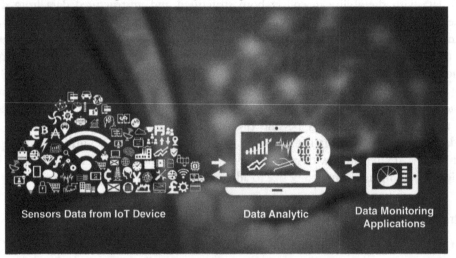

Source: Zapp2Photo/Shutterstock.com.

Artificial Intelligence Machine Learning and Computer Vision

Today's computer vision has the goal of helping computers see. It uses artificial intelligence and machine learning to digitize imagery for analysis. Tasks that come easily for humans are a challenge for computer vision. A human easily understands that a car in the distance moving toward them appears larger as it gets closer. Computers need to be taught that the change in size does not indicate several different cars. I recall the early days of my supporting a computer vision startup. The engineers were excited that they had taught their program to detect a bare arm reaching for a controller. It worked fine until someone wore a long-sleeve shirt. The program ignored the arm because it had not been taught to consider an arm with clothing.

The manufacturing use cases for computer vision are varied and continue to grow. As the affordability and ease of installation continue to improve, it is reasonable to predict the demise of "dumb" cameras, even for consumer uses. Some of the more popular computer vision (CV) applications include:

- **Human/Machine Interaction.** CV is able to capture how operators interact with their equipment. Unlike manufacturing shop floor control software programs, CV captures the actual efficiency of operators (the number of units produced against a standard) and utilization of equipment (the hours in operation against the total available shift hours). All this is accomplished without entering data or scanning barcodes. CV can also identify the actual percentage of time an operator is at their workstation versus away from it. This is also accomplished without entering data or scanning barcodes, so there are no data entry errors.
- **Anomaly Detection.** CV is able to flag such anomalies as forklift and truck drivers speeding, people standing in restricted areas, an excessive number of workers congregating in a work center, or conveyors running without materials or parts after an established maximum time.
- **Defect reduction.** CV is able to automate visual inspection to complement or replace human inspectors of parts, assemblies, and packaging. This overcomes inconsistencies from inspector to inspector and from shift to shift. Multiple smart cameras capturing 100 images per second can inspect even the most complex assemblies in a few seconds.
- **Barcode and Label Scanning.** This is one of the simple use cases for CV eliminating human error and inconsistencies.
- **Safety and Security Violations.** CV is able to help organizations maintain social distancing and mask requirements during the COVID-19 global pandemic. In some cases, CV has helped to prevent serious injuries when operators clearly violated safety requirements by standing behind trucks in busy terminals or not wearing their safety harnesses and hard hats in restricted areas.

Networking for Mobile-Edge Computing

Smart Manufacturing is powered by mobile computing. Without mobile technology, Smart Manufacturing would not be practical. Mobile devices are the platforms by which manufacturing workers and managers can connect easily to the Cloud. The IIoT generates massive amounts of data with connected devices. By combining mobile's ability to provide networks with the data generated by the IIoT, manufacturers have powerful new sources of information to improve operations and eliminate paper-based practices.

Mobile communications has been with us for decades. The first mobile communication was designated as 0G and generally thought to start with the car phone, introduced in 1946 by the Bell System. Beginning in the 1980s, the first generation of wireless analog cellular phones was introduced. These are part of the 1G generation. Starting in 1991 in Finland, the first commercially available digital cellular phones were introduced, creating the 2G generation. Beginning in 2009, 4G was commercially launched in Sweden and Norway. In the United States the launch was in 2010. Then in 2019, 5G technology was launched almost simultaneously in South Korea and the United States, and the rollout is expanding now around the world. Finally, 6G is expected to launch commercially sometime after 2030.

5G has provided manufacturing with the high bandwidth, low latency, and high reliability that are critical to many mobile computing applications. In the past these applications required fixed-line connections. 5G technology will be key to increasing flexibility, shortening lead times, and lowering costs on the factory floor. While 6G is still years away, some early estimates predict 10 to 100 time increases in speed. This will continue to drive down costs while increasing the capabilities of all types of mobile computing devices.

Edge computing refers to the location of a computer relative to sensors feeding it data. Exhibit 1.10 is a graphic showing the process of gathering data from smart sensors and transmitting it to Edge computers for real-time alerts and actions and then to the Cloud for analysis.[19]

EXHIBIT 1.10   Edge computing

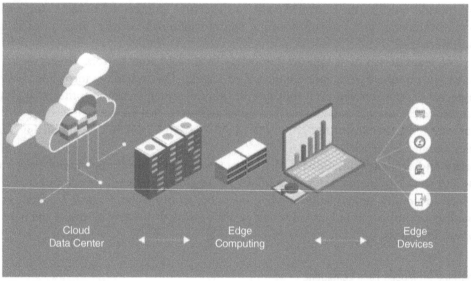

Source: Zzins/Shutterstock.com.

Additive Manufacturing and 3D Printing

Additive manufacturing, often referred to as 3D printing, plays an important role in Smart Manufacturing. It is especially useful for small production runs and rapid prototyping, which helps shorten the times required for new product introduction (NPI) while lowering development costs. It also allows for greater product customization by eliminating the long setup times required with traditional production processes.

3D printing makes objects according to a 3D digital model. 3D printers add material layer by layer, creating objects ranging from simple to highly complex. Originally limited to plastics, 3D can now use a variety of metals to produce parts quickly and inexpensively. As this technology advances, 3D printing can move beyond prototyping and short runs to support agile manufacturing, allowing consumers to expect multiple design changes over short periods of time.

Exhibit 1.11 shows a 3D printer creating a complex object.[20]

EXHIBIT 1.11   A 3D printer at work

Source: Asharkyu/Shutterstock.com.

Robotics

Robotics will play a central role in Smart Manufacturing and will grow in importance as its costs continue to decline. Many workers fear that robots will replace them. The truth is that robots will help to save manufacturing jobs by keeping domestic producers competitive with foreign competitors who enjoy lower-cost labor. In many cases robots are performing dangerous or repetitive tasks so boring that nobody wants to do them. This frees workers to focus on higher-value and more interesting activities.

Some of the benefits of manufacturing robotics include:

- Improved efficiencies throughout all phases of production
- The ability to operate on a 24/7 basis and in a lights-out environment
- Greater agility to react quickly to product and process changes
- Increased ability to bring back domestic manufacturing that was previously offshored
- Help in addressing chronic labor shortages, especially for boring, repetitive, or dangerous tasks
- Improvements in quality and reliability
- ROIs achieved in a few months in many cases

Exhibit 1.12 shows a small robotic arm with a mechanical hand.[21]

EXHIBIT 1.12   Robotic arm with mechanical hand

Source: A. Suslov/Shutterstock.com.

Smart Technology to Improve Life on the Factory Floor

Over the past 50 years the number of manufacturing jobs in the United States has dropped from 20 million to 12 million. Much of this has been attributed to automation, according to the Center for Business and Economic Research at Ball State University. Of course, offshoring plays a major role as well.[22] Unfortunately, this has given Smart Manufacturing a bad reputation among American factory workers.

What is missed in the criticism is the great number of Smart Manufacturing tools that help make life better for factory workers. Inexpensive smart sensors monitoring production equipment and tools can spot problems and prevent plant shutdowns. Such disruptions reduce revenue and customer satisfaction, potentially leading to loss of wages and layoffs. IIoT sensors and computer vision can also help to flag safety and security violations reducing accidents and injuries.

Adding smart sensors along the entire production line provides a wealth of valuable information that can foster continuous improvement and Lean Six Sigma programs. The new data sources do not necessarily make equipment more powerful, but they do make humans more knowledgeable. This leads to better decision making.

Summary: The Advantages of Smart Manufacturing

Whether you call it Smart Manufacturing or Industry 4.0, a major transformation of physical operations is underway in which a complete digital twin of people, materials, and equipment will be created on factory and warehouse floors, in distribution centers, and in ports and terminals. The following are some of the benefits that can be expected.

Improved Quality and Safety

The cumulative effect of IIoT sensors, computer vision, robotics, smart machines, data analytics, and so forth will help to reduce errors and defects by eliminating the bulk of inefficient human interactions. It will also help to standardize and automate processes around best practices. These same technologies will help flag unsafe activities on factory floors and in distribution centers.

Improved Margins

Robust new data sources are creating a digital twin of operations so manufacturers can determine optimal areas for automation. In the past, incomplete data did not accurately capture physical operations, leading to investing in new equipment that was not needed or investing in the wrong types of equipment.

Improved Cycle Times

Lean has always striven for continuous improvement to reduce manufacturing cycle times. Smart Manufacturing identifies new opportunities to reduce cycle times. AI makes machines intelligent, robots shorten production times, and computer vision identifies production bottlenecks. The combined effect is to shorten cycle times without sacrificing quality or safety.

High Efficiency with Well-Defined Smart Factory Processes[23]

Lean also strives to replace traditional material systems that pushed production through factories based on forecasts with a pull system based on customer orders. This is only possible after the great majority of waste is removed from the entire manufacturing cycle. What results is the ability to make to customer orders, even if the lot size is only one. For 35 years in manufacturing this seemed an impossible dream in the factories I supported. With Smart Manufacturing, this dream is now achievable.

Sample Questions

1. What are the major differences between Smart Manufacturing and Industry 4.0?
   a. Industry 4.0 is only used in Europe.
   b. Smart Manufacturing is only used in the United States.
   c. Smart Manufacturing generates more data than Industry 4.0.
   d. They are different names for the same thing.
2. A major change from the First Industrial Revolution to the Second was the addition of
   a. Steam power
   b. Water power
   c. Electric power
   d. Computers
3. A major change from the Second Industrial Revolution to the Third was the addition of
   a. Computers
   b. Electric power
   c. The assembly line
   d. Modern highways

4. The Fourth Industrial Revolution, or Smart Manufacturing,
   a. Creates a digital twin of physical operations
   b. Relies heavily on mobile computing
   c. Deploys many types of IIoT sensors
   d. All of the above
5. Computer vision helps to improve safety by
   a. Creating alerts to safety violations in real time
   b. Requiring managers to review hours of video
   c. Preventing workers from violating rules
   d. Automating tasks
6. The benefits of robotics include
   a. Improved efficiencies throughout all phases of production
   b. Ability to operate on a 24/7 basis and in a lights-out environment
   c. Greater agility in reacting to product and process changes
   d. Increased ability to bring back domestic manufacturing that was previously offshored
   e. All of the above
7. Computer vision
   a. Can only analyze vehicles
   b. Does not typically require artificial intelligence (AI)
   c. Digitizes physical operations to improve efficiencies and safety
   d. Is too expensive for most organizations
8. The role of Big Data analytics in Smart Manufacturing
   a. Is critical because of the many new sources of data created
   b. Is of minor importance
   c. Requires PhDs to analyze the data
   d. Can only be used by large organizations
9. What is Smart Manufacturing's (SM's) role in creating high-paying jobs?
   a. SM is a jobs killer.
   b. SM automates low-skilled, mundane, and dangerous jobs.
   c. SM creates high-paying jobs by helping to keep American manufacturing competitive.
   d. SM has no role.
   e. b and c
10. How does Smart Manufacturing improve quality?
    a. It uses robotics to eliminate errors and standardize operations.
    b. It digitizes all process steps for analysis – there is no need to use small sample sizes.
    c. It uses computer vision to quickly accept or reject parts.
    d. All of the above

Notes

1. Monopoly 919/Shutterstock.com. Smart factory which uses futuristic technology. *Shutterstock* ID: 1826784071 (accessed June 1, 2021).
2. Schwab, K. (December 12, 2015). The Fourth Industrial Revolution. *Foreign Affairs.* https://www. foreignaffairs.com/articles/2015-12-12/fourth-industrial-revolution.

3. Ivezic, N., Kulvatunyou, B., and Srinivasan, V. (May 2014). On architecting and composing through-life engineering information services to enable Smart Manufacturing. *Science Digest* 22: 45–52. https://www.sciencedirect.com/science/article/pii/S2212827114008178.

4. Lu, Y., Xu, X., and Wang, L. (July 2020). Smart manufacturing process and system automation – A critical review of the standards and envisioned scenarios. *Journal of Manufacturing Systems* 56: 312–325. https://www.semanticscholar.org/paper/Smart-manufacturing-process-and-system-automation-%E2%80%93-Lu-Xu/51a385ed6729864a74f71731193e29680f487eeb.

5. Hilbert, M., and López, P. (April 2011). The world's technological capacity to store, communicate, and compute information (PDF). *Science* 332 (6025): 60–65. https://science.sciencemag.org/content/332/6025/60.

6. Denker, O./Shutterstock.com. Blacksmith at work outside his shop in a medieval European town, 3D render. Shutterstock ID: 1809699526 (accessed June 17, 2021).

7. Everett Collection/Shutterstock.com. Machines making cotton thread by performing mechanical versions of carding drawing and roving in a mill in Lancashire England ca 1835. Engraving with modern watercolor. Shutterstock ID: 237232108 (accessed June 15, 2021).

8. US Census Bureau. https://www.census.gov/prod/2002pubs/censr-4.pdf.

9. Everett Collection/Shutterstock.com. Workers building cars in factory. Shutterstock ID: 696639136 (accessed June 14, 2021).

10. Taylor, W. (1911). *Shop Management*. New York: Harper and Brothers.

11. Purdue University Engineering. https://engineering.purdue.edu/Engr/Research/GilbrethFellowships.

12. Gorodenkoff. Factory worker in a hard hat is using a laptop computer with an engineering software. Shutterstock ID: 702079537 (accessed June 15, 2021).

13. CESMI – The Smart Manufacturing Institute. https://www.cesmii.org (accessed June 14, 2021).

14. Bonner, M. (March 2, 2017). What is Industry 4.0 and what does it mean for my manufacturing? *Saint Clair Systems Blog*. https://blog.viscosity.com/blog/what-is-industry-4.0-and-what-does-it-mean-for-my-manufacturing.

15. NDOELJINDOE/Shutterstock.com. Portrait Asian engineer smiling look on you [sic] camera. Shutterstock ID: 137604548 (accessed June 17, 2021).

16. Military Standard 105e. Quality Assurance Solutions. https://www.quality-assurance-solutions.com/mil-std-105e.html (accessed June 17, 2021).

17. Edwards, E. Different types of Internet of Things (IoT) sensors. *Thomas Industry Insights*. https://www.thomasnet.com/articles/instruments-controls/types-of-internet-of-things-iot-sensors/ (accessed June 11, 2021).

18. Zapp2Photo/Shutterstock.com. Internet of things (IoT) data analytic concept. Infographic of cloud, Wi-Fi, data analytics, data monitoring application and texts with blur man suit holding tablet abstract background. Shutterstock ID: 515607865 (accessed June 17, 2021).

19. Zzins/Shutterstock.com. Edge computing Vector Pictogram. Service delivery computing offload IOT management storage and caching system. Shutterstock ID: 1856928346 (accessed June 17, 2021).

20. Asharkyu/Shutterstock.com. Modern 3D printer printing figure close-up macro. Shutterstock ID: 676361872 (accessed June 17, 2021).

21. Suslov, A. Robotic arm 3d on white background. Mechanical hand. Industrial robot manipulator. Shutterstock ID 746139496 (accessed June 17, 2021).

22. Savić, A. (2017). Using smart technology to aid factory workers. *The Atlantic*. https://www.theatlantic.com/sponsored/vmware-2017/human-ai-collaboration/1721/.

23. Futurism Technologies. (August 31, 2020). Understanding the advantages of smart factory in a simple way. https://www.futurismtechnologies.com/blog/understanding-the-advantages-of-smart-factory-in-a-simple-way/.

CHAPTER 2

Lean Six Sigma in the Age of Smart Manufacturing

Anthony Tarantino, PhD

Introduction

The history of Lean and Six Sigma are worth understanding in light of the waves of criticism both have received over the years. The criticism of Lean intensified during the COVID-19 pandemic when so-called "Lean" inventory practices were blamed for shortages of virtually every critical commodity. We argue in this chapter that much of the criticism of Lean and Six Sigma is not warranted. To the contrary, we will demonstrate that they are more valuable than ever with the advent of Smart Technologies. We will also give some background on how the criticism came about. The next chapter is focused on a deeper dive into the many continuous improvement tools that are empowered by Smart Technologies.

Lean is a philosophy that advocates for continuous improvement and the elimination of waste of all kinds. Six Sigma is a data-driven framework for solving complex problems where the customer is known and the solution is unknown. While the terms have been used separately with success for decades, combining them makes sense, because this overcomes weaknesses in both approaches.

Lean lacks Six Sigma's proven framework to run continuous improvement projects and initiatives. Six Sigma suffers from its end-of-project approach that hopes improvements will sustain themselves after the project is completed. By combining both, Six Sigma's process improvements and error reductions can be embedded into Lean's continuous improvement philosophy.

A simple way to combine the two approaches is by using value stream maps, which we detail in the next chapter. Value stream maps, a classic Lean tool, are a great way to highlight the largest bottleneck or constraint in a process. Once the bottleneck is identified, Six Sigma's DMAIC framework (Define-Measure-Analyze-Improve-Control) can be applied to improve an existing process. For a new process or a process so broken that it needs to be replaced,

Six Sigma's DMADV (Define-Measure-Analyze-Design-Verify) can be applied. DMADV is also known as Design for Six Sigma.

The next step is critical to convert a Six Sigma project into Lean's continuous improvement philosophy, in which the journey never ends: once the existing bottleneck is eliminated or substantially reduced, rerun the value stream map. The bottleneck will always move, sometimes to unexpected areas of the process. The next Six Sigma project should tackle the new bottleneck, creating a true Lean Six Sigma approach of continuous improvement.

The History of Lean – American Assembly Lines

In reading the history of the Toyota Production System, you will find complementary references to Henry Ford, who is credited with creating very efficient mass production of cars in the 1920s. The process was successful in creating quality products that the average American of the 1920s could afford.

The assembly line predates Henry Ford by many years. Ford credited the meat packing industry for the concept of workers remaining in a fixed location while work was moved along a line to them. The first automotive assembly lines were created in 1901 by Ransom Olds, whose Oldsmobile cars were popular throughout the twentieth century.[1]

The genius of the assembly line was that each unit of work was designed to fit the speed that the line was run. Ford was instrumental in streamlining the assembly line with his own version of Lean. The complete assembly process was reduced to 93 minutes with a car completed every six minutes. Because workers remained in one location, Ford was able to greatly reduce the rate of injuries and accidents. Greatly improved efficiency meant that the average Ford worker could buy a new car with only four months of their salary.[2]

Exhibit 2.1 shows a modern assembly line with an overhead conveyor system.[3]

EXHIBIT 2.1   A car assembly line

Source: Shutterstock.com.

While Ford was able to revolutionize manufacturing in the 1920s with his own version of Lean, American manufacturing practices had some major drawbacks that began to emerge in the 1960s. The major American car manufacturers were slow to understand the voice of their customers, who would come to demand a wide variety of features and options. Ford was famous for his quip that customers could have any color car they wanted as long as it was black.

America's major car manufacturers were also slow to respond to customer demands for greater quality and reliability. Levels of quality and standards of customer service declined in the 1960s and 1970s. American manufacturers ignored the growing threat from Japanese manufacturers that offered high-quality products at lower prices.

The History of Lean – Toyota Embraces Deming and Piggly Wiggly

The origins of Japan's high-quality and reliable cars can be traced back to Dr. W. Edwards Deming, who helped to bring about Japan's economic revival after the devastation of World War II. Starting in 1950, Deming championed Japan's improved product designs, improved product quality, and robust statistical methods for product testing, including statistical process controls.

The other major and legionary event of the 1950s came about by accident when Toyota engineers came to the United States to visit Ford's River Rouge Factory production lines. While the engineers enjoyed their visit to Ford, they were more impressed by what they witnessed at the local Piggly Wiggley grocery store. Piggly Wiggly is credited with inventing the modern grocery store with self-service, shopping baskets, and checkout stands. In the 1950s Piggly Wiggly maintained 4,000 SKUs (stock keeping units), which was four times higher than the SKUs its competitors carried. The Toyota visitors were most impressed with the *just-in-time (JIT)* inventory, which offered fresh produce, meats, and dairy products. This was unlike anything they had in Japan with its small markets and limited selections.

Exhibit 2.2 shows one of Piggly Wiggly's early stores.[4]

EXHIBIT 2.2   An early Piggly Wiggly store

Source: Piggly Wiggly.

The cumulative effect of automobile assembly lines, Dr. Deming, and Piggly Wiggly grocery stores forged the foundations of the Toyota Production System (TPS). Toyota revolutionized manufacturing by building to customer demand and continuously attacking waste of all types to improve quality and process times. Toyota's continuous flow manufacturing made traditional batch manufacturing obsolete in the automotive industry. In order to compete, American manufacturers were compelled to embrace the new system called JIT, and, later, Lean.

The Toyota Production System: The Birthplace of Lean

Originally known as just-in-time, or JIT, the Toyota Production System is a socio-technical system, a method of complex organizational work design recognizing the interaction between technology and people in the workplace. It is also a management system for organizing automotive logistics and manufacturing. It was the brainchild of Toyota's founder, Sakichi Toyoda, his son, Kiichiro Toyoda, and their senior engineer, Taiichi Ohno. It evolved over the 1950s and through the 1970s.[5]

Exhibit 2.3 shows a photo of Taiichi Ohno.[6]

EXHIBIT 2.3   Toyota senior engineer Taiichi Ohno

Source: Wikipedia.

Taiichi Ohno was very impressed with the stories of Piggly Wiggly but pondered how to implement that concept into manufacturing. A supermarket customer takes the items they desire off the shelves. Store clerks then restock the shelves with exactly the amount of product needed to fill up the open space. Ohno writes: "Similarly, a work-center that needed parts would go to a 'store shelf' (the inventory storage point) for the particular part and 'buy' (withdraw) the quantity it needed, and the 'shelf' would be 'restocked' by the work-center that produced the part, making only enough to replace the inventory that had been withdrawn."[7]

There is a strong argument that Taiichi Ohno did more to advance manufacturing in the twentieth century than anyone else. Interestingly, his ideas were very unpopular when first introduced at Toyota. The reasons were obvious: he challenged the conventional wisdom of running large batches of production to sales forecasts, stocking plenty of just-in-case inventory, and factory workers rarely challenging management decisions.

The traditional push approach to manufacturing builds lots of components and finished goods to a forecast. It was not considered practical to build smaller lot sizes to customer orders, due to long setup and processing times. While there were industrial engineering efforts to automate production, the process never achieved the levels of excellence that Ohno envisioned for Toyota.

Ohno argued that the push system, with its long processing times and large lot sizes, was inherently wasteful. Rather than piece-meal efforts at process improvements, Ohno introduced a culture of continuous improvement that attacked waste of all types. One of his greatest achievements was tackling long setup times.

As with the Piggly Wiggly inspiration, Ohno first learned of setup time reductions during a 1955 visit to Danly Manufacturing in Pennsylvania. Upon his return to Japan, Ohno applied a structured approach he learned from the US military called ECRS: Eliminate, Combine, Rearrange, and Simplify. The decades-long effort produced dramatic results in setup time reductions – from hours to 15 minutes in the 1960s, to three minutes in the 1970s, and 180 seconds by the 1990s.[8] The technique became known as the Single Minute Exchange of Dies, or SMED, and was pioneered by Shigeo Shingo in the 1970s.[9] Through massive setup time reductions and close working relationships with key suppliers, Toyota was able to transition from a push system to a just-in-time pull system.

EXHIBIT 2.4   Push versus pull systems

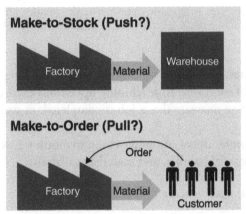

Source: Courtesy of All About Lean.

Exhibit 2.4 shows the fundamental difference between traditional material planning systems that "pushed" materials into manufacturing based on forecasts, and the Toyota/Lean system that "pulled" materials through manufacturing based on actual customer orders.[10]

There are three major goals in the Toyota Production System (TPS). Ironically, many current advocates and critics of Lean have only focused on the first goal, the elimination of waste, or *muda* in Japanese. The second goal is to design out overburden, or *muri*, which means that the system is flexible to accommodate changes in supply and demand. The third goal is to eliminate inconsistency, or *mura*.

With Toyota, waste is not limited to simply eliminating excess inventory. Exhibit 2.5 shows eight types of waste that Lean attacks on a continuous basis.[11]

EXHIBIT 2.5 The Eight Wastes of Lean

THE 8 WASTES

DEFECTS	OVERPRODUCTION	WAITING	NON-UTILIZED POTENTIAL
Efforts caused by rework, scrap & incorrect information	Production that is more than needed or before it is needed	Wasted time waiting for the next step in the process	Not using people's talents, skills & knowledge to their full potential
TRANSPORT	INVENTORY	MOTION	EXTRA PROCESSING
Unnecessary movements of products & materials	Excess products & material not being processed	Unnecessary movements by people (eg. walking)	Unnecessary processing or activities in the process that do not add value

www.haldanconsulting.com

Source: Courtesy of Nawras Skhmot.

One of the biggest wastes, or *muda*, is waiting. In many cases long wait times are caused by factors other than material shortages. In machining operations these could be operators tending multiple machines being unaware that a job is completed, or waiting for tool changes, or waiting for an inspector.

Exhibit 2.6 shows a complex use of computer vision to monitor machine status via an andon light, the operator, the material handler, and material availability.[12]

EXHIBIT 2.6 Using computer vision to monitor machine status

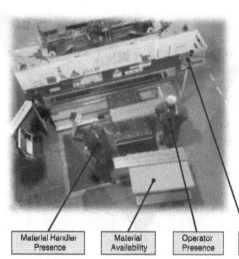

Computer vision can help eliminate constraints without new equipment by monitoring the status of:

- Operator
- Material Handler
- Material
- Andon Light

| Material Handler Presence | Material Availability | Operator Presence | Andon Light Signaling Job Status |

Source: Courtesy of Atollogy.

Another major innovation associated with Toyota and Lean was the introduction of kanbans. Kanbans predate Toyota and were used by the British in World War II for aircraft production. Known as the two-bin system, kanbans are an ideal means to manage inventory in a JIT/pull system, by aligning inventory levels with actual consumption. When customer demand consumes the contents of a kanban bin, the empty bin alerts a supplier to produce and replenish the bin. In this way, nothing is produced that is not needed.[13] The beauty of the kanban system is its simplicity. In its purest form, it is a simple visual system requiring no computer programs. The visual system can be a card, a bin, or other containers.

Kanbans can also be automated with the use of electronic messaging of consumption. One of the most common automations is the use of barcode scanners to mark inventory movement throughout the various stages of production. All major retailers use a form of electronic kanbans by scanning consumption at the point of sale. This creates a signal in real time to replenish inventories for sold items. This also eliminates errors by eliminating the need for manual data entry.

Exhibit 2.7 shows three examples of kanban bins. The first is a two-bin system for use within a facility, the second is a two-bin system used with a supplier, and the third is a three-bin system which shows the path from supplier to storeroom to factory floor.[14]

EXHIBIT 2.7 Three kanban bins

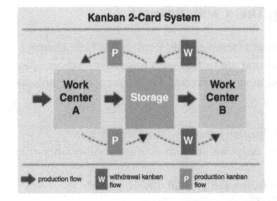

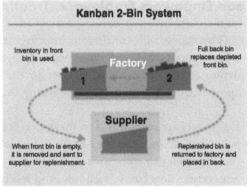

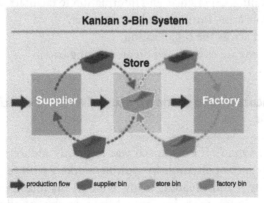

Source: Courtesy of Creative Safety Supply.

Exhibit 2.8 shows two uses of computer vision: monitoring kanban bin inventory levels, and monitoring kanban pallets. When bin or pallet levels hit a designated amount, an alert is sent that will replenish the inventory.

EXHIBIT 2.8   Monitoring kanban bin inventory levels and pallets

Smart Manufacturing KanBans – Using AI and Computer Vision

- Monitor stocking locations
    - Monitor both staging and consumption locations
    - Generate alerts for critically low or stockout
- Benefits
    - Replace operator-initiated replenishment signal
    - Link staging and consumption inventories

Monitoring Floor Locations

Source: Courtesy of Atollogy, www.Atollogy.com.

Lean Empowers Employees, Treating Them with Respect

Lean's obsession with continuous improvement as a never-ending journey has little chance of success without dedicated, energized employees that are treated with respect and whose opinions are considered a valuable asset to the company's success. The elements of employee empowerment are shown in the following list:[15]

- Employees are empowered to make decisions formerly made by their supervisors.
- Self-directed work teams make most decisions, only escalating to their supervisors when they cannot come to a consensus.
- Supervisors now coach, facilitate, and resolve issues, rather than directing work.
- Employees have the training, experience, and information to facilitate process improvements.
- When problems or errors arise, learning replaces blaming.
- Best practices are shared across the organization.
- Training and development are encouraged on an ongoing basis.

Resilient Supply Chain Management: How Toyota Fared During the COVID-19 Pandemic

While a major benefit of Ohno's JIT approach was to greatly reduce inventory, the primary driver was to eliminate wastes of all kinds by treating workers with respect and empowering them to work intelligently. Unfortunately, many Western manufacturers became enthralled with the prospect of major inventory reductions without developing all the needed disciplines that make those reductions possible.

In 1989, I experienced this problem firsthand when a medical device manufacturer in South Carolina decided to implement JIT by moving all their inventory to the shop floor. Unfortunately, they did so with poor bill-of-material and inventory accuracy. The results were disastrous, creating major inventory write-offs quarter after quarter. It took several months and a shakeup in management ranks to resolve the problems and prevent the company from being delisted from the stock exchange.

During the COVID-19 pandemic, Lean became a convenient target for global commodity shortages. While Lean has resulted in major reductions in inventory levels, it never advocated offshoring strategic suppliers to chase the lowest labor rates, a practice known as labor arbitrage. Lean, as envisioned by Ohno and Toyota, built strong supplier relationships, using local suppliers whenever possible. More recently, Toyota implemented a program to understand its second- and third-tier suppliers. This program allowed Toyota to avoid much of the automotive chip shortage that plagued the industry during the COVID-19 pandemic.[16]

The History of Six Sigma: Bill Smith and Jack Welch

Bill Smith joined Motorola in 1987 after 35 years of experience in engineering and quality assurance. Motorola's leadership was shocked by the poor quality of the new pager product they were rolling out. The leadership recognized the need for a more structured approach to data collection and analysis applying statistical methods.[17] While Smith is acknowledged as the father of Six Sigma, there was very much a senior management team effort to remake Motorola's quality culture and embrace Japan's philosophy of continuous improvement. Allied Sigma, now part of Honeywell, was also an early champion of Six Sigma.

Jack Welch was the legendary CEO of General Electric from 1981 to 2000 and the most famous champion of Six Sigma. Welch followed the lead of Motorola and Allied Sigma to embrace Japan's continuous improvement philosophy and its efforts to improve productivity. GE began its Six Sigma programs in 1995, reporting $12 billion in cost savings over the next five years.

Welch made Six Sigma certification a management priority, requiring most managers to achieve a black belt status or see their careers decline. Six Sigma follows a belt system similar to the martial arts, typically starting with a green belt and progressing to a black belt. Black belts are required to actively work projects and train new black belts. GE also hired master black belts to train and mentor black belts.

There is controversy around Welch and his approach to Six Sigma. Six Sigma's strength is its proven framework to use data analysis to solve complex problems. It is not a strategic management or change management system. While Six Sigma can help reduce defects in a product line, it will not help determine whether the product line should be discontinued. These decisions are typically made at a strategic and enterprise level, above the work level of Six Sigma.

Another criticism of Six Sigma as applied by GE was the onerous and time-consuming nature of green and black belt projects. The projects were so painful that many of those receiving their belts hesitated to apply the framework again. It is a valid success metric to measure the number of green belts who continued to used Six Sigma after receiving their belt. I ran into this problem while supporting a major high-tech firm over a seven-year period. It was very rare to find a green belt willing to apply Six Sigma a second time. Their typical response was that they appreciated learning the tools and processes but felt that following the framework with tollgate reviews unnecessarily delayed projects.

Six Sigma's DMAIC Framework to Fix an Existing Process

DMAIC is an abbreviation that stands for the five phases of a Six Sigma project: Define, Measure, Analyze, Improve, and Control. To be successful, all five steps are followed, typically progressing one step to the next after passing a tollgate review process.

1. **Define.** The first phase of a Six Sigma project is to define a problem, define the proposed solution, determine how success will be measured, identify the resources needed, and determine the amount of time that will be required to complete the project. This should be captured in a project charter, described later in this chapter. Key to Six Sigma's approach is determining the customer for the problem being addressed. Known in Six Sigma parlance as the voice of the customer (VOC), the customer is the person or organization next in a process, not necessarily a company's paying customer. Notice that the Define phase does not attempt to describe the solution. Six Sigma should only be used when the customer is known, and the solution is not known.

2. **Measure.** The second phase is to measure the problem and goal specification. Put simply, the Measure phase is how you define success – how you know you have solved the problem. It should include developing a data collection plan for the process, collecting data from a variety of sources to determine defect types and defect metrics, and then comparing data to feedback from your customer to determine the nature of the problem.

3. **Analyze.** This is the phase where the bulk of the work is done on any Six Sigma project. The goal is to identify and validate all potential root causes of the problem. A good way to begin route cause analysis is to use simple qualitative tools such as Five Whys and Fishbone (Ishikawa) diagrams. Six Sigma applies the principle of causation to look at all problems as governed by the law of cause and effect, that is, to identify how process inputs (Xs) affect process outputs (Ys). The data are then analyzed to discover how large the contribution of each root cause, X, is to the project metric, Y.

   Detailed processes or value stream maps are also good ways to spot bottlenecks or constraints in a process. The easiest way to define the bottlenecks is to compare the duration of the process step versus the work time. The larger the difference, the larger the bottleneck. In the next chapter we discuss a wide variety of tools, from the basic to the complex. The key is to only use the tools you need, otherwise the project will be needlessly delayed.

4. **Improve.** The purpose of the Improve phase is to identify, test, and implement potential solutions to the problem. There are a variety of tools for the Improve phase, such as simple brainstorming sessions, Failure Mode and Effects Analysis (FMEA), and Quality Function Deployment (QFD). There are also complex tools, such as design of experiments (DOE), that may require expert help.

5. **Control.** Historically this has been the most challenging phase of Six Sigma. The purpose is to hand over improvements to customers so that they embrace the new process, which is challenging because Six Sigma typically follows a project with a beginning and end. Once the project ends, teams move on to work other tasks. Rarely is there budgeting to monitor and audit the implemented improvements to verify their "stickiness."

The DMAIC Framework Using Smart Technologies

Smart Technologies offer the means to improve all phases of the DMAIC framework by creating a digital twin of physical operations, automating data collection, and using Big Data analytical tools.

Exhibit 2.9 provides examples of how Smart Manufacturing technologies fundamentally change the way Six Sigma projects are run, improving the outcomes while reducing the effort.

EXHIBIT 2.9 Smart Manufacturing technologies change how Six Sigma projects are run

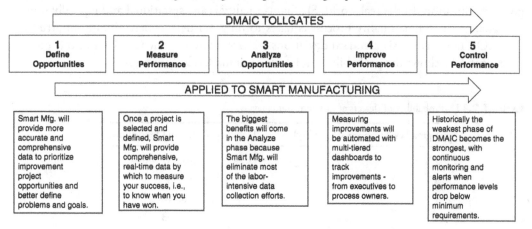

Six Sigma's DMADV Framework to Design a New Process

The great majority of the training most Six Sigma professionals receive is in the DMAIC framework, which is designed to reduce defects and improve processes. DMAIC is less effective in creating new processes where none existed before or in cases where the existing process is too broken to be fixed. This is where Design for Six Sigma, or DMADV (Design, Measure, Analyze, Design, and Verify) should be used. There are two good tools used in DMADV projects to assist in product designs: the Pugh Matrix and Quality Function Deployment, also known as the House of Quality. We will cover both these tools in the next chapter.

Exhibit 2.10 shows the decision process between DMAIC and Design for Six Sigma.

EXHIBIT 2.10 Choosing between DMAIC and Design for Six Sigma

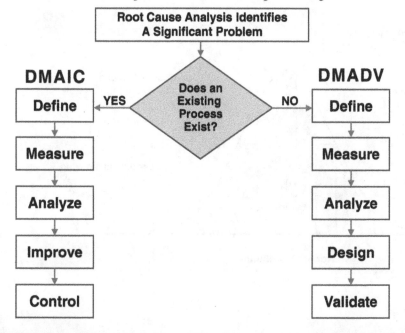

The Statistics Behind Six Sigma

The term *Six Sigma* comes from statistics and specifically refers to normal distribution. In manufacturing, it refers to the ability to produce a very high proportion of output within a given specification. In the real world, Six Sigma quality is an aspirational and typically unrealistic goal. A Six Sigma quality process equates to only 3.4 defects per million opportunities (DPMO). In statistics, six standard deviations are represented by the Greek letter σ (sigma).

Exhibit 2.11 shows the probability of defects at sigma levels 2 through 6.

EXHIBIT 2.11   Probability of defects at different sigma levels

Sigma Level (Process Capability)	Defects per Million Opportunities
2	308,537
3	66,807
4	6,210
5	233
6	3.4

Normal distribution, or Gaussian distribution, is one of the most important statistical distributions used in Six Sigma. Sachin Naik describes a normal distribution as "the data distribution that you get when the data is clustered around the center (mean) of the data and extends towards both sides almost symmetrically. This means that the maximum number of data points are at the center of the range of your data set as compared to both ends of the range."[18]

Exhibits 2.12 and 2.13 provide examples of a normal distribution followed by a normal distribution with standard deviations of 1.0, 1.5, and 2.[19]

EXHIBIT 2.12   A normal distribution curve

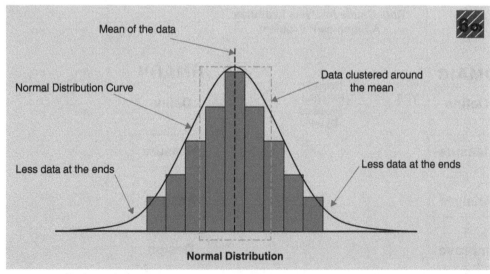

Source: Courtesy of Sachin Naik.

EXHIBIT 2.13   Normal distribution curves with three different standard deviations

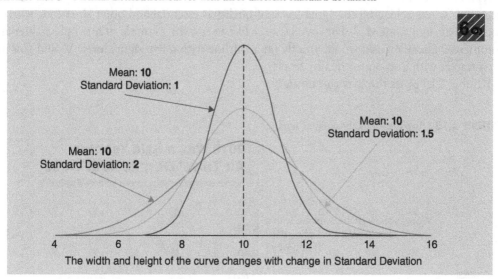

Source: Courtesy of Sachin Naik.

Examples of normal distribution, or bell-shaped, curves are very common in everyday life. For example, bell-shaped curves can be seen in academic test scores, in which the bulk of students receive a score of C with small numbers receiving a score of B or D, and even smaller numbers receiving a score of A or F. Other examples of bell-shaped curves can be found in people's:

- Weights
- Heights
- Blood pressure
- Salaries

In a typical bell-shaped curve, the percentage of data points or observations that fall between a certain number of standard deviations from the mean are as follows:

- 68% of data falls within one standard deviation of the mean.
- 95% of data falls within two standard deviations of the mean.
- 99.7% of data falls within three standard deviations of the mean.

Exhibit 2.14 shows the distribution of observations for one, two, and three deviations.[20]

EXHIBIT 2.14   Distribution of observations

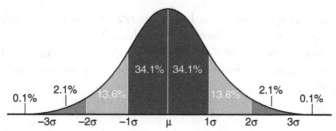

Source: From Statistics How To.

As mentioned earlier, Six Sigma is an aspirational and not necessarily realistic goal. There are situations where achieving six-sigma levels of quality is essential and other situations where it is unrealistic and unneeded. For my classes, I like to use the example of a simple stapler and commercial aircraft. I ask, would you fly on an airline with a five-sigma level? Would you pay for a stapler with a six-sigma quality level?

Exhibit 2.15 poses the two questions.

EXHIBIT 2.15   Two questions about sigma levels

Would you fly on an airplane with a five-sigma level of quality? If not, why?

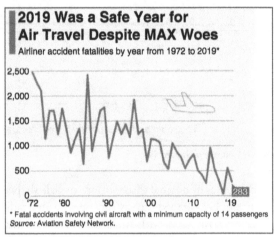

2019 Was a Safe Year for Air Travel Despite MAX Woes
Airliner accident fatalities by year from 1972 to 2019*

* Fatal accidents involving civil aircraft with a minimum capacity of 14 passengers
Source: Aviation Safety Network.

Would you pay for a six-sigma quality level stapler, i.e., one with only 3.4 jammed staples in one million uses or opportunities? If not, why?

View on Amazon

Swingline is essentially the staple king. You have probably seen this classic iteration of a stapler thousands of times. They are the OG of staplers and produce very high-quality products. This one is definitely a bit more heavy-duty than the last and can staple up to 20 sheets.

So still nothing too crazy, but that will cover a lot more of your needs than 10 sheets and it's made out of metal so it just feels sturdier. It also contains a specialized inner rail to help prevent jams.

Most of the students get the point and answer "no" to both questions. In 2018 there was one fatal accident for every 5.58 million flights. This equates to a sigma level of 6.57. If the sigma level dropped to five, there would be 1,000 fatal accidents for every 5.58 million flights.

As for the stapler, my experience is that I get a jam nearly every time I staple more than a few documents. This equates to a sigma level of about three. With an average stapler cost of $20 for a three-sigma quality level, a six-sigma quality level stapler would most likely cost several hundred or thousands of dollars. Even then, it would be some feat for it to only jam 3.4 times for every 1,000,000 uses.

Six Sigma Professionals in the Age of Smart Manufacturing

A major improvement over earlier quality improvement programs was the professionalization of Six Sigma practitioners. Before Six Sigma, quality management programs were run by folks from manufacturing who may have received little formal training and definitely received no certifications. Professionalization came in the way of formal training and progressive levels of belt certifications. Professionalization also included using tollgate reviews for each phase of DMAIC or DMADV projects. Besides training for belts, Six Sigma also provides training for champions who sponsor projects. The belt certifications typically follow this format:

1. **Yellow Belt** – Taking very basic training covering one or two days, passing a short quiz.
2. **Green Belt** – Taking one week of classroom training, completing one project, passing a written exam.
3. **Black Belt** – Completing a green belt certification, mentoring green belts, taking two weeks of training, completing multiple projects over two to three years, passing a written exam.
4. **Master Black Belt** – Completing a black belt certification, mentoring several green and black belts, completing five black belt projects, developing, maintaining, and revising Six Sigma curriculum, and delivering classroom training.

Exhibit 2.16 shows the belt hierarchy.[21] While there are no hard estimates of the number of Six Sigma professionals, LinkedIn Sales Navigator shows over 3 million people listing "Six Sigma" in their profile, 720,000 listing "Six Sigma green belt," and 320,000 listing "Six Sigma black belt."

EXHIBIT 2.16   Six Sigma belts

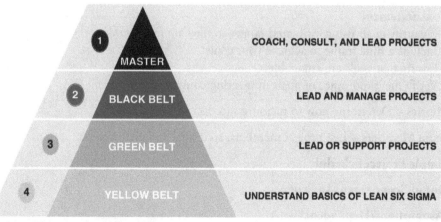

Source: Courtesy of the Six Sigma Global Institute.

Smart Manufacturing provides challenges to and opportunities for Six Sigma professionals. The challenge comes from grasping the many new tools that will help automate the data collection process and learning more advanced analytical tools than spreadsheets. The opportunities come in their enhanced value using Smart Technologies. It sounds counterintuitive, but Six Sigma professionals will be able to shorten analysis times while substantially increasing the amounts of data analyzed.

Six Sigma Project Charters and SMART Goals

Six Sigma follows a structured project management framework. Successful projects have in common a good project charter that is treated as a living document, that is, it is updated as the project progresses. Maintaining a project charter and publishing it on a regular basis is an ideal way to keep all interested parties aware of a project's status. This may seem obvious, but I have seen many projects fail because key stakeholders could never agree on the problem, the goal, and the timing.

By their very nature, Six Sigma projects are difficult in that a solution is not known, requiring data analysis using the scientific method to devise a solution. The process will typically take a minimum of a few months for even a green-belt-level project to several months for black-belt-level projects. It will also require the concerted efforts of a cross-functional team. Without a robust project charter, it is nearly impossible to keep all project members and stakeholders up to date as to its status.

While we will be discussing Lean and Six Sigma project charters, the following examples will apply to most projects. Some of the basic elements of a project charter should include the following:

Problem Statement
- Quantify the problem.
- State how the problem hurts your organization.

Goal Statement
- Quantify goals using estimated ranges so they are measurable.
- Include a time frame to achieve your goals.
- Ensure that goals are attainable.
- Verify the importance of goals in meeting company objectives.

Metrics – Determine how to measure success.

Team Members – List project members, sponsors, and champions.

Simple Project Schedule

Out of Scope versus In Scope

Assumptions/Dependencies

Of these charter elements, the most important are the problem and goal statements. It is important to state problems and goals in quantifiable terms. It is helpful to use ranges when describing goals rather than boxing yourself in with a specific target quantity and date. Updating the problem and goal statements as the project progresses is also valuable. Very few successful projects end without some changes to their original problem and goal statements.

Exhibit 2.17 is an example of a project charter with time-bound, quantifiable goals.

EXHIBIT 2.17   Example project charter

<u>**Problem Statement**</u>

Currently there are **38,612 days per year** associated with preparing and attending meetings for _____ group of 1,100 FTEs, or over **six (6) hours per day in meetings.** There is no root cause analysis as to why meetings consume such a large portion of the workweek.

<u>**Goal Statement**</u>

A root cause analysis to identify the major opportunities to achieve a **20% to 30% reduction in meetings** (in both number and average meeting length), plus more focused communications will **generate savings of over 7,000 days per year by July 31, 202X.**

<u>**Metrics and KPIs**</u>

Meeting root cause analysis = Monthly rate of meetings attended as measured by Outlook calendar
Meeting efficiency = Participation rate (active participants as a percentage)

<u>**Constraints:**</u> None

<u>**Assumptions:**</u> VOC interviews available and questionnaires completed per schedule

<u>**Scope**</u>
In Scope: All meetings and communications for the Product Ops
Out of Scope: Meetings and communications outside of Product Ops

<u>**Project Team**</u>	<u>**Project Timeline**</u>	
1. Project Champion:	Define/Measure Tollgate	~May 31, 202X
2. Project Sponsor:	Analyze Tollgate	~June 30, 202X
3. Project Manager and LSBB:	Improve Tollgate	~July 31, 202X
4. Project Members:	Control Tollgate	~August 31, 202X

Rather than a subjective problem statement complaining of terrible quality, a quantifiable problem statement captures the magnitude of the problem, that is, 5% of routers are returned, costing the company $3 million per year. This also establishes the metric for the project's goals.

It may be a bit confusing to write about SMART goals in a Smart Manufacturing textbook, but it is well worth remembering this acronym. SMART goals are those that are **S**pecific, **M**easurable, **A**chievable, **R**elevant, and **T**ime-bound.

- **Specific.** Rather than saying "I want more production," you could say, "I want to increase production by 10% in the fall quarter. Try using action verbs to focus on what specifically the team needs to do, what the objectives are, what needs to be achieved, who is responsible, and what steps are needed to achieve the goal.
- **Measurable.** How do you measure success in quantifiable terms? Using a range for goals is preferable and including stretch goals is acceptable.
- **Achievable.** It is fine to have stretch goals, but goals must be realistic. It is better to under-promise and overdeliver than the other way around. Many projects fail due to a scope that is too broad. Therefore, it may make sense to take on smaller projects, especially when a new team is coming together for the first time.
- **Relevant.** It makes little sense to solve a problem that is not mission-critical to your organization. Lean and Six Sigma projects consume valuable resources and should only be applied to solve major problems.

- **Time-bound.** Setting time frames for your project is critical to its success. There is an old saying that applies: "a project delayed is a failed project." In larger organizations there are typically constraints to moving quickly on any project. The longer the delays, the less chance for success for the simple reason that keeping teams focused over several months is challenging. Just finding available times to meet with project champions, executive sponsors, and your team members can be a challenge. Therefore, publishing conservative timelines may be necessary.

Lean and Six Sigma Uses of the Scientific Method

Lean and Sigma, along with other quality and process improvement programs, use variations of the scientific method to improve processes. The variations come in the tools and techniques they deploy and their focus. Dr. Jaap van Ede describes the scientific method as follows:

1. Observe processes in an objective manner.
2. Next, formulate an interesting question. For example: What can be done better in the eyes of the customer? Defining the problem is perhaps the most important and difficult part of the process.
3. Then, formulate a hypothesis that may answer your question. True scientists develop a model, as a simplified representation of reality. This model should not only describe the observations but should also predict what happens in the new and improved situation (e.g., what the output will be of a production line after an adaptation). This principle is called falsifiability. A hypothesis is only interesting if it has a predictive value and can therefore also turn out to be incorrect.
4. Next, conduct experiments to test your predictions.
5. Now check whether the hypothesis was correct and if the process in question is improved. Depending on the results, modify the hypothesis as needed. Then go back to the first step.[22]

Summary: Six Sigma's Marriage to Lean

After 40 years, Six Sigma has gained a mixed reputation. For our purposes, Six Sigma is a good framework in which to use data analysis to solve complex problems that are critical to your organization's success. The DMAIC or Design for Six Sigma frameworks have proven their effectiveness over the years, generating hard financial savings. With the addition of Smart Technologies generating massive amounts of data from a wide variety of sources, Lean Six Sigma has a new lease on life.

The problems for Six Sigma began when advocates claimed that it was much more than a project management framework for solving complex problems. Frank Wyatt, in his 2018 article, writes that Six Sigma has "not lived up to the remainder of its wide claims as a stand-alone program for strategy, change management, leadership development, and as a quality and continuous improvement strategy, these weaknesses primarily being traced to Six Sigma's minimal/poor consideration of the human, behavioral, and team-participative aspects of creating and driving sustainable change."[23]

There is a growing consensus that Six Sigma remains the ideal means to reduce defects and variation in manufacturing and distribution operations. There is also consensus that the marriage of Lean and Six Sigma makes a lot of sense as a means to overcome Lean's lack of structure and lack of data and Six Sigma's lack of focus on waste reduction beyond defects.[24] Another benefit of combining Lean and Six Sigma is converting the project mindset of Six Sigma to the continuous improvement journey embedded in Lean's philosophy. Because of its project framework, Six Sigma has a limited ability to transform factory-floor culture. On the other hand, Lean only works when every factory worker embraces a culture of continuous improvement and continuing education. It is now very common to see references to Lean Six Sigma and there have been a diminishing number of references to Six Sigma or Lean alone.

Organizations that have embraced Six Sigma over the past 30 years have reported major cost reductions:

- In the 1990s Allied Signal (now Honeywell) saved $800 million by implementing Six Sigma programs championed by Bill Smith.[25]
- Under Jack Welch's leadership, General Electric began implementing Six Sigma in 1995. After five years, GE reported savings of $12 billion.[26]
- In 2005 Motorola attributed over $17 billion in savings to Six Sigma.[27]
- In 2007, the US Army reported savings of over $2 billion over a year-long period.[28]
- In 2015, GM reported $3 billion in savings by reductions in waste and variation.[29]
- In 2019, 3M reported annual savings of about $1 billion with over 65,000 employees trained in Lean Six Sigma.[30]

Because of the continuous improvement nature of Lean initiatives, there is not as much documentation around cost savings as in traditional Six Sigma projects. Adding Lean to traditional Six Sigma projects helps to end the start-and-stop mentality in which a team is created, works a project, and then disbands. Lean's continuous improvement philosophy would look for a team that would work a problem, solve the problem, and then go on to the next most important problem the company faces. In other words, they no longer work on an ad hoc basis. With their belt certifications, they would ideally be working a series of projects on a full-time basis.

Sample Questions

1. What are advantages from combining Lean and Six Sigma?
   a. Overcoming the project limitations of Six Sigma
   b. Overcoming Lean's lack of a formal framework like DMAIC
   c. Overcoming Lean's lack of continuous improvement
   d. a and b
   e. a, b, and c
2. SMART stands for
   a. Specific, Measurable, Attainable, Relevant, Technical
   b. Specific, Manageable, Attainable, Relevant, Time-Bound
   c. Specific, Measurable, Attainable, Realistic, Time-Bound
   d. Specific, Measurable, Attainable, Relevant, Time-Bound

3. A six-sigma level of quality means there are ___ errors out of 1 million opportunities.
   a. 34
   b. 3.4
   c. .34
   d. 340
4. A fair criticism of Six Sigma is that it should *not* be thought of as a
   a. Change management strategy
   b. Leadership development program
   c. Continuous improvement program
   d. All of the above
5. What are some differences between DMAIC and DMADV?
   a. DMAIC is ideal when designing a new process.
   b. DMADV is ideal for improving an existing process.
   c. DMAIC is a continuous improvement program.
   d. DMAIC was created by GE.
6. Criticisms of Lean as contributing to commodity shortages during the COVID-19 pandemic are
   a. Mostly justified
   b. A convenient excuse for poor sourcing strategies
   c. Not supported given Toyota's ability to avoid major shortages
   d. Fully justified because of Taiichi Ohno advocated offshoring of Toyota's suppliers
   e. B and C above
7. Lean strives to eliminate all types of waste, or *muda*, except in
   a. Transportation
   b. Defects
   c. Overproduction
   d. Training
   e. Transportation
8. Kanbans
   a. Can be in the form of bins, cards, or electronic signals
   b. Are an ideal way to pull inventory through production
   c. Help to prevent overproduction
   d. Can work as a simple visual system
   e. All of the above
9. Six Sigma is a framework for
   a. Data-driven problem solving where the customer is known, and the solution is not known
   b. Continuous improvement
   c. Change management
   d. Data-driven problem solving where the customer is not known, and the solution is known
10. Lean is
    a. A continuous improvement philosophy
    b. A framework for data-driven problem solving
    c. A system that pushes materials through production
    d. A program that only works in manufacturing

Notes

1. Domm, R. (2009). *Michigan Yesterday & Today*. McGregor, MN: Voyageur Press.
2. Ibid.
3. Shutterstock.com. (July 2017). Assembly line of LADA cars B0 platform in automobile factory. Shutterstock ID: 728066197.
4. Piggly Wiggly. https://www.pigglywiggly.com/about-us (accessed August 9, 2021).
5. Ohno, T. (1988). *Toyota Production System*. New York: Productivity Press.
6. Taiichi Ohno. Wikipedia. https://en.wikipedia.org/wiki/File:Taiichi_Ohno.jpeg.
7. Ohno, T. (1988). *Just-In-Time for Today and Tomorrow*. New York: Productivity Press.
8. The history of quick change over (SMED). (March 2014). *All About Lean*. https://www.allaboutlean.com/smed-history/.
9. Shingo, S. (1985). *A Revolution in Manufacturing: The SMED System*. New York: Productivity Press.
10. Push vs Pull wrongly based on make-to-order make-to-stock. *All About Lean*. https://www.allaboutlean.com/push-pull/make-to-oder-stock/ (accessed June 2, 2021).
11. Skhmot, N. (August 5, 2017). The 8 wastes of lean. *The Lean War*. https://theleanway.net/The-8-Wastes-of-Lean.
12. Atollogy. www.Atollogy.com.
13. Ohno, *Toyota Production System*.
14. Kanban bins. Creative Safety Supply. https://www.creativesafetysupply.com/articles/kanban/ (accessed June 5, 2021).
15. A Lean Journey. (December 2012). Top 10 principles of employee empowerment. http://www.aleanjourney.com/2012/12/top-10-principles-of-employee.html.
16. Davis, R. (April 7, 2021). How Toyota steered clear of the chip shortage mess. *Bloomberg/Businessweek*. https://www.bloomberg.com/news/articles/2021-04-07/how-toyota-s-supply-chain-helped-it-weather-the-chip-shortage.
17. Schroeder, R. (2006). *Six Sigma: The Breakthrough Management Strategy Revolutionizing the World's Top Corporations*. Sydney: Currency.
18. Naik, S. (January 28, 2020). Normal distribution for Lean Six Sigma. *Lean Six Sigma Simplified* https://www.lsssimplified.com/normal-distribution-for-lean-six-sigma/.
19. Ibid.
20. Normal distributions (bell curve): definition, word problems. *Statistics How To*. https://www.statisticshowto.com/probability-and-statistics/normal-distributions/ (accessed June 8, 2021).
21. Six Sigma Global Institute. Six Sigma belts explained. https://www.6sigmacertificationonline.com/six-sigma-belts/ (accessed June 15, 2021).
22. van Ede, J. (June 19, 2017). Scientific method core of all improvement methods. *Business-Improvement EU* https://www.business-improvement.eu/worldclass/scientific_method_process_improvement.php.
23. Wyatt, F. (February 8, 2018). The demise of Six Sigma: The right-sizing of a problem-solving methodology. Medium. https://medium.com/business-process-management-software-comparisons/the-demise-of-six-sigma-the-right-sizing-of-a-problem-solving-methodology-4e49b4442bf7.
24. Ibid.
25. JM. (February 13, 2015). Saving from the start with Lean Six Sigma. *Six Sigma Daily*. https://www.sixsigmadaily.com/saving-start-lean-six-sigma/.
26. Six Sigma case study: General Electric (May 22, 2017). *SixSigma*. https://www.6sigma.us/ge/six-sigma-case-study-general-electric/.
27. About Motorola University (December 22, 2005). https://web.archive.org/web/20051222081924/http://www.motorola.com/content/0,,3071-5801,00.html.
28. Leipold, D. (June 4, 2007). Lean Six Sigma Efforts Near $2 Billion in Savings. *US Army*. https://www.army.mil/article/3446/lean_six_sigma_efforts_near_2_billion_in_savings.

29. Tramontana, D. (June 19, 2019). The impact of Six Sigma in the automotive industry. *QAD Blog* https://www.qad.com/blog/2019/01/the-impact-of-six-sigma-in-the-automotive-industry.
30. Scotty. (March 28, 2019). 3M looks to continue to bolster manufacturing with Lean Six Sigma. Pyzedek Institute. https://www.pyzdekinstitute.com/blog/news-blog/3m-looks-to-continue-to-bolster-manufacturing-with-lean-six-sigma.html.

CHAPTER 3

Continuous Improvement Tools for Smart Manufacturing

Anthony Tarantino, PhD

Introduction

Continuous improvement techniques and tools date back centuries. Examples include Qin dynasty emperor Shi Huangdi using interchangeable parts to make crossbows in the first century BCE, Guttenberg's printing press in 1400 that made mass education possible, the Gilbreths' introduction of process charts in 1900 to improve throughput, and Ford's perfection of the assembly line that made cars affordable for millions of Americans in the 1920s. During the Cold War (1950s–1980s) continuous improvement programs became a major focus of the scientific community and the defense industry and have continued to grow in use and acceptance ever since.

W. Edward Deming and Taiichi Ohno championed a major change in process improvement starting in the 1960s. Rather than the traditional break-and-fix mentality in which process improvement efforts were reactive in nature and project-based, continuous improvement became an enterprise-wide philosophy that accepts that improvement is a never-ending journey requiring management commitments, staffing, and budgets.

This chapter highlights many of the more widely used tools and techniques that have proven their effectiveness over the decades. With the advent of Smart Technologies, each of these tools has become more effective in delivering tangible benefits, doing so with less effort and in a shorter time frame.

I have used the large majority of these tools in my projects and in classroom exercises. They are effective in small or large organizations, when used manually or using automated tools. Some of the advanced tools require a focused team effort, but feedback from my students has been consistent that it was worth the effort. The tools and techniques we cover are shown in the following list. Tools 1–12 are listed in a rough order of use in many continuous improvement projects. Tools 13–19 are classic Lean tools. Finally, tools 20–23 are more advanced and require more effort, but are the ones I have found essential in tackling the most challenging problems and opportunities.

Popular Process Improvement Tools and Techniques
1. Voice of the Customer using Smart Technologies
2. Voice of the Customer (VOC) using Net Promoter Score (NPS) surveys
3. Voice of the Customer surveys and interviews using the Delphi Technique
4. Voice of the Customer using the Kano Model to understand customer loyalty factors
5. Brainstorming with an affinity diagram to organize many ideas into common themes
6. Critical to Quality (CTQ) to convert the VOC into measurable and actionable objectives
7. Benchmarking of peer organizations for best practices
8. Process maps, the most basic process improvement tool
9. Value stream maps to identify waste and bottlenecks
10. Root cause analysis using a fishbone diagram with a risk matrix
11. Root cause analysis using Five Whys
12. Pareto chart to prioritize efforts
13. Kanban pull system to use only those materials needed to fulfill orders
14. Poka-Yoke to error-proof operations
15. Five S to maintain a well-organized and spotless shop floor
16. Heijunka to level production, balancing supply and demand
17. Plan-Do-Check-Act (PDCA) to drive continuous improvements
18. Kaizen to drive continuous improvement
19. Setup time reductions using the Single-Minute Exchange of Dies (SMED)
20. Gage repeatability and reproducibility to reduce process variability
21. Failure modes and effects analysis (FMEA) to solve complex problems
22. Pugh Matrix to design optimal products and services
23. Quality Function Deployment (QFD), balancing the VOC with product and service offerings

Voice of the Customer in the Age of Smart Manufacturing

One of the major strengths of the Six Sigma framework is its insistence on identifying the customer for any project and thoroughly understanding their requirements. Known as the Voice of the Customer (VOC), the customer is the person or group next in the process you are attempting to improve. Hearing the customer's voice means meeting their preferences and expectations for a product or service being addressed or the process of your project. There are two types of customers: those internal to your organization and those external to your organization, that is, the paying customer.

The goal of any successful continuous improvement project or initiative, whether it is using a Six Sigma framework or not, is to convert the subjective and emotional nature of the VOC into more tangible and actionable requirements using Critical to Quality (CTQ) tools, covered later in the chapter.

A customer's actual requirements may not be included in their stated VOC comments. For example, the VOC may state that customer service is unacceptable, but the measurable requirements to improving customer service may include picking up incoming calls within 15 seconds and resolving over 90% of service issues on the first call.

There is also an issue with customers demanding conflicting objectives. For example, a car buyer wants a low price but also wants to own the fastest car on the road – that is, sports car tastes on a subcompact car budget. The Quality Function Deployment (QFD), covered later in the chapter, is a good tool to address conflicting customer wants.

The Six Sigma Institute lists the following as traditional sources of the VOC:[1]

- **Surveys.** Questionnaires are sent to current and potential customers. While cost effective, they suffer from low response rates.
- **Interviews.** Personal meetings are conducted with current and potential customers using a standardized list of questions. These discussions should always be treated as confidential. (The confidentiality issue will be covered more fully later in the Delphi technique.)
- **Focus Group.** A group of current customers or potential customers are brought together to meet in person to record their needs and wants.
- **Suggestions.** Solicit feedback from all current and potential customers and provide the means for suggestions to be made anonymously.
- **Observations.** The concept of management by walking around applies here. Whenever possible, randomly visit the customers to observe their suppliers, their processes, and their outputs to their customers.

Exhibit 3.1 shows how verbatim VOC is converted into needs and into more specific, measurable requirements.

EXHIBIT 3.1   VOC converted into needs and requirements

Verbatim	Need	Requirement
"I want the pizza that I ordered."	Right pizza to right person	Accuracy
"I want my pizza when you said it would be here."	Pizza delivered on time as promised to customer	Timeliness
"I want my delivery person to be friendly."	Pizza delivery person is polite	Complaints
"I'm not going to pay a lot for this pizza."	Price is equal to or less than all other pizza providers	Price

Source: Six Sigma Institute.

The VOC is often full of emotions. One common problem is that customers may not speak with one voice, presenting a challenge on deciding what customer pain points to focus efforts on. Customer statements need to be restated into fact-based, performance requirements that continuous improvement efforts can focus on. It is also important to not take the VOC too literally. While customers obviously feel the pain, they may not know the cause of or cure for it. Customers may be describing a symptom or effect of the problem and not the problem itself.

Smart Technologies offer a fundamental improvement in capturing the customer's voice, the importance of which cannot be understated. According to Gartner research, 89% of businesses will compete mainly on customer experience, versus 36% in earlier surveys.[2] According to Bain & Company, companies that excel at customer experience grow revenues 4–8% above their competition's.[3] Smart Manufacturing sources for VOC will automate and simplify the process. These are some of them:[4]

- Customer interviews via videoconference
- Online customer surveys
- Live chat data
- Call center data
- Social media
- Website analytics
- Online customer reviews
- Feedback forms analyzed with artificial intelligence
- Focus groups via video conference
- Online NPS surveys
- The Delphi Technique, using video conferences and video calls

Ian Luck, writing for *Customer Gauge*, argues that the Internet of Things (IoT) will have a major impact on customer service and customer experience. "As a customer, you'll be contributing towards an environment that constantly learns about your buying behavior and your peers, from seamless shopping experiences to lightning-quick solutions from customer service hotlines. And as a business, you'll have an enormous amount of data that helps you consistently provide a better experience for your customers—driving revenue and ramping up your internal measures like Net Promoter Score®."[5]

With the global number of IoT devices forecast to nearly triple from 8 billion in 2020 to 25 billion devices,[6] nearly every physical activity will be digitized, generating usage statistics in real time.

Voice of the Customer Using Net Promoter Score

Net Promoter Score (NPS) is a streamlined survey technique to gauge customer loyalty to companies, products, and services. NPSs possess only one basic question of customer loyalty: "Would you recommend _____ to a friend or to a colleague?" Scores are measured on an 11-point scale, with zero being the lowest score and 10 the highest. The 11-point scale is used to allow for a neutral score of 5 in the middle.

NPS was developed by Bain & Company in 2003 and has grown in popularity, with two-thirds of the Fortune 1000 now using some form of NPS.[7] The rise in popularity is based on its simplicity and its ease of use. It is possible to complete a NPS survey on a smart phone in less than a minute, which is why response rates are so much higher than traditional surveys composed of several questions.

NPS sets a high bar to receive a good customer loyalty rating. Qualtrics describes the three-tiered scale as follows:
- **Promoters** respond with a score of 9 or 10 and are typically loyal and enthusiastic customers.
- **Passives** respond with a score of 7 or 8. They are satisfied with your service but not happy enough to be considered promoters.
- **Detractors** respond with a score of 0 to 6. These are unhappy customers who are unlikely to buy from you again and may even discourage others from buying from you.[8]

Calculating an NPS score is simple: subtract the percentage of detractors from the percentage of promoters. For example, if 20% of respondents are detractors, 20% are passives, and 60% are promoters, your NPS score would be 60 – 20 = 40. Exhibit 3.2 shows the NPS scale.[9]

EXHIBIT 3.2   The NPS scale

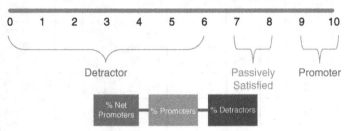

Source: Team Support.

Another reason for the success and popularity of NPS is its use of emoticons or emoji, which is an easy way to overcome language barriers and confusion about the top and bottom of the scale. An early successful application of emoji was by medical professionals when asking patients their level of pain. The use of the graphical images avoided any language issues and worked well with children as their "owie" scale. For business it is a measure of a company's pain when their customers dislike or hate their products and services and of a company's success in winning their loyalty.

Exhibit 3.3 shows the 11 emoji used in a typical NPS scoring system.

EXHIBIT 3.3   The 11 emoji in an NPS scoring system

My experience with NPS started when we introduced it as a painless survey technique to replace an unpopular extended survey used internally and externally to measure software license renewal user experience. Using the extended survey, response rates were as low as 1% or 2%. Using only three questions, we received a response rate of over 19% of the 660 surveys issued using an online survey tool. We did provide two optional questions so respondents could offer a comment and if they wanted to be interviewed. Only the first question was required. Unfortunately for the organization being surveyed, its internal customers were far from satisfied, but they very freely offered advice in how the process could be improved. Using the NPS we were able to interview survey respondents who were the most concerned with the poor service levels and develop a get-well plan.

Exhibit 3.4 is the summary of the survey of license renewal user experience.

EXHIBIT 3.4   License renewal user experience survey results

Survey Group and Leader	Mail List Distribution	Response Rate		Highest Score	Lowest Score	Avg Score	Satisfied		Neutral		Unsatisfied		Std Dev
							Count	Prct	Count	Prct	Count	Prct	
John Negron - Security	259	33	13%	10	0	5.8	6	18%	14	42%	13	39%	3.0
Susan Mitchell softwa	68	32	47%	10	0	5.1	1	3%	14	44%	17	53%	2.6
Karen Manning - Grou	46	4	9%	7	1	3.0	0	0%	1	25%	3	75%	2.7
Karen Manning - Grou	72	13	18%	8	1	2.7	0	0%	1	8%	12	92%	1.9
Karen Manning - Grou	143	27	19%	8	0	3.5	0	0%	3	11%	24	89%	1.9
Doug Merio	21	4	19%	8	1	4.0	0	0%	1	25%	3	75%	3.2
Marcy Blair - PSDM	51	14	27%	7	0	3.1	0	0%	6	43%	8	57%	2.7
Totals and Average	660	127	19.2%	8.29	0.43	4.39	7	6%	37	29%	83	65%	2.71

Voice of the Customer Using the Delphi Technique

The Delphi Technique had its origins in the Cold War, developed by the RAND Corporation and the US Air Force. Its goal was to reach a consensus on complex problems, but to do so in a manner that combined the best outputs from confidential interviews and from team meetings. The name came from the Oracle of Delphi, who was famous with the ancient Greeks for divining the future and consulted before beginning any new undertaking.

Anyone who has attended or facilitated team brainstorming meetings has seen that typically a few strong and outspoken personalities will dominate the conversation as to what solutions should be pursued. This is especially the case if the team includes both subordinates and their supervisors. Challenging one's boss in an open group meeting can be a career-limiting exercise.

The problem with these types of meetings is that they may promote a groupthink in which the dominate personalities force a consensus and minority opinions are discarded or not even expressed.

Groupthink is a type of behavior in which the group reaches a perceived consensus based on the majority view but does not critically think or test an alternative hypothesis. Groupthink is especially common in more homogeneous and cohesive groups, especially ones led by a directional leader. Groupthink discourages minority opinions, especially those that challenge the beliefs of the group. History is filled with examples of groupthink leading to disastrous consequences in which contrary opinions were unwelcomed or punished.

The Delphi Technique is an effective countermeasure to groupthink. The process can work like this:

1. A facilitator provides participants with initial questionnaires or, ideally, conducts individual interviews. The identities of participants are kept confidential from other members of the group.
2. The facilitator then reviews questionnaire answers and comments from individual interviews, filtering out any irrelevant information and identifying common themes.

3. Based on the results of the first phase, the facilitator next creates questionnaires or interview questions for the second round of the process.
4. At this stage, the facilitator may choose to bring the entire group together to review all the results. It may be helpful for the facilitator to rephrase and summarize various responses to help maintain the anonymity of the participants. Participants are encouraged to change their minds as they learn from other participants, which is a major departure from other brainstorming techniques.

Patty Mulder, writing for *Tools Hero*, describes four environments in which the Delphi Technique can be effective in overcoming organizational barriers in solving problems and improving processes:[10]

1. In organizations suffering from poor informal and formal communication, the Delphi's anonymous technique can prevent personal contact, avoiding personality conflicts and arguments. Depending on the skill of the facilitator, it may even improve respect for each other's opinions.
2. In situations where a problem can be solved by bringing together a team of subject matter experts, Delphi encourages everyone's subjective opinions to be captured.
3. In situations where there is excessive subjectivity in suggested solutions, it may help to bring in external subject matter experts to work with an anonymous group, who independently express their views on the problem.
4. In situations where there is a lack of objective data and where inputs from all stakeholders are needed, Delphi can be very effective.

What follows is an example from Expert Program Manager showing how the Delphi Technique can expose an opportunity not found using traditional surveys. It also shows how Delphi can bring about a strong consensus.[11]

In our example, a program manager wants to achieve the most successful product launch in recent memory and do so with a strong consensus from all internal stakeholders. The program manager uses the Delphi Technique to create a survey form for the first round of questions. Respondents can assign a score from 1 to 5, with 5 being the most important. Each respondent's responses are kept anonymous. For this example, the names are shown for clarity, but would be kept confidential in actual practice.

Exhibit 3.5 shows the first-round survey results including their standard deviation.

EXHIBIT 3.5 First-round survey results

Goal	Andy	Alice	James	Peter	Mean	Standard Deviation
Improve development team productivity	1	5	3	2	2.75	1.7078
Provide tiered product pricing	3	4	1	3	2.75	1.2583
Increase the size of the sales team	4	3	4	4	3.75	0.5
Respond rapidly to customer feedback	2	2	2	5	2.75	1.5
Other	5		5			

Source: From Expert Program Management.

The four options the product manager listed received a mixed response. The "Other" option received two strong votes from Andy and James. In the comments section of the survey, Andy and James suggested launching the new product in China. For the next round of questions, the China launch is added to the list of options.

With the addition of the China launch as an option, the voting changes dramatically. Not only does the China launch receive the highest vote, but it receives the lowest standard deviation, demonstrating a strong consensus.

Exhibit 3.6 shows the results of the second-round questionnaire.

EXHIBIT 3.6   Second-round survey results

Goal	Andy	Alice	James	Peter	Mean	Standard Deviation
Improve development team productivity	1	5	3	1	2.5	1.9149
Provide tiered product pricing	3	3	1	2	2.25	0.9574
Increase the size of the sales team	4	2	4	3	3.25	0.9574
Respond rapidly to customer feedback	2	1	2	5	2.5	1.7321
Launch in China	5	4	5	4	4.5	0.5774

Source: From Expert Program Management.

When using the Delphi Technique the ideal outcome is a high median score accompanied by a low standard deviation for the selected option or options. Without low standard deviations, there is no strong consensus.

Using Smart Technologies, the time and effort needed to conduct a Delphi Technique can be substantially decreased. Calls using video conferencing will provide a positive and confidential environment in which participants feel comfortable sharing insights. While onsite team meetings are always preferred, video conferencing is a good alternative. Finally, online survey tools make it simple to distribute questionnaires.

I have used some form of the Delphi Technique in nearly all my continuous improvement projects over the years. It is a great way to capture participants' confidential and candid insights and then use them within a team brainstorming process to come to a strong consensus. My green belt students have also reported the effectiveness of the Delphi Technique. One example proves its value. A large computer hardware manufacturer was experiencing quality issues with its primary contract manufacturer. Meeting with them as a group produced only politically correct responses without any substantive suggestions. The student decided to meet with team members individually, away from their supervisor. The suggestions were surprisingly blunt in exposing inadequate training and vague work instructions. Because the interviews were confidential, and the student was clever in summarizing suggestions in his own words, the participants had no fear of retaliation.

Voice of the Customer Using the Kano Model

The Kano Model is named after Dr. Noriaki Kano, a quality management professor at Tokyo University. In the 1980s Professor Kano developed his model to help understand customer success and loyalty factors. The Kano Model divides product and service features in satisfaction categories along two axes, assuming customer perceptions change over time and that what was initially a delighter will later become an expected feature.

The Kano Model gauges customer satisfaction along a two-axis model. The *x*-axis shows whether the feature a product feature is present or absent. The *y*-axis shows how customers are expected to respond to the product feature, that is, whether they are delighted or dissatisfied. The Kano Model then breaks out product feature categories: **must-have**, **performance**, **delighter**, **indifferent**, and **reverse**.[12]

Exhibit 3.7 shows an example of the Kano Model with its *x*-axis and *y*-axis.

EXHIBIT 3.7   Example Kano Model

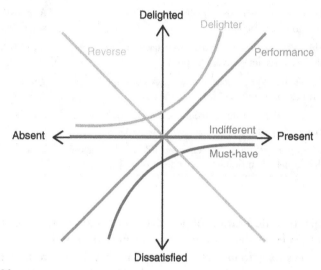

Source: Expert Program Management.

The Kano Model typically breaks product and service features into the following categories:[13]

- **Expected/Must-Have Features.** These are features considered essential and must always be included. For a car it could be a steering wheel. For a computer it could be a screen. These are the customer's basic requirements. Their presence will not delight customers in any way; they are simply neutral.
- **Linear/Performance.** These are features where an increase in their quality will directly increase customer satisfaction. For a car it could be shorter stopping distance. For a personal computer it could be longer battery life. These are known as performance features in that they improve product performance.
- **Exciters/Delighters.** These are features that increase satisfaction levels. Because they are not necessarily expected, their absence is a neutral. For a car it could be a leather-wrapped steering wheel. For a personal computer it could be a free extended warranty.
- **Indifferent/Don't Care.** These are features that have little impact on customer satisfaction. For a car it could be the type of jack in the trunk. For a personal computer it could be the color of packaging material.
- **Reverse/Negative.** These are features customers view as a negative or as a product flaw. They will annoy and frustrate customers when they exist and make them happy when they do not

exist. For a car it could be a navigation system that is too complicated to use. For a personal computer it could be the lack of a warning before it shuts down from low power.

- **Questionable/Mistake.** These are typically caused by misunderstanding or confusion over a survey question.

Exhibit 3.8 describes the six Kano Model categories with examples for each for a car purchase.

EXHIBIT 3.8   Kano Model of a car

Category	Description	Example
Expected	Must-have or basic expectation.	A car with a steering wheel
Linear	Things whose increase in quality directly increases consumer satisfaction.	A car with responsive braking and turning
Exciter	Things whose presence increases satisfaction, but whose absence does not elicit any response.	A car with a butterfly door
Indifferent	Things that do not increase or decrease consumer satisfaction in any significant manner.	The color of the car's A/C vent
Reverse	A feature that consumers (could be a minority, could be a majority) react to opposite to your expectations.	A car with an electronic parking brake
Questionable	When you get conflicting responses, this usually suggests that somebody misunderstood a question in your survey.	A car with less cargo capacity

Source: From What is SixSigma.net.

Smart Technologies provide a variety of new data sources to determine customer satisfaction levels. These include social media, customer reviews posted to online consumer catalogs such as Amazon, and online survey tools such as Survey Monkey. Richer sources of customer sentiment will increase the accuracy and usefulness of the Kano Model.

There are also exciting advancements in using machine learning for creating Kano Models. Writing for ASME's *Journal of Mechanical Design*, Feng Zhou, Jackie Ayoub, Qianli Xu, and X. Jessie Yang advocate for the use of machine learning in developing the next generation of Kano Model. "We propose a machine-learning approach to customer needs analysis for product ecosystems by examining a large amount of online user-generated product reviews within a product ecosystem . . . We applied a rule-based sentiment analysis method to predict not only the sentiment of the reviews but also their sentiment intensity values. Finally, we categorized customer needs related to different topics extracted using an analytic Kano model based on the dissatisfaction-satisfaction pair from the sentiment analysis."[14]

Affinity Diagrams to Organize Many Ideas into Common Themes

The affinity diagram is a popular brainstorming tool used to organize and summarize many ideas around some common themes. While the concept of grouping common ideas together dates back centuries, Jiro Kawakita devised a structured approach in the 1960s that is simple to use.[15]

An affinity diagram can be effective when several improvement ideas are being offered, or when customers are making several complaints about their major pain points. Affinity diagrams can help in creating a project's problem statement and its goal statement. The idea is to take the many and disparate ideas or complaints and group them into a few common themes. It is not unusual for the same idea or problem to be stated in a variety of ways.

One visual way to conduct the exercise is to use cards or sticky notes placed on a whiteboard or wall. Each card represents an input. During the initial phases, it is critical to accept all inputs and not to judge them. Typically a facilitator will meet with stakeholders to begin the grouping process.

Exhibit 3.9 shows how the grouping process works. The first phase is to capture as many as ideas as possible from all stakeholders. The source of the ideas may be kept anonymous if stakeholders are hesitant to participate. Next the ideas are grouped. Finally the groups are given a name for the common ideas they represent.[16]

EXHIBIT 3.9 Affinity diagram grouping

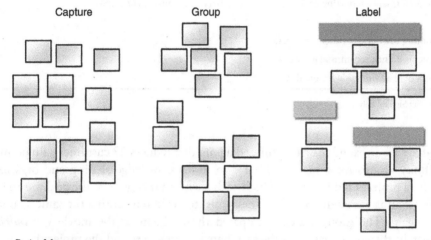

Source: From ProjectManagement.com.

The affinity diagram can be used at any stage of a continuous improvement project. Six Sigma provides the following list of affinity diagram advantages as a brainstorming technique:

- Consensus is reached with less conflict
- Communication is clear and easily understandable
- Every team member contributes
- Ideas are more easily organized into potential solutions[17]

The following affinity diagram example demonstrates its use in winning acceptance in adopting the Six Sigma methodology. The members quickly jotted down a dozen challenges. Exhibit 3.10 is a list of the dozen challenges the team suggested.[18]

EXHIBIT 3.10 An affinity diagram list of challenges

• Cost accounting discourages other measures.	• Performance measures discourage cooperation.
• Culture does not encourage quality at the source.	• Poor cooperation among departments.
• Distrust of "new initiatives."	• Poor opinion of team-based projects.
• Inadequate performance-reporting tools.	• Reward systems do not accommodate teams.
• No current process champions.	• Supervisors resistant to required time to train.
• Operators not well trained in quality.	• Suppliers not held accountable for quality.

Source: From Six Sigma Daily.

Exhibit 3.11 shows how the team organized the dozen challenges into three common themes.

EXHIBIT 3.11 Challenges organized into themes

Management	Training
Poor cooperation among departments.	Operators not well trained in quality.
Performance measures discourage cooperation.	Supervisors resistant to required time to train.
	Culture does not encourage quality at the source.
Poor opinion of team-based projects.	Distrust of "new initiatives."
Inadequate performance reporting tools.	No current process champions.
Systems	
Cost accounting discourages other measures.	
Reward systems do not accommodate teams.	
Suppliers not held accountable for quality.	

Source: From Six Sigma Daily.

Exhibit 3.12 is an example of an affinity diagram that reduces 28 customer pain points down to six affinity groups for a major initiative to foster knowledge sharing and organizational learning for a project I led for a global high-tech client. In this project, the goal was to improve knowledge sharing across dozens of business units to avoid reinventing the same solution over and over again. The group leaders compared the problem to the movie *Groundhog Day*. Unfortunately, there were few incentives to share best practices and the project failed.

EXHIBIT 3.12 An affinity chart that reduces pain points to affinity groups

Pain Point	Cited Causes
Lack of quantitative and qualitative measures	• No review process to determine the scalability of learnings and improvements • No means to validate performance claims made by the business units • No annual review of systemic learning's obsolescence and perishability • Inconsistent quality control and validation in the customer complaint database • Poor understanding and review of the repeatability and scalability of a learning
Inconsistent process	• Lack of ground rules (process documentation) for sharing learnings • Lack of root cause analysis, best practice, and playbook templates • Lack of metadata classifications that work beyond the customer and business unit • Customer complaint prioritization system fails to differentiate catastrophic outages • No template for the product improvement suggestions • Lack of standardized template for customer outages requiring root cause analysis • Lack of customer win and loss analysis for critical customer deals • Lack of triggering mechanism to determine time to share learnings • No effective means to share disruptive innovation beyond the business unit

EXHIBIT 3.12 An affinity chart that reduces pain points to affinity groups (*Continued*)

Pain Point	Cited Causes
Inconsistent policy and governance	• Lack of ground rules (process documentation) for sharing learnings • No procedure to require creation and sharing of root cause analyses • No policy to mandate annual program reviews • Few rules in the sharing of sensitive information outside the business unit
Counterculture	• Unrealistic to expect employee incentive program changes to reward systemic learning • Punitive measures that punish failure to use lessons learned are counterproductive • Few managers require periodic program review and updates • Some directors believe the status quo supports systemic learnings
Lack of metrics	• Lack of success metrics
Fear of sharing	• Services sells intellectual capital and fears internal competition • Advanced services fears providing intellectual property to partners who sell to clients at a lower price • Business units fear sharing that exposes people, process, and technology shortcomings

Source: From Six Sigma Daily.

This example demonstrates the advantage of organizing the common themes into affinity groups. It is far easier to address issues when they fall into three categories, requiring three plans of attack.

Critical to Quality to Convert the VOC to Measurable Objectives

Understanding the Voice of the Customer (VOC) is the starting point for successful continuous improvement projects. As discussed earlier, the VOC is subjective, full of emotion, and can include conflicting wants. As such it is difficult to quantify and measure. Critical to Quality (CTQ) is the tool of choice to convert the VOC into measurable and actionable requirements. These requirements will be how the success of any continuous improvement should be measured.

Exhibit 3.13 lists the steps to convert the VOC to actionable and measurable CTQs.[19]

EXHIBIT 3.13 Flowcvhart for VOC to CTQ

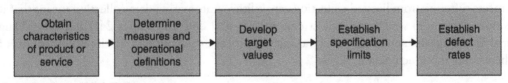

Source: J. DeLayne Stroud, iSixSigma.com.

Stroud defines the operational steps of the process as follows:

• **Product or Service Characteristics:** A few words or a short phrase describing the aspects of product or service. For example, what is the total processing time for major car repairs?
• **Definitions of Quantifiable Measures:** A definition of how the product or service is to be quantified or measured. There can be several areas to define, but it is important to only

measure the significant few that are truly critical to its quality. For example, what is the unit of measure to cover repair time? If the car is kept overnight, are the total hours based on a 24-hour clock or an 8-hour workday clock?

- **Product or Service Target Values:** What is the performance goal for a product or service? For products or services without variation, this is the goal that would always be met. For the car repair example, all repairs would be finished within eight hours of dropoff.
- **Limitations of Specifications:** What is an acceptable variation in meeting customer requirements? For our car repair example, it might be a 15-minute tolerance to meet the eight-hour turnaround goal.
- **Rate of Defects:** For the car repair, what is an acceptable rate of defects that the repair service is willing to accept, and is this defect level acceptable to their customers? This could be the rate of customer complaints per competed service repair, or low scores on customer satisfaction surveys.

Exhibit 3.14 is an example of a Critical to Quality tree converting the VOC to measurable specifications.

EXHIBIT 3.14   CTQ tree

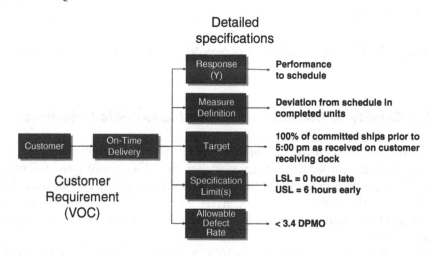

One of the major advantages in bringing Smart Technologies to the CTQ process is the improvement in data quality. Data gathering has traditionally been a manual, labor-intensive process suffering from limited collection capabilities. Automating data collection means that it is now possible to watch and collect data for physical operations on a 24/7 basis. Big Data analytics simplifies and streamlines the analysis process, overcoming traditional complaints of there being too much data to digest.

Types of Data

Data types can be qualitative or quantitative. Quantifiable data can be either continuous or discrete. Quantifiable continuous data is best for data analysis because it can be captured at many different points. Examples of continuous data include temperature, size, time, length, and weight. Another advantage of continuous data is that it be broken down into smaller units of measure, for example, feet to inches, pounds to ounces.

Qualitative data, also known as attribute data, is binary in nature, for example, go or no go, yes or no, good or bad. Attribute data is not useful in data analysis and should be converted to some type of discrete data, in order to be counted. See the following summary of data types.[20]

Attribute Data (Qualitative): Binary with only two values (0, 1)
- Yes/No
- Go/No go
- Pass/Fail

Variable Data (Quantitative)
Discrete (Count) Data: Categorized in a classification and based on counts
- Number of defects
- Number of defective units
- Number of customer returns

Continuous Data: Measured on a continuum with meaningful subdivisions
- Time, pressure, conveyor speed, material feed rate
- Money
- Pressure
- Conveyor speed
- Material feed rate

A common trap in many continuous improvement projects is attempting to measure too many goals. A simple test is to ask whether a metric tells if you have won or lost. In some cases two related, codependent metrics may be needed. For instance, the goal of lowering product costs should be tied to no reduction in quality.

Another trap many continuous improvement projects fall into is to set unrealistic goals. It is best to provide a range of benefits. For instance, a goal of lowering product costs by 5% to 8%, with a stretch goal of 10%, is more realistic than setting a high bar at 10 percent. Project goals should not be set in concrete and updated as the project progresses. Initial goal estimates are just that, estimates. It is only after completing a thorough analysis will a more realistic target be known.

In a later section, we discuss Quality Function Deployment (QFD), commonly referred to as the House of Quality (HOQ). QFD is a powerful tool to convert CTQ outputs into critical parameters of success, score each CTQ against product and service design, and also compare you to your competitors.

Benchmarking

Ironically, many organizations faced with solving major problems do not bother to benchmark their peers to determine whether a solution already exists. Companies in the same industry verticals will undoubtedly face the same operational challenges and try similar solutions to address them.

Benchmarking other industries for best practices also may make sense. For instance, if benchmarking employee onboarding, a function that is not unique to a given industry, it makes sense to benchmark organizations known for their best practices in onboarding.

Benchmarking is a way of discovering what is the best performance being achieved – whether in a particular company, by a competitor, or in an entirely different industry. This information can then be used to identify gaps in an organization's processes in order to achieve a competitive advantage.

Here are some of the reasons benchmarking makes sense:

- Your problem is unlikely to be unique.
- Others have struggled with and found solutions to similar problems.
- Reinventing knowledge is costly.
- You can significantly shorten the problem-solving and improvement process by standing on the shoulders of others.
- Potential solutions may be identified during benchmarking.

Here are some tips for benchmarking peer organizations and competitors:

- Identify a process that is likely to have similar problems to yours. It may come from an organization outside of your industry.
- Do not be shy about requesting to see another company's process. If a company believes they are world-class, they may want to show it off.
- Prepare for the benchmark study and have a prioritized list of areas you would like to observe and questions you would like to ask.
- Scale the benchmarking effort with the project. Your effort should grow with the magnitude of the problem you are attempting to solve.

While benchmarking may be one of the most powerful tools used in a continuous improvement project, it is far from the easiest. It requires specialized skills and practices that some may find unethical, such as role playing in what is known as a mystery shopper. For example, a mystery shopper would contact direct competitors posing as a potential customer. The mystery shopper would then gather data around competitors' service levels.

The good news in conducting benchmarking is the existence of several sources of publicly available data. The growth of social media, online customer reviews, business journals, and investigative reporters all help the benchmarking process. Ironically, the use of publicly available data also dominates the activities of government intelligence agencies, undermining their cloak-and-dagger reputations.

For larger organizations, internal benchmarking can provide valuable insights not available from company reporting. This is especially true for organizations with multiple facilities and multiple business lines. There can be a wide range of service levels, productivity, safety compliance, and employee turnover from location to location or product line to product line.

J. DeLayne Stroud, writing in iSixSigma, describes three types of benchmarks:[21]

1. **Internal benchmarking.** While this may appear to be the simplest method of benchmarking, many organizations operate in silos with little cooperation or sharing across divisions or business units. Internal benchmarking is the means to discover hidden best practices within an organization. It can also be the means to identify parts of the organization that are not operating at desired levels as set by their leadership.

2. **Competitive benchmarking.** This is the means to measure your relative position in the marketplace and to identify industry leaders.
3. **Strategic benchmarking.** This is the means used to identify world-class organizations and their best practices. Often, world-class operations exist outside of your industry and these benchmarks are obtained from outside industries.

Exhibit 3.15 is a benchmarking example of seven peer organizations around their percentage of the day consumed by meetings, emailing, and instant messaging. In this case there is no best practice for most of their peers, with little quiet time available to work.

EXHIBIT 3.15 Benchmarking example

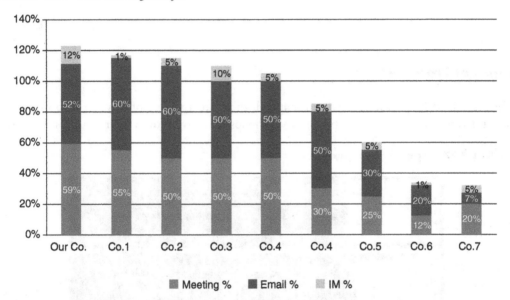

Process Maps

Process maps are the most widely used tool in continuous improvement projects and programs for the simple reason that it is difficult to improve a process without knowing the process steps and the process owners. Process maps, also known as flowcharts, were introduced in the early twentieth century by Frank and Lillian Gilbreth, pioneers in industrial engineering.

Process maps are defined as graphical displays of steps, events, and operations that constitute a process, a structured approach, focused on improving processes to deliver the highest quality and value of products and service to the customer. Process maps can identify value-added and non-value-added steps in the business. They provide a common understanding of the process and help identify key inputs and outputs. Process maps also identify outputs for capability and measurement analysis.

It may be helpful to start the mapping process with a high-level process map that includes only the core steps in a process. This helps in establishing process boundaries and scope and gaining buy-in from stakeholders and leadership.

Exhibit 3.16 is an example of a high-level process map with six steps in the hiring process.

EXHIBIT 3.16 Hiring process map

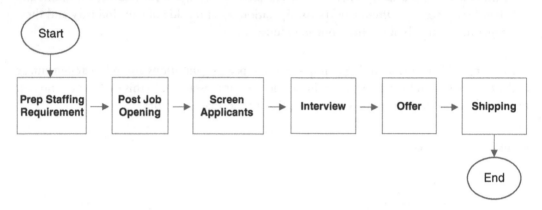

Types of Process Maps

Once the high-level process has been mapped, there are a variety of ways to map in greater detail. Exhibit 3.17 shows six types of process maps that have proven their value for decades.

EXHIBIT 3.17 Types of process maps

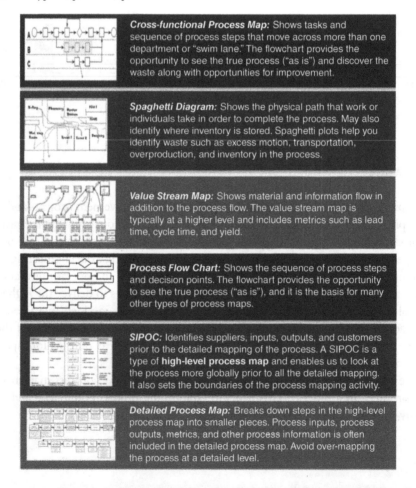

SIPOC

SIPOC is a high-level process map that ties stakeholders to a process. SIPOC stands for **S**upplier, **I**nput, **P**rocess, **O**utput, and **C**ustomer. SIPOC is a good way to demonstrate project improvements by comparing the current-state SIPOC with the projected future-state SIPOC. A SIPOC graphically does the following:

- Correlates supplier-input-process-outputs-customer
- Documents, at a high level, a process from suppliers to customers
- Identifies both internal and external stakeholders
- Records notable requirements of the process inputs and/or outputs
- Discovers potential impacts to suppliers and customers

Exhibit 3.18 is a graphical display of a SIPOC with definitions for each step.

EXHIBIT 3.18   SIPOC steps and definitions

S	Supplier	Provides the inputs to the process
I	Input	Material or data that is transformed in the process step
P	Process	The series of activities that provide an output to the customer
O	Output	Material or data that results from the process step
C	Customer	Receives the outputs of the process

Exhibit 3.19 lists the five steps of a SIPOC. Notice that with this process, you start in the middle and first move to the right and then move to the left.[22]

EXHIBIT 3.19   The five steps of a SIPOC

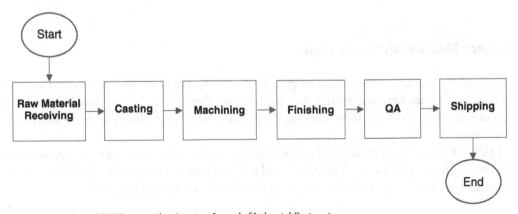

Source: From Mohamed K. Hassan in the *American Journal of Industrial Engineering.*

1. Define a high-level process, typically not more than 10 steps.
2. Identify outputs generated from each step.
3. Identify all customers who consume the process outputs.
4. Identify inputs transformed in each step.
5. Identify all suppliers who provide inputs to the process.

Process Maps with Decision Points

Process maps should include any relevant decision points with a flow for a "yes" or "no" response. Exhibit 3.20 is an example of a process map with a decision point, usually symbolized with a diamond shape.

EXHIBIT 3.20   Process map with a decision point

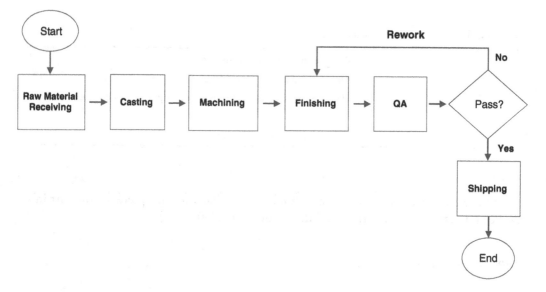

Process Maps with Swim Lanes

Using swim lanes is a good means of graphically displaying participants in a process and their dependencies between groups and individuals. Exhibit 3.21 is a process map using functional swim lanes to show the handoffs from one functional group or individual to another.

Exhibit 3.22 shows an example of a more complex process that crosses several departments or groups. This map is for an IT help desk and includes nine decision points. Without the use of swim lanes, streamlining a complex process like this becomes challenging.[23]

EXHIBIT 3.21 Process map using swim lanes

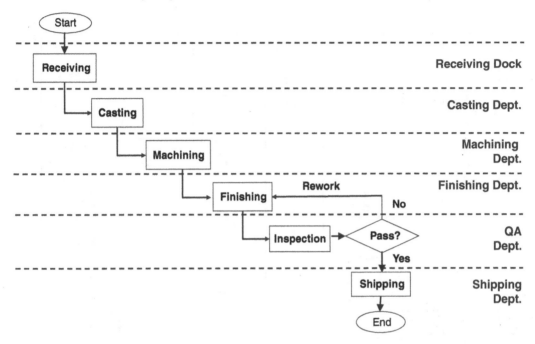

Limited Data Collection and the Hawthorne Effect Impacting Process Mapping

Traditional process maps are created by conducting interviews, holding brainstorming meetings, and through personal observations. The limitations come in process variations such as:

- **Shifts.** Swing and night shifts are often less effective due to a variety of issues including sleep deprivation, reduced supervision, and higher employee turnover.
- **Individual Workers.** There can be major differences based on experience.
- **Days of the Week.** Mondays and Fridays are known to be less effective.
- **Seasons of the Year.** Seasonality has been well documented.

There is another issue with process improvement efforts in general that impacts process and value stream mapping. It is known as the Hawthorne effect, which was first described in the 1950s based on experiments in the 1920s and 1930s at Western Electric's Hawthorne Works in Cicero, Illinois. It refers to the type of reactivity in which individuals modify their behavior in response to their awareness of being observed. Process improvement efforts suffer from the Hawthorne effect, which translates into improvements diminishing once the project ends and observers depart. It is human nature for folks to want to do better when they are being observed and being measured.

EXHIBIT 3.22 Process map with nine decision points

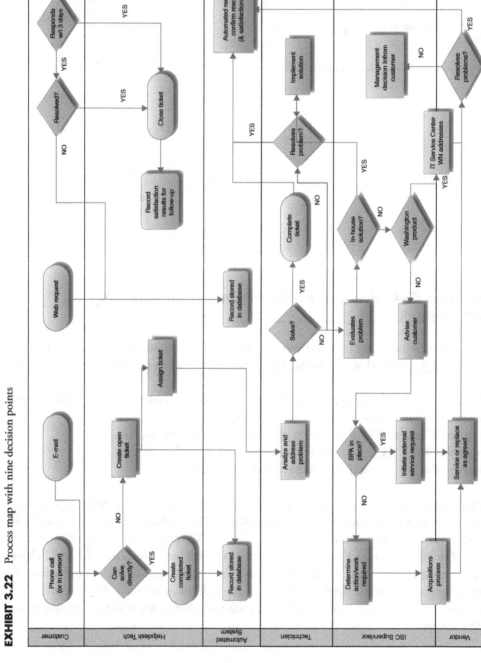

Source: From Cacoo.

Smart Technologies offer a solution to both limited data collection and the Hawthorne effect. Exhibit 3.23 compares traditional manual methods of process mapping with its limited data collection against automated process mapping with data collection on a 24/7 basis.

EXHIBIT 3.23   Traditional versus automated process mapping

TRADITONAL VALUE STREAM MAP   SMART MFG VALUE STREAM MAP

Manually monitor an area of interest with sample sizes as small as 1%.   Automatically monitor an area of interest capturing 100% of transactions over extended duration.

   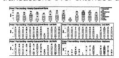

Labor Intensive Data Gathering Struggles to Capture Actual Processes   Automated Data Gathering Generates Analytics & Dashboards

Suffers from the Hawthorne Effect      Helps to Defeat the Hawthorne Effect

One final point about the value of process maps. It has always surprised me about how poorly many organizations understand processes that are important to their operations. In project after project we would begin with whiteboard sessions with experienced stakeholders, those who follow the process in question as part of their daily activity. In group meetings, it was not unusual to find major differences of opinions on how the process worked or for members to change their minds or simply admit they were unsure of the details.

Value Stream Maps to Eliminate Waste

A process map can be converted to a value stream map by adding two elements of time: the cycle time for the process to complete, and the actual work time during the process. Lean's overriding goal is the elimination of waste of kinds, and the simplest way to capture waste is to compare cycle time against work time. Work time is considered value-added. The balance of cycle time is considered waste. I like to use a measure of waste instead of attempting to break out value-added versus non-value-added time.

Exhibit 3.24 shows a simple example of a value stream map. This example is unfortunately typical of many manufacturing processes, in that the process is very wasteful: 3.6 hours of work time across five process steps, with a total cycle time of 20 hours. This equates to a process with about 80% waste. It may seem extreme, but this example was not unusual before Lean and Six Sigma continuous improvement programs became popular.

EXHIBIT 3.24   Simple value stream map

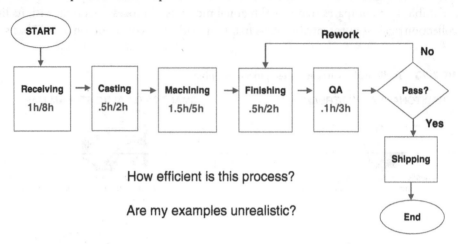

In our example, the receiving department is the largest bottleneck with the greatest opportunity for improvement. Once the bottleneck has been eliminated or greatly reduced, the value stream map process should be repeated. The bottleneck will move, but, interestingly, it may not move where you would expect it to. Once the receiving bottleneck is addressed, it will be interesting to see if non-value added time increases in the casting department, which is the department downstream to receiving. This is why Lean advocates a continuous improvement process – each new improvement will expose other, possibly hidden, bottlenecks.

Value-Added Activity versus Non-Value-Added Activity

The overriding focus of Lean is eliminating waste of all types and in all parts of an organization. This requires an organization to categorize all activities as either value-added or non-value-added. There are also activities that may not be value-added but necessary. Mike Beels, writing for *Michigan Manufacturing Technology Center*, details the differences.[24]

- **Value-Added Activities.** These must meet all three of these criteria:
  1. All work or activities that your customer will pay for
  2. All work or activities that physically transform a product or service
  3. Work or activities completed correctly, the first time
- **Non-Value-Added Activities.** These require work or activities but do not add value to products and services.
- **Necessary Non-Value-Added Activities.** These are more difficult to classify and can include meeting workplace safety or regulatory requirements, and processing engineering change orders.

On the manufacturing floor, there are many opportunities to reduce non-value-added activities. They include reducing the distance between work centers, reducing setup times, and converting from batch processing to making only what the customer has ordered.

Once you leave the manufacturing floor, distinguishing between value-added and non-value-added activities becomes more challenging. It is essential to be sensitive when making declarations as to what is waste outside of the shop floor. No one wants to be told that their efforts are not value-added or are wasteful. Many companies have outsourced many aspects of manufacturing. There are many activities required to manage outsourced manufacturing and classifying their value-add is not a simple matter.

I have found it effective to avoid making value-added versus non-value-added declarations and instead focus on reducing the delta between total cycle time and total work or activity time. This was very helpful when working with manufacturing engineering, product testing, supply chain, and quality assurance. Using a strict definition, inspecting products is not considered value-added. Imagine starting a continuous improvement product in quality assurance by declaring their inspections a waste of time.

Root Cause Analysis Using a Fishbone Diagram and Risk Matrix

The fishbone diagram was developed by Dr. Kaoru Ishikawa in the 1960s as a simple graphical method to search for the root causes of an effect. It is also known as the Ishikawa or Cause-and-Effect Diagram. When completed the diagram resembles fishbones, hence its name. (It also makes it easy to remember the name.)

Exhibit 3.25 shows the cause and the effect bones of the diagram.

EXHIBIT 3.25   Fishbone diagram

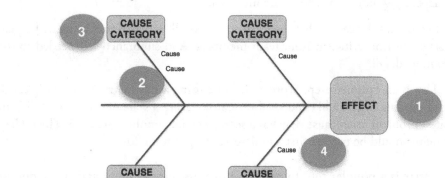

This version of a fishbone diagram follows a four-step process. The fourth step is the addition of a Likelihood and Severity Matrix to quantify the risks/opportunities listed on the diagram.

Exhibit 3.26 shows the basic structure of the fishbone with the problem or effect as the head of the fish, the causes as its ribs.[25]

EXHIBIT 3.26   Basic structure of a fishbone diagram

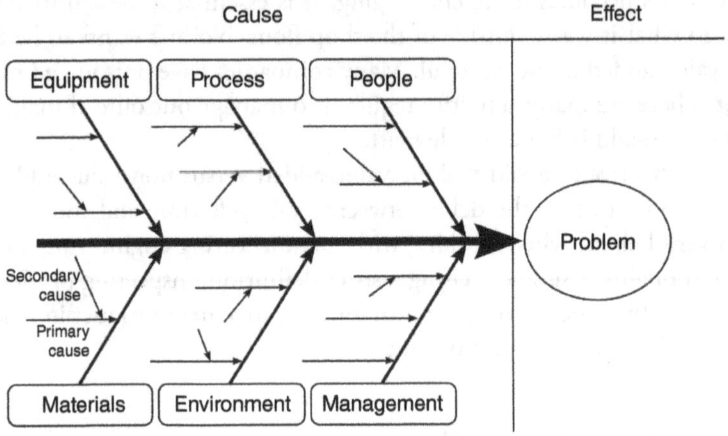

Source: From Lean Six Sigma Definition.

These are the four steps in creating a fishbone diagram:

Step 1. Identify the process effect to be analyzed as part of the continuous improvement effort. This becomes the symbol on the right of the diagram, the head of the fish.

Step 2. Next add the ribs to the fish containing the causes of the effect. The ribs should be divided into categories. The classic categories include materials, people, equipment, method, measurement, and environment. These categories are only suggestions and will change depending on the causes you are investigating.

Step 3. Identify all the causes that the team suggests. All causes should be treated equally at this stage with no criticism from team members. Add subbranches if needed to drill down to more detail.

Step 4. This is an optional step. Score each cause using a risk matrix that multiplies the severity of the cause by the likelihood of the cause. Using a scale of 1 to 5, with 5 being the greatest problem, each cause now has a score. The maximum score is 25. The highest-scoring items should be tackled first and deserve the greatest effort.

The risk matrix is a popular tool to prioritize process improvement and risk reduction efforts. The risk matrix is easy to use and works well for team brainstorming sessions. It is not a quantitative tool, so it is subject to subjectivity among its participants. Exhibit 3.27 shows the matrix as a heat map, with green as the lowest risk and red as the greatest risk.[26]

Exhibit 3.28 shows an example of a simple fishbone exercise to find the causes for a car not starting.[27]

EXHIBIT 3.27   Risk matrix as a heat map

			Impact		
		Negligible	Minor	Moderate	Significant	Severe
Very Likely		Low Med	Medium	Med Hi	High	High
Likely		Low	Low Med	Medium	Med Hi	High
Possible		Low	Low Med	Medium	Med Hi	Med Hi
Unlikely		Low	Low Med	Low Med	Medium	Med Hi
Very Unlikely		Low	Low	Low Med	Medium	Medium

(Likelihood axis runs vertically on left side)

Source: Dan Boers for ARMS Reliability.

EXHIBIT 3.28   Example of a fishbone exercise

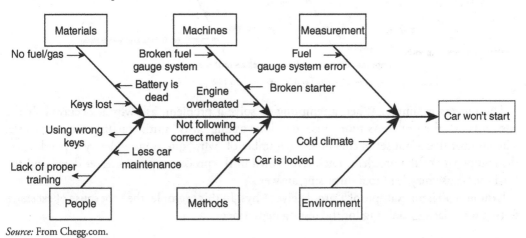

Source: From Chegg.com.

Exhibit 3.29 shows an example of a complex fishbone exercise to find the causes for inefficient supply chain management.[28]

Root Cause Analysis Using the Five Whys

The Five Whys method of root cause analysis was developed by Sakichi Toyota as part of the Toyota Production System. It has been widely adopted in both Lean and Six Sigma programs.

On the surface, Five Whys seems like a simplistic technique with limited value, but it is effective and easy to use in drilling down to root causes for less complex problems. It works because it eliminates symptoms of a problem, one layer at a time, until it reaches the final "why," which is the root cause.

EXHIBIT 3.29 Complex fishbone exercise

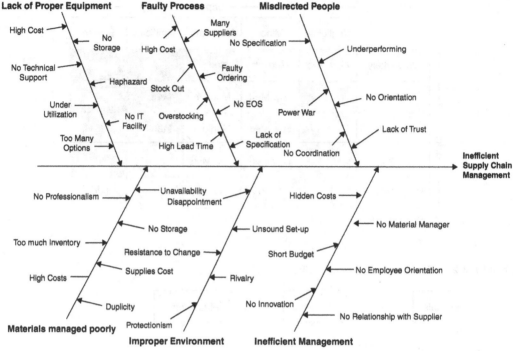

Source: Tarun Kanti Bosean, in *International Journal of Managing Value and Supply Chains.*

The process is simple. When attempting to solve a problem, ask why it occurred. If the answer is not the problem's root cause, continue asking "why" until reaching the root cause. The number five is not set in concrete; the number of "why" questions will vary depending on the complexity of the problem. Each level down in the process exposes greater detail and the final root cause may lead to a surprising answer.

Exhibit 3.30 is an example of a simple Five Whys. In this example, the root cause of excessive testing was a lack of training for the testing department.

EXHIBIT 3.30 A simple Five Whys

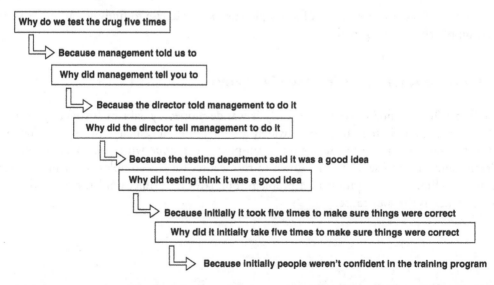

Exhibit 3.31 is a more complex example of a Five Whys. Notice that the root cause of the initial problem turned out to be completely different from what might have been expected. It also turned out to be a process problem, not a technical problem.

EXHIBIT 3.31   A complex Five Whys

<u>**Why we did not send the newsletter on time**</u>

Why didn't we send the newsletter on time? Updates were not implemented until the deadline.

Why were the updates not implemented on time? Because the developers were still working on the new features.

Why were the developers still working on the new features? One of the new developers didn't know the procedures.

Why was the new developer unfamiliar with all procedures? He was not trained properly.

Why was he not trained properly? Because the CTO believes that new employees don't need thorough training and they should learn while working.

Changes Coming to Root Cause Analysis with Smart Technologies

Fishbone diagrams and the Five Whys have been the most commonly used tools in manufacturing for root cause analysis (RCA) for many years. They are considered a good brainstorming technique but primarily a manual process with little automation. That is changing with Smart Technologies using artificial intelligence (AI), specifically machine learning. For example, AI can help formulate predictions as to equipment performance and preventative maintenance requirements.

Automating RCA will help replace subjective inputs to both fishbones and Five Whys with data-driven inputs that are automatically generated. The new data will enable predictive modeling and anomaly detection, which are not possible when only relying on expert options.

Pareto Chart

The Pareto principle is also known as the 80/20 rule and is based on the law of the vital few and the trivial many. The concept was developed in 1906 by Vilfredo Pareto, an Italian economist, who discovered that about 80% of effects come about 20% of the causes. Following Pareto's logic would suggest that 80% of sales are generated by the top 20% of customers, or that 80% of on-hand inventory is made up of 80% of stock keeping units (SKUs). It is remarkable how often the 80/20 rule applies in manufacturing operations.

Joseph M. Juran is credited with adapting Pareto's concept into a popular tool. It is popular because it visually captures the vital few that should be addressed in most continuous improvement efforts. Beyond Lean and Six Sigma, Pareto charts are also one of the seven basic quality assurance tools.

A Pareto chart is typically used as a combination bar graph and line graph. The line is in ascending order, while the bar is in descending order. Each bar represents one cause or factor.

Exhibit 3.32 shows a Pareto chart of late arrivals.[29]

EXHIBIT 3.32   Pareto chart of late arrivals by reported cause

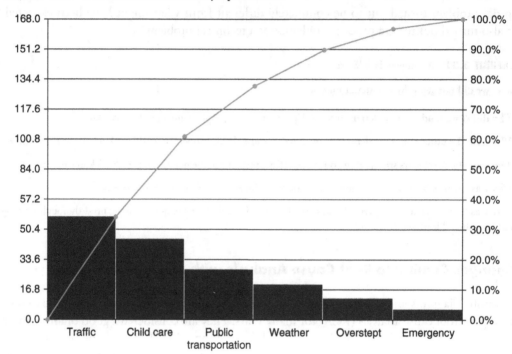

Source: From Six Sigma Daily.

Pareto charts can be generated using a variety of tools:

Spreadsheet programs
- Apache OpenOffice/LibreOffice Calc
- Microsoft Excel

Visualization tools
- ThoughtSpot
- Tableau Software

Secialized statistical tools
- Mini Tab
- JMP
- QI Macros

Exhibit 3.33 is a Pareto chart created using QI Macros that shows that 70% of all bank service calls are for only two types of inquiries. For simplicity's sake several other responses are grouped together as "other." Remember, the goal is to only focus on the vital few, not the trivial many.

Exhibit 3.34 is a Pareto chart created using QI Macros that shows the types of manufacturing defects found in a carton production line. In this example three issues are responsible for two-thirds of all defects.

The Pareto chart makes a very compelling argument for determining the top priorities of a project. Without distinguishing between the critical few and the trivial many, projects can fail by attempting to tackle too many problems, what is sometimes referred to as trying to boil the ocean.

EXHIBIT 3.33 Bank service call Pareto chart

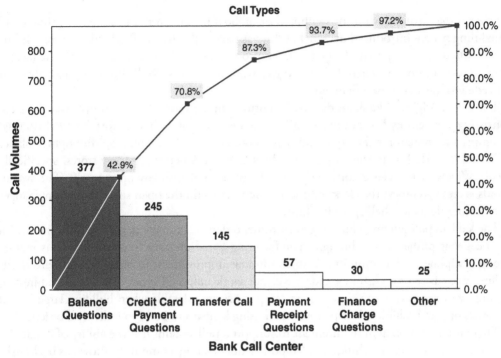

Source: From QI Macros.

EXHIBIT 3.34 Manufacturing defect Pareto chart

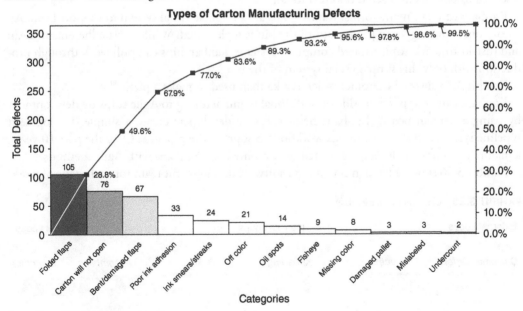

Source: From QI Macros.

Kanban Pull System

Kanban is the Japanese word for signboard or billboard. Kanban is a Lean, just-in-time (JIT) scheduling system introduced by Taiichi Ohno as part of the Toyota Production System in the 1950s. Ohno used a system of cards to track production as it was pulled through the factory floor to meet customer demand. This reduces work-in-process (WIP) inventory to only what is needed to meet customer demand.

The actual origins of kanbans date back centuries in the form of a very simple visual system consisting of an empty box or bin to indicate the need to replenish inventory. The British used an empty box system for the rapid and mass production of their iconic Spitfire fighter planes in World War II. Toyota was inspired by the success of American grocery stores, specifically Piggly Wiggly, where clerks constantly scan shelves to replenish any open spaces. The system avoids excess inventory; the clerk only adds inventory to fill the open space. Based on changes in customer demand, shelf space is adjusted.

The kanban pull system of building to customer orders was a major departure from traditional practices that pushed work through manufacturing based on sales forecasts. Kanbans require strong disciplines and a commitment to continuous improvement in order to work effectively. Pulling work based on customer orders will not work unless all types of waste are addressed, starting with excessive setup times. Long setup times are tolerable when building large lots of materials but are prohibitively expensive when using kanbans to build to individual orders.

There is only one system requirement to support a pull system. It is the ability of a material planning system to use a technique called post-deduct issuing, commonly known as backflushing, to relieve items on a bill of material (BOM). Naturally, this only works with 100% BOM accuracy and timely transactions. Deploying a kanban, just-in-time system without all the essential disciplines in place is doomed to fail.

Kanbans can take many forms. Kanban bins are a popular visual system to support Lean. An empty bin is the trigger to produce until the bin is replenished. While most adherents of Lean and just-in-time use sophisticated computer systems, kanban bins can pull work through production with only this simple visual system of the bin itself.

Exhibit 3.35 shows a customer order as a kanban used to make a pizza.[30]

Kanbans can be applied outside of traditional manufacturing to Agile software development by using a kanban board. Kanbans helps software developers create a simple development pipeline system and the continuous workflow to support the pipeline. Over the past 10 years, a majority of software developers started to combine Lean kanbans with Agile methods.[31]

Exhibit 3.36 shows a kanban board for a software development team for each stage of work.

EXHIBIT 3.35 Customer order kanban

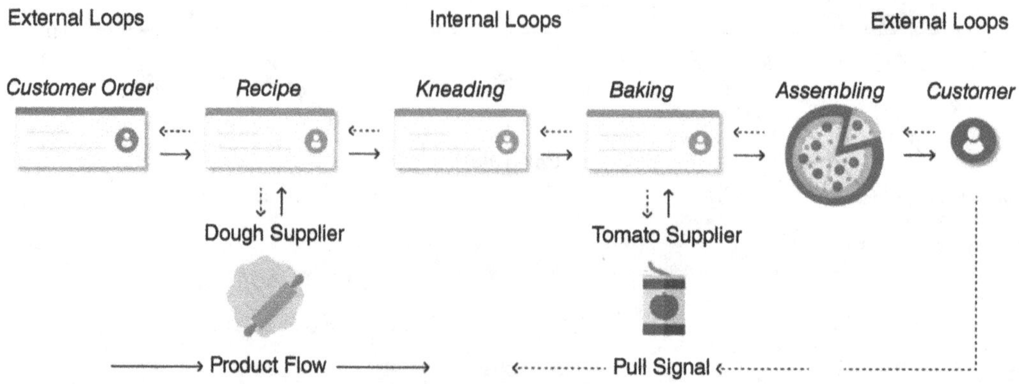

Source: From Kanbanize.

EXHIBIT 3.36 Example kanban board

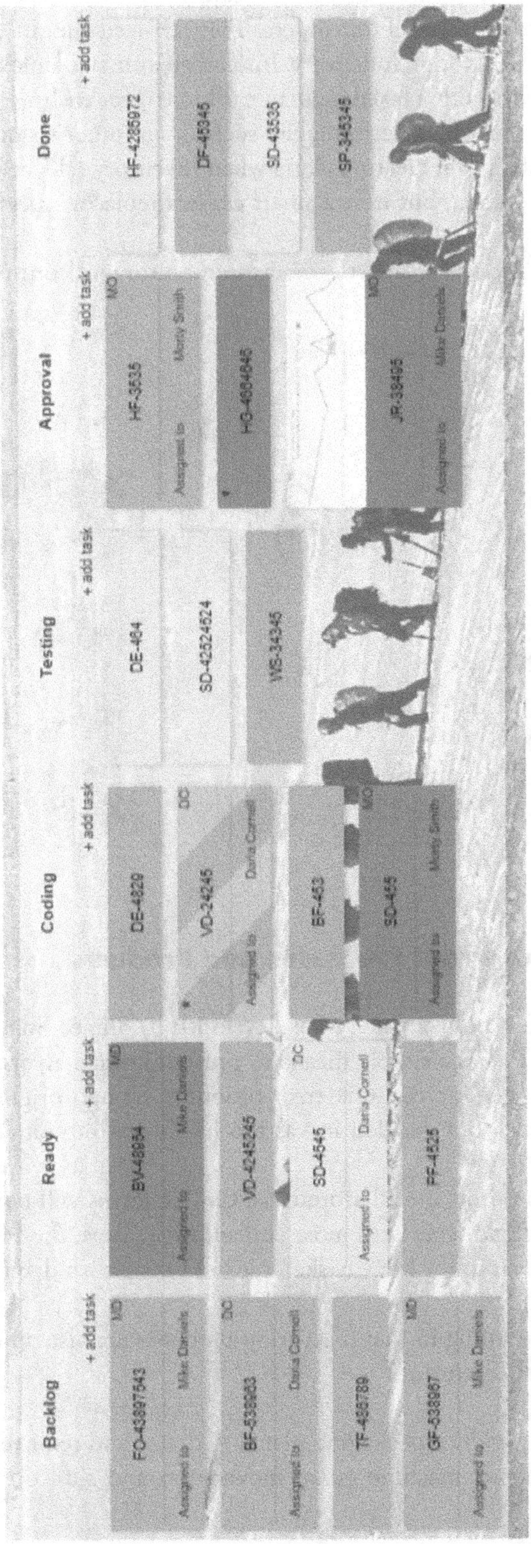

Source: Kanban Tool.

Kanbans can also be electronic. Early in my career, we used a fax machine to send kanban signals to a local supplier to fulfill our orders. This replaced the time and money of creating individual purchase orders for each delivery. Instead an annual blanket purchase order covered contractual requirements with a faxed kanban card to trigger each release.

Today kanbans can use Smart Technologies such as computer vision to monitor the level of inventory in each kanban bin, creating alerts when inventory falls below a designated level. It is still a simple visual system, but now a smart camera replaces a person looking at dozens of kanban bins.

Exhibit 3.37 shows the use of smart cameras to monitor and control kanbans.

EXHIBIT 3.37 Using smart cameras with kanbans

Poka-Yoke to Error-Proof Processes and Products

Poka-yoke (pronounced "poh ka yoke") was developed by Shigeo Shingo as part of the Toyota Production System (TPS) as a simple means to prevent human errors, defects, and variations in a process. Its more generic name is error-proofing. Shingo originally wanted to call the method "idiot proofing," but decided it was too harsh, settling on error-proofing as a more acceptable name.

Examples of error-proofing are all around us. Online forms will not allow a user to submit the form if a required field is missed, most coffeemakers move the fill spout when the top is closed so the water goes into the brew basket, most ATMs return debit cards before dispensing cash, and so on.

Examples of error-proofing in manufacturing processes are also abundant, as shown in the following list from Lean Factories.[32]

- Magnets are used in produce-packaging plants to detect and remove metal contaminants.
- Interlock switches detect machine guard movements and shut off the machine when the guard is lifted.

- Factory light curtains detect when a worker is too near a running machine, switching off the machine to avoid injuries.
- Power guards on machines with moving parts cannot be opened until parts stopped, preventing accidents.
- Machines are set up so that both hands must operate them, preventing a hand getting caught.
- Personal protection articles used in food production are blue in color, so they are easy to detect if they fall into the food.

Poka-yoke, as envisioned by Shingo, aimed to automate manual processes that were prone to errors. Smart Technologies add several opportunities to error-proof manufacturing processes. Computer vision using Edge computing can detect a defect or a safety violation within a second or two and trigger a warning, stop a machine, kick a bad part off the production line, and so on. IOT sensors can detect unacceptable changes in temperature, vehicles moving into restricted areas, forklifts speeding in a warehouse, and so on.

Michael Schrage, writing in the *Harvard Business Review*, notes that "Shingo looked for the simplest, cheapest, and surest way to eliminate foreseeable process errors. To make sure an assembler uses three screws, for example, package the screws in groups of three. The package is a poka-yoke device. Obvious? Perhaps. But 'obvious' is often an underutilized and underappreciated asset . . . and, more importantly, constantly looking for creative ways to minimize mistakes pushes people to rethink the process. That's healthy. It invites innovation."[33]

RNA Automation lists six principles of mistake-proofing that make for a continuous process as Shingo and Toyota envisioned:[34]

1. **Elimination.** Redesign processes and products to eliminate any possibility of an error.
2. **Prevention.** Design and build processes and products that make it virtually impossible to make mistakes.
3. **Replacement.** Replace a process or product with one that is more consistent and reliable.
4. **Facilitation.** Combine process steps to simplify and shorten the work.
5. **Detection.** If unable to prevent errors, identify them before they go to the next process step.
6. **Mitigation.** When unable to completely prevent errors, develop the means to minimize their impact.

Five S

Taiichi Ohno, and Toyota executives Sakichi Toyoda and his son Kiichiro, developed the 5S good housekeeping methodology as part of the Toyota Production System in the 1950s and 1960s. Their visits to the Ford production line revealed a great deal of waste, including workers waiting for assemblies, leading to layoffs. They also noticed that employees were not encouraged or empowered to proactively suggest process improvements.[35]

Toyota implemented 5S with the goal of organizing the production floor to make and keep it standardized, clean, safe, and efficient. The five "s" words to achieve this goal are *sort, set, shine, standardize,* and *sustain* (*seiri, seiton, seiso, seiketsu,* and *shitsuke* in Japanese).

Exhibit 3.38 shows a graphical representation of the 5S methodology.[36]

EXHIBIT 3.38 The 5Ss

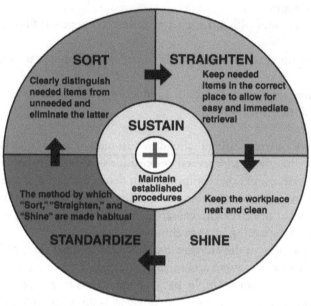

Source: Paul Akers, in *Beyond Lean*.

Matt Wrye describes the traditional method of implementing 5S:[37]

Sort. First workers are encouraged to take the initiative to sort everything out in the workplace in an organized and uncluttered manner, removing anything not needed for production. Seldom-used items are typically moved out of the immediate area. When completed, the workplace is optimized for production with safety hazards removed.

Set. Next workers are encouraged to follow the old adage, "A place for everything and everything in its place." Items should be moved to the location where they are consumed in the process. The goal is that they be located as close as possible to the work to avoid wasted motion. This also makes for a safer workplace.

Shine. Now that messy clutter is eliminated and items are stored in an organized manner, the next step is to clean the workplace and to always keep it clean. This includes machines and other equipment, tools, storage bins, or racks. It is hard for workers to be proud of their workmanship in a dirty workplace. A clean, well-organized, and efficient workplace makes for happier and more motivated workers.

Standardize. This is the step to sustain the sort, set, and shine improvements that have been made. The improvements now become standard operating procedure.

Sustain. We described the Hawthorne effect earlier in the chapter – the tendency for improvements to degrade and potentially vanish once oversight and measurements are ended. This final step in the process is to perpetuate the improvements. Toyota's leadership realized that the only way to achieve this was for workers and their supervisors to take ownership of 5S transformations they helped to make possible. Remember, all these improvements will ideally come from the workers themselves, not from outside consultants or because of harsh management mandates.

As with the control phase of DMAIC Six Sigma projects, sustaining improvements will always be the largest challenge to the 5S methodology. Smart Technologies, especially the use of smart cameras using deep learning algorithms, offer a low touch and automated means to sustain 5S improvements.

Exhibit 3.39 shows the use of computer vision AI to enforce 5S improvements.

EXHIBIT 3.39 Computer vision AI and 5S

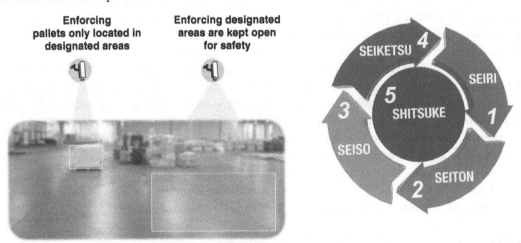

The COVID-19 pandemic could be said to have added a sixth "s" to the 5S methodology. Sort, set, and shine all advocate a safe workplace. With the pandemic, computer vision systems have been installed that flag violations of social distancing requirements.

Exhibit 3.40 shows five levels of social distancing, from a safe work center to a severe violation. Not only is this a violation of social distancing, but it also violates other efficiency and safety guidelines with so many workers so close to working machines.

EXHIBIT 3.40 Five levels of social distancing

Heijunka

Part of the Toyota Production System, Heijunka, or leveling in Japanese, is a popular Lean Six Sigma tool to equalize production levels while meeting customer demand without creating excess inventory. It reduces load and helps manage changes effectively. With this technique, short segments of standardized work are used to meet customer demands.

Heijunka is quite different from traditional line balancing in which standardized batches of production are used to achieve a steady and a level flow of materials. Traditional material requirements planning (MRP) systems build inventory to a forecast by pushing batches of production through the factory. This may balance production, but it creates excess inventory, because production is not matched to customer order requirements.

Lean production, as developed by Toyota, works in a very different and more efficient manner. Lean attacks all types of waste, including excessive setup times, excessive movement, poor quality, and so on. By eliminating waste, Lean is able to build to customer orders in a pull system. The goal is to balance the distribution of production volume and product mix evenly over time. By doing so uneven customer pull is converted into a predictable, stable, and even production process.[38]

Heijunka's production leveling controls the workload rate entering production. It cannot control demand based on customer orders in Lean/JIT systems.

Exhibit 3.41 shows the unevenness of a Lean/JIT production line simply pulling orders through production and a production line with level loading using Heijunka.[39]

EXHIBIT 3.41   Lean/JIT production line with and without leveling

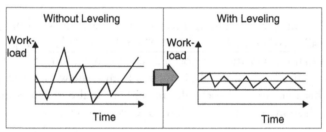

Source: From Shmula.com.

Exhibit 3.42 shows traditional, non-Heijunka, level loading of production balancing the production of three items – A, B, and C. While production is smoothed and stabilized using traditional batch level loading and sales forecasting, it does not meet customer order demand for A, B, and C.

EXHIBIT 3.42   This is *not* Heijunka

Hours	Tuesday	Wednesday	Thursday	Friday
1	A	B	C	A
2	B	C	A	B
3	C	A	B	A
4	A	B	C	B
5	B	C	A	A
6	C	A	B	B
7	A	B	C	A

Source: From Shmula.com.

Exhibit 3.43 shows Heijunka level loading of production while meeting customer order demand for the three items, A, B, and C. Production is smoothed and stabilized without sacrificing customer requirements.

EXHIBIT 3.43   This is Heijunka

Hours	Tuesday	Wednesday	Thursday	Friday
1	A	A	B	C
2	A	A	B	C
3	A	A	B	C
4	A	A	B	C
5	A	B	B	C
6	A	B	B	C
7	A	B	B	C

Source: From Shmula.com.

Heijunka can also be used outside manufacturing in such areas as software development, especially for game developers. Heijunka is used as a specialized Kanban board to support a sequential development process. Its goals are to create a predictable workflow and a parallel completion of work. In this way all team members are kept busy.

Exhibit 3.44 shows a Heijunka kanban board.[40]

Plan-Do-Check-Act

The history of Plan-Do-Check-Act (PDCA) goes back to 1939 with the Shewhart Cycle, which advocated a three-step scientific process to improve operations. The scientific method of hypothesis, experiment, and evaluation is modified for operations to specification, production, and inspection. In 1950, W. Edwards Deming modified the Shewhart Cycle to stress the constant interaction among the four steps of design, production, sales, and research. It was known as the Deming Wheel or Deming's Circle. Toyota used the work of Shewhart and Deming to create a four-step cycle they called PDCA, for **Plan, Do, Check, Act**.

PDCA's four-step process shares similar characteristics with Six Sigma's five-step DMAIC and Design for Six Sigma frameworks. They all follow the basic scientific process steps:

- Defining and planning an investigation to be conducted, agreeing on the measurements
- Performing a data analysis, testing the process, and checking the results
- Implementing the improvements that the data analysis and testing suggests
- Sustaining the improvements made

PDCA has evolved into a popular process improvement tool because it is simple and effective. The PDSA cycle is agnostic and can be applied to any industry as an iterative design and management method. It becomes a continuous improvement tool when iterative loops are followed to improve operations. Once the most immediate problem is solved, you move on to the next most immediate problem.

EXHIBIT 3.44 Heijunka kanban board

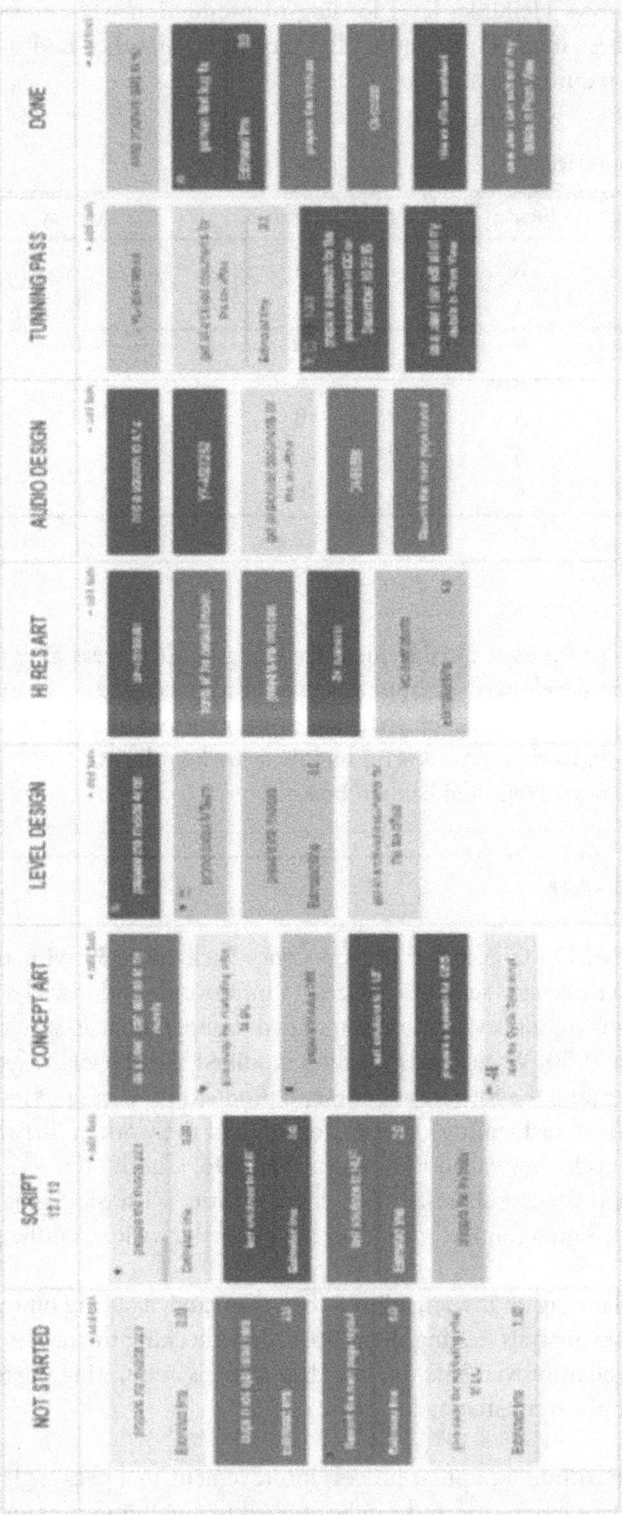

Source: From Kanban Tool.

Exhibit 3.45 shows iterative steps of PDCA.[41]

EXHIBIT 3.45   Steps of PDCA

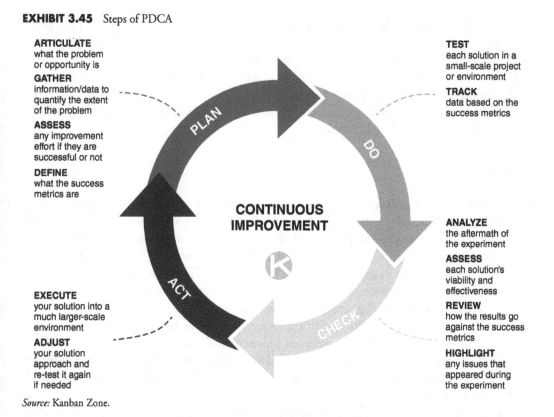

ARTICULATE
what the problem
or opportunity is

GATHER
information/data to
quantify the extent
of the problem

ASSESS
any improvement
effort if they are
successful or not

DEFINE
what the success
metrics are

TEST
each solution in a
small-scale project
or environment

TRACK
data based on the
success metrics

PLAN

DO

CONTINUOUS
IMPROVEMENT

ANALYZE
the aftermath of
the experiment

ASSESS
each solution's
viability and
effectiveness

REVIEW
how the results go
against the success
metrics

HIGHLIGHT
any issues that
appeared during
the experiment

ACT

CHECK

EXECUTE
your solution into a
much larger-scale
environment

ADJUST
your solution
approach and
re-test it again
if needed

Source: Kanban Zone.

The four-step PDCA process typically involves the following:[42]

Plan. This is where a problem or an opportunity is defined with a problem and goal statement quantifying the magnitude of the problem and establishing the metrics to measure improvements. A project charter should be created to show the resources needed, basic timelines of the effort, data sources for the analysis, and any constraints to the project.

Do. This is where each potential solution is tested and evaluated. The most promising solutions may need to be tested further and in greater detail.

Check. This is where the results of the testing are analyzed as to the solution's effectiveness and viability. This is also where the solution is compared against the success metrics established in the Plan phase.

Act. This is where the selected solution is implemented, assuming it is considered viable and worth the effort.

The following tips to help ensure the success of a PDCA initiative are from Bright Hub PM.[43]

• The PDCA team investigates the Voice of the Customer with interviews, surveys, social media, and so on. The customer may be internal or external to your organization.

- The project or initiative leader shows good communication and facilitation skills at all levels of management. They also need to be a good motivator of team members to keep them engaged.
- The project or initiative leader impresses corporate sponsors of the seriousness of the problem being attacked. They also demonstrate a good return on investment (ROI) in terms of cost reductions, safety improvements, or greater customer satisfaction.
- The team benchmarks other organizations to identify best practices inside and outside your industry that can help shape your solution.
- The team implements their improvements on a small test basis, adjusting the solution based on the results before scaling the solution.

Kaizen

Kaizen is the Japanese term for change for better, or improvement. It was developed in Japan after World War II based on the work of Walter Shewhart and W. Edwards Deming, which Toyota applied to its Lean philosophy and PDCA methodology of continuous improvement.[44]

As envisioned by Toyota, Kaizen's philosophy can be combined with PDCA or other process improvement frameworks to drive continuous improvement. The Kaizen way is to take gradual, commonsense steps forward in a never-ending process of saving money, improving quality, reducing accidents, or increasing customer satisfaction.

Kaizen is not designed to drive major process improvement projects or other radical changes. This is a major difference between Japanese and Western thinking. In the West, major and abrupt change is often the preferred method to drive big operational improvements. The problem comes in making them stick, avoiding the Hawthorne-type of backsliding discussed earlier.

Major change efforts also suffer from rapidly changing priorities. In running or supporting over 30 process improvement projects, it was not unusual to be pulled out of one crisis-inspired project to take on the next crisis or grand idea to transform the organization. It came to be known internally as the chasing the shiny ball syndrome. Unfortunately, grand ideas and crisis-inspired projects do little to create a continuous improvement culture.

Kaizen is often associated with a system of events lasting from two to five days. While Kaizen events are called "Kaikaku" in Japanese for radical change, the events typically take on smaller problems that can be resolved in less than a week. As a comparison, even lower-level Lean Six Sigma projects are rarely completed in less than a few months and major projects can take several months to complete.

To be successful, a Kaizen event should have these qualities:

- The event is preceded by a period of detailed preparation.
- The event is well prepared, with a very specific scope and achievable goal.
- All resources are fully dedicated and involved with the event; no outside work is permitted.
- Like a Six Sigma project, the solution is not known but the customer is known.
- The problem is an immediate and pressing one.
- The event is followed by an implementation period of a few weeks.
- The Kaizen ends with the implementation.

There is a good argument for combining Six Sigma with Kaizen. Six Sigma will teach Kaizen leaders and participants the disciplines and tools to define problems and solutions, find the best metrics to measure improvements, and the basics of root cause analysis. Participating in a Kaizen event is a great way for new Six Sigma practitioners to start implementing process improvement before they take on major Lean Six Sigma projects.

Following a seven-step Kaizen cycle can help make process improvement a continuous reality.[45]

1. **Get people involved.** Engage with people closest to the processes that are impacted by the Kaizen. By treating them with respect, as subject matter experts, they will be open to sharing their candid views and make relevant suggestions. This also helps them buy in to the changes under discussion.
2. **Discover problems.** Talk to everyone involved with the processes that are part of the problem under discussion. Combining confidential individual interviews and brainstorming group meetings is a good combination to get candid views and achieve a consensus.
3. **Create solutions.** The same people who help uncover the problems can help create the solutions to those problems. Strive for a consensus as to the optimal solution.
4. **Test solutions.** Implement the chosen solutions with Kaizen participants and as many stakeholders as possible.
5. **Analyze results.** On a periodic basis, check the progress and work to keep the participants motivated and engaged. Determine whether the solution is worth adopting.
6. **Adopt the solution.** If the solution is acceptable, adopt it organization-wide.
7. **Plan for the future.** Look at the opportunities for related improvements the Kaizen exposed.

Setup Time Reduction Using Single Minute Exchange of Dies

Programs to reduce setup times date back to the early twentieth century with the work of Frank Gilbreth who, along with his wife Lillian, was a pioneer in industrial engineering. He addressed many areas to improve manufacturing efficiencies and working conditions. In his book *Motion Study*, published in 1911, he describes methods to reduce setup times. While Henry Ford sought to reduce setup times, there was no structured approach and concerted effort to reduce setup times until Toyota made it part of the Toyota Production System beginning in 1950.[46]

Like all other auto manufacturers, Toyota suffered from extended machine setup or changeover times that could last from two to eight hours. Long lead times made it impractical to run smaller production lots to meet customer demand. While the big American manufacturers such as General Motors made over one million of their Chevrolet cars per year throughout the 1950s, Toyota was only making a few thousand cars annually.

Manufacturers typically used a popular formula to calculate the economic order quantities (EOQ) for items they purchased and items the made. The EOQ dates back to the early twentieth century and attempts to balance ordering or setup costs, carrying costs, and annual demand. Due to long setup times, the EOQ formula would suggest making very large lot sizes.

Exhibit 3.46 is the EOQ formula used for manufactured items.

EXHIBIT 3.46   EOQ formula

$$Q = \sqrt{\frac{2DK}{h}}$$

where

Q = optimal order quantity
D = annual demand quantity
K = fixed cost per order or setup cost
h = annual holding/carrying cost per unit

While the EOQ formula can be a good guide to optimal production quantities, EOQ is not an absolute. There has been an ongoing debate as to how to calculate carrying and setup costs. Holding inventories for extended periods presents risks of potential obsolescence and ties up cash and valuable warehouse space. Large batch quantities were a natural response to long setup times, creating excessive inventory, but ironically not necessarily meeting customer requirements.

On a trip to the United States, Taiichi Ohno was impressed with the rapid setup times achieved by Danly for their stamping presses. As a consequence, Toyota purchased several Danly presses for their Motomachi facility. Toyota also began a long-term, structured program to reduce changeover or setup times following a framework from the US military called ECRS, which stands for **E**liminate, **C**ombine, **R**earrange, and **S**implify.

The results of the Toyota program were dramatic. It took years of effort, but setup times of four to eight hours were slashed to 15 minutes in the 1960s, and then to three minutes in the 1970s. With such short setup times, the EOQ formulas now suggested much smaller production quantities, enabling Toyota to build to customer requirements.

The Toyota program came to be known as the *Single Minute Die Exchange*, or *SMED*, and consists of seven steps that Shigeo Shingo taught to American manufacturers. While not universally applicable to all types of setups, the biggest takeaway is to attack setup times on a continuous, iterative basis. The seven steps are as follows:[47]

1. Observe every detail of the current setup process.
2. Separate internal from external setup activities. (Internal activities can only be performed when operations are stopped. External activities can be performed while operations are running.)
3. Convert as many internal activities to external activities as possible.
4. Streamline and simplify the remaining internal activities.
5. Streamline and simplify the external activities in a way that they resemble internal activities.
6. Document all the procedural changes.
7. Repeat the entire process with the goal cutting times by 20% to 30% with each iteration.

While not mandatory, Smart Technologies can play a role by using computer vision to analyze the current state setup process. Start by measuring the actual time working at the setup area versus the time away from the setup area. It is not unusual to find long periods without any actual setup activities near the machine being changed. I recall watching a setup time video early in my Lean training of an American outboard motor manufacturer. We watched a time-lapsed video showing that the setup person spent much of their time going for tools and parts and very little time directly working on the machine. Just by moving tools close to the machine and organizing them in their order of use on a mobile cart will bring about a quick win in reduction times.

Maybe the best advice in attacking changeover or setup times is to use your imagination to rethink the entire process. For example, it takes about 90 minutes to change four tires on a passenger car. It takes only a few seconds to change four tires on a Formula One race car. This required developing specialized tools and training highly skilled pit crews who follow a precise choreography.

Two final points: While Toyota's focus was on metal fabricators, SMED can be applied to all types of physical operations. Second, "single minute" should not be taken literally, as the goal is to substantially cut setup times to respond to customer requirements, reduce work in process (WIP) inventory, and increase machine utilization rates.

Gage Repeatability and Reproducibility (Gage R&R)

A measurement system is used in the process of associating numbers with physical quantities and phenomena. It can include weight, distance, length, temperature, luminosity, pressure, electric current, and so on. To be accurate, a measurement system must contain accurate and reliable standards of mass and length and agreed-on units. Gage R&R stands for gage repeatability and reproducibility. Gage R&R is a method to assess a measurement system's repeatability and reproducibility:

- A process is **repeatable** if the same person working in the same conditions can repeat a process at other times such as other days of the week, seasons of the year, and so on.
- A process is **reproducible** if another person or group can reproduce a process working in the same conditions, but not necessarily in the same facility or department.

In manufacturing, Gage R&R is the tool of choice to analyze variations of a worker from one day to the next, and variations between various shifts, and variations from one facility to the next. On the surface, working conditions may appear identical between first and second shifts, but history suggests that second shifts suffer from higher employee turnover and reduced levels of supervision. There can be significant changes between two facilities even when using identical equipment run by workers who received the same training and had the same level of experience.

Exhibit 3.47 shows a Gage R&R example of variances across operators and across shifts.

Variation in measurements can be caused by a variety of factors, not all of them obvious. Besides variations caused by different operators and different equipment, variations in the

EXHIBIT 3.47 Gage R&R example

Gage Repeatability and Reproducibility

Parts/Operation (Std=100)

Day Shift	Day	Operator 1	Operator 2	Operator 3	Day Avg	Std Dvd
	Mon	97	96	91	94.7	2.6
	Tue	98	96	94	96.1	1.6
	Wed	102	100	98	100.0	1.6
	Thu	103	101	99	101.0	1.7
	Fri	96	94	92	94.1	1.6
Standard Deviation		2.8	2.6	3.1	SHIFT AVERAGE	97.1
Average		99.2	97.4	94.8		

Swing Shift	Day	Operator 4	Operator 5	Operator 6	Day Avg	Std Dvd
	Mon	88	84	81	84.5	2.8
	Tue	91	91	87	89.8	1.7
	Wed	95	91	88	91.3	3.0
	Thu	98	94	86	92.7	5.0
	Fri	78	75	76	76.3	1.3
Standard Deviation		6.9	6.9	4.5	SHIFT AVERAGE	86.9
Average		90.0	87.1	83.6		

Night Shift	Day	Operator 7	Operator 8	Operator 9	Day Avg	Std Dvd
	Mon	80	76	72	76.1	3.2
	Tue	84	80	76	79.9	3.3
	Wed	86	82	78	81.8	3.4
	Thu	90	86	81	85.6	3.6
	Fri	93	88	84	88.4	3.7
Standard Deviation		4.5	4.3	4.1	SHIFT AVERAGE	82.3
Average		86.6	82.3	78.2		

DAY AVERAGE				
Mon	Tue	Wed	Thu	Fri
85.1	88.6	91.0	93.1	86.3

DAY AVERAGES

SHIFT AVERAGE		
DAY	SWING	NIGHT
97.1	86.9	82.3

SHIFT AVERAGES

working environment or calibration equipment can play a role. A simple example is the decline in productivity as temperatures and humidity levels on the shop floor reach uncomfortable levels. Temperature and humidity changes can also change materials being worked on. Most challenging are variations that occur over extended periods of time such as seasons of the year or between facilities in different regions of the world.

The math behind Gage R&R can be challenging, but there are free Excel templates and many sample sheets to help beginners navigate the process. Dr. Jody Muelaner offers a comprehensive guide, free templates, and examples to follow.[48] QI Macros offers free trials to try out their Excel add-on, which includes instructions, examples, and templates.[49]

Gage R&R can be one of the most powerful tools to improve processes and reduce defects by analyzing all potential sources of variations, but the required data gathering can be challenging, sometimes limiting its effectiveness. For example, what if the major variations in a process occur because of seasonal changes or changes when operating in different regions or countries?

Smart Technologies offer cost-effective and low-touch tools to automate the great majority of data gathering. Rather than an industrial engineer or project participant physically watching operations, or data analysis entering production data from different operators and machines, IOT sensors and AI-based computer vision using smart cameras can record all relevant operators and machines over extended periods of time and operating in different regions or countries.

Failure Modes and Effects Analysis (FMEA) to Solve Complex Problems

Failure modes and effects analysis (FMEA) was one of the first systematic techniques for failure analysis. It was developed by reliability engineers in the 1950s to study problems that might arise from malfunctions of military systems. FMEA involves reviewing as many components, assemblies, subsystems, and processes as possible to identify failure modes (the way something fails), their causes, and their effects. For each component, the failure modes and their resulting effects on the rest of the system are recorded in a specific FMEA worksheet.

FEMA has evolved over the decades and has become one of the most popular and effective means to tackle complex problems, the ones that defy simple root cause analysis. FMEA works because it creates the means to evaluate every potential cause of a failure, a defect, or an ineffective process. It then calculates each mode as to its potential severity if it occurs and the likelihood of it occurring.

FMEA was first used in 1949 by the defense department and by the defense industry. In the 1960s NASA adopted it for the space program. Responding to serious quality issues with its Pinto model cars, Ford Motors introduced FMEA to the auto industry in the 1970s. Today FMEA is used in all industries.

FMEAs can be divided into two categories:

1. **Design FMEA:** When developing new products and services, design FMEA is arguably the most valuable tool to evaluate how they may malfunction, suffer from a limited product lifecycle, or create safety concerns. As to when to start a design FMEA, the answer is to start as soon as possible. Don't wait until you have all the needed information because that may never happen.
2. **Process FMEA:** Once a product or service is in use, process FMEA is very effective in identifying potential failures, quality and reliability issues, customer concerns, complaints, or safety concerns. As with design FMEAs, start process FMEAs as soon as practical. FMEA is an iterative process that will evolve as more data becomes available.

The proper and timely use of both types of FMEA can prevent major disasters. It is far better to catch product design issues with a design FMEA than to miss the design flaw and have to fix the problem with a process FMEA. In the case of the Boeing 737 Max, there has been criticism that a much greater use of design FMEAs using Boeing's advanced modeling software programs could have averted the two fatal crashes that occurred over a five-month period, killing 346 people. Boeing paid out $2.5 billion in settlements but refused to admit that its obsession on profits diminished safety.[50]

FMEAs are not easy nor are they fast to conduct. They require cross-functional teams of subject matter experts and may run over multiple months. But FMEA has proved to be very effective in its ability to shape product or service designs and resolve problems that arise once in production. Requiring no specialized software, FMEAs can be used in all sizes of organizations.

FMEA creates a *Risk Priority Number (RPN)* as follows:

$$RPN = Severity\ Rating \times Occurrence\ Rating \times Detection\ Rating$$

- **Severity.** A rating corresponding to the seriousness of an effect of a potential failure mode. On a scale of 1 to 5, a 1 rating means it has no effect on the customer, and a 5 rating means it has a hazardous effect.
- **Occurrence.** A rating corresponding to the likelihood that the failure will occur. On a scale of 1 to 5, a 1 rating means failure is very unlikely, and a 5 rating means failure is certain.
- **Detection.** A rating corresponding to the ability to detect the failure before it occurs. On a scale of 1 to 5, a 1 rating means detection is certain, and a 5 rating means detection is very unlikely.

Exhibit 3.48 shows an FMEA process flow starting with a process map to identify potential failures.

EXHIBIT 3.48 FMEA process flow

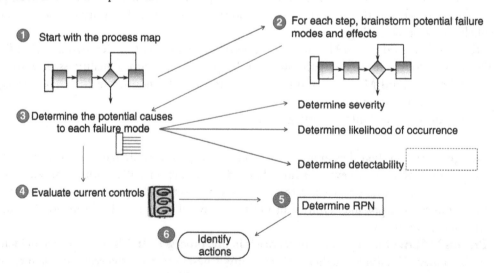

The FMEA process is typically performed using an Excel analysis sheet that calculates the RPN.

Exhibit 3.49 is an example of a FMEA analysis sheet.

EXHIBIT 3.49 FMEA analysis sheet

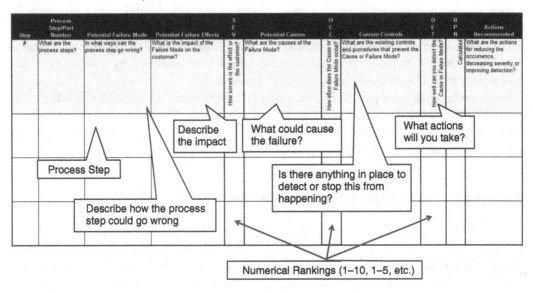

Exhibit 3.50 is an example of a FMEA worksheet used to evaluate the risks surrounding outsourcing. Notice that two items present the largest risks by far.

EXHIBIT 3.50 FMEA worksheet

Risk	Opportunity	Probability	Severity	Risk priority number
Cost				
Unforeseen vendor selection cost	2	4	2	16
Unforeseen transition cost	2	4	2	16
Unforeseen management cost	4	4	3	48
Lead Time				
Delay in production start-up	2	4	4	32
Delay in manufacturing process	5	3	2	30
Delay in transportation of goods	4	2	2	16
Quality				
Minor cosmetic/finishing defect	5	4	1	20
Major cosmetic/finishing defect	5	2	2	20
Component will not fit with mating parts–requiring rework	5	2	4	40
Structural defect–function failure	5	1	5	25

Exhibit 3.51 shows an example of a manufacturing FMEA of the failure modes and effects of bottles on a palletizing conveyor and the recommendations to reduce the RPN scores. Notice the recommendations are manual and do not include any automation. While the RPNs are reduced, they are still fairly high.[51] In this example, running the FMEA again makes sense to explore additional improvement opportunities.

Very few product and service design issues can be solved with a simple root cause analysis. Once in production, few defects or failures are from a solitary cause. For these reasons FMEA will continue to be the tool of choice to solve complex problems. While FMEAs require significant effort, the results justify the commitment. I have been surprised by how many of my students chose FMEA in their projects, even at a green belt level.

Smart Technologies will play an increasingly vital role in performing FMEAs. While FMEA is considered a qualitative tool, it requires massive data sets to effectively analyze all the potential modes and effects of failure. The quality of FMEAs will grow as Smart Technologies are able to digitize all aspects of physical operations.

Pugh Matrix to Design New Processes and Products

The Pugh Matrix is a decision-making tool to formally compare new concepts (processes, services, products) based on customer needs and functional criteria. While many of the tools covered in this chapter can be used to improve existing processes, the Pugh Matrix is a great choice for design engineers and product managers to help design the optimal product or service to serve customer needs.

The Pugh Matrix is sometimes used in combination with Quality Function Deployment (QFD), covered in the next section, as the primary tools for Design for Six Sigma projects. Of the two tools, the Pugh Matrix is easier and faster to navigate. Most Six Sigma training is focused on improving existing processes or eliminating defects in existing products using the DMAIC framework. Therefore, most of them are not exposed to the Pugh Matrix.

The Pugh Matrix was introduced in the 1980s and 1990s by Stuart Pugh, a professor and head of the design division at the University of Strathclyde in Glasgow. The Pugh Matrix was part of Pugh's concept for a total design that he describes as ". . . the systematic activity necessary, from the identification of the market/user need to the selling of the successful product to satisfy that need – an activity that encompasses product, process, people and organization."[52]

The Pugh Matrix or similar versions of it has been widely used because of its simplicity and not being very mathematically intensive. Ironically, the easiest-to-use Pugh has a history of suggesting the same solutions as those derived from mathematically and labor-intensive solutions.[53] It is known by a variety of names:[54]

- Pugh method
- Pugh analysis
- Decision matrix method
- Decision matrix
- Decision grid
- Selection grid
- Selection matrix
- Problem matrix

EXHIBIT 3.51 A manufacturing FMEA

Process Step	Potential Failure Mode	Potential Failure Effects	SEV	Potential Causes	OCC	Current Process Controls	DET	RPN	Actions Recommended	Response	Actions Taken	SEV	OCC	DET	RPN
Palletizing Conveyor	Bottles slip, crash and break on the conveyor while traversing	Production output delayed	4	Lubricate pressure low	4	Pressure control value in place	2	32							
				Lubricant jet blocked	6	Manual checking by operator	8	192							
		Breakage cost	4	Lubricate pressure low	4	Pressure control value in place	2	32							
			4	No collection space around conveyor	9	Design limitations – no space	10	360	Employee additional staff on	Brian	Complete	4	7	10	280
		Injury	10	Employees not wearing gloves	5	Educated, provided gloves. Supervisor to check manually	9	450	Adherence	Kimberly	Complete	10	3	9	270
			10	Lubricant jet blocked	6	Manual checking by operator	9	540	Hourly inspection by maintenance	Brian	Complete	10	6	6	360
			10	Lubricate pressure low	4	Pressure control value in place	2	80							

Source: From Nilakanta Srinavasan.

- Problem selection matrix
- Problem selection grid
- Solution matrix
- Criteria rating form
- Criteria-based matrix
- Opportunity analysis

The Pugh Matrix works well to quickly identify the strengths and weaknesses for a group of potential solutions with the goal of preserving the strength and correcting the weaknesses. By doing this, the optimal solution will emerge, hopefully improving over the initial design ideas.

To be effective, the Pugh Matrix should be run as a cross-functional team effort. Besides the obvious involvement of design engineering and product management, adding quality assurance, marketing, and manufacturing can add valuable insights. Adding the Delphi Technique, with its combination of confidential interviews or surveys and team meetings, may also be helpful.

The process starts by creating a set of design criteria and then creating a group of potential candidate designs. One of these criteria is set as a baseline and may be an existing design. The baseline is compared to the new design options as to whether they are better, worse, or the same. The numbers of times "better" and "worse" appeared for each design are then displayed, but not summed up. Each criterion is given a weighted value of importance, making Pugh a weighted decision matrix.

The following five-step process to use a Pugh Matrix is from the *MSG Management Study Guide*.[55]

> **Step 1 – List the Criteria in a Vertical List.** Begin by listing the criteria that will be used for the evaluation on the left side of a vertical list on either a whiteboard or spreadsheet.
>
> **Step 2 – Select the Datum or Baseline.** The next step is to select the datum or baseline for the initial most feasible solution. Each potential solution will be compared to the datum.
>
> **Step 3 – List the Various Solutions Horizontally.** The next step is to list all the alternative solutions horizontally, creating a matrix with criteria on the vertical axis and the solutions on the horizontal axis.
>
> **Step 4 – Score the Pugh Matrix.** One method of scoring the matrix is by using "+", "–", or "S."
> - "+" means a solution is better compared to the datum.
> - "–" means a solution is worse compared to the datum.
> - "S" means a solution is the same compared to the datum.
>
> **Step 5 – Aggregate the Scores.** The final step is to count the number of "+" and "–" scores for each solution. Scores for each criterion may be weighted.

Since the datum is always assigned a "S" score for all of its criteria, its total score is zero. Any solution with a score over zero is superior to the datum or baseline. Any solution less than zero is worse than the datum or baseline.

Exhibit 3.52 shows an industry example of a Pugh Matrix in which the scores are weighted.[56]

Exhibit 3.53 shows a simple Pugh Matrix exercise using a whiteboard. In this example the digital scrapbook is clearly the superior choice.[57]

EXHIBIT 3.52

Evaluation Criteria	Improvement 1	Improvement 2	Improvement 3	Improvement 4	Improvement 5	Importance Rating
Easy replication	Baseline	(−)	(−)	S	(+)	3
Cost of implementation	Baseline	(−)	(+)	(−)	(+)	9
Speed of implementation	Baseline	(−)	S	S	()	3
Consistent with core competencies	Baseline	(+)	(−)	(+)	(+)	1
Improve customer satisfaction	Baseline	(−)	(−)	(+)	(−)	3
Increase revenue	Baseline	(−)	(+)	(+)	(+)	3
Minimal training needed	Baseline	S	(+)	(−)	S	1
Can use existing infrastructure	Baseline	(+)	(+)	(+)	(+)	9
TOTAL +		4	4	4	5	
TOTAL −		5	3	2	2	
Net value		−1	1	2	3	
Weighted Total +		10	22	16	25	
Weighted Total −		21	7	10	6	
Final Weighted Value of Improvement		*−11*	*15*	*6*	*19*	

Source: From iSixSigma.

EXHIBIT 3.53 Simple Pugh Matrix exercise

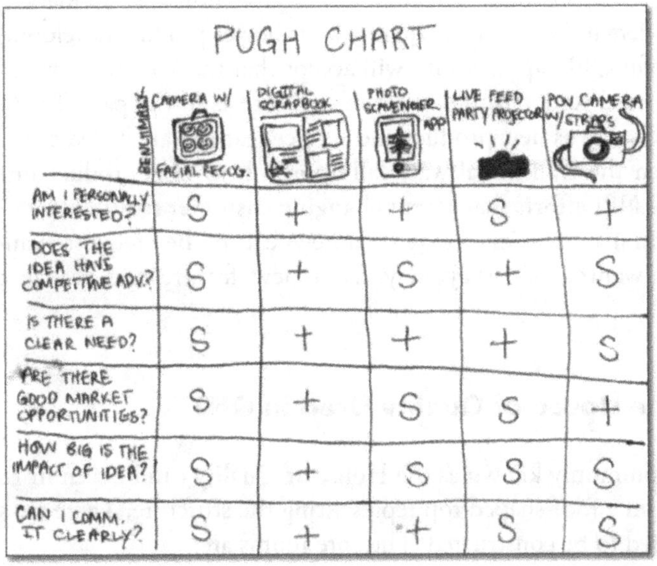

PUGH CHART

	CAMERA W/ FACIAL RECOG.	DIGITAL SCRAPBOOK	PHOTO SCAVENGER APP	LIVE FEED PARTY PROJECTOR	POV CAMERA W/ STRAPS
AM I PERSONALLY INTERESTED?	S	+	+	S	+
DOES THE IDEA HAVE COMPETITIVE ADV.?	S	+	S	+	S
IS THERE A CLEAR NEED?	S	+	+	+	S
ARE THERE GOOD MARKET OPPORTUNITIES?	S	+	S	S	+
HOW BIG IS THE IMPACT OF IDEA?	S	+	S	S	S
CAN I COMM. IT CLEARLY?	S	+	+	S	S

Source: From Yogesh Prasad.

Quality Function Deployment (House of Quality)

History of QFD

Quality Function Deployment (QFD) is a graphical tool that brings together the Voice of the Customer (customer needs) and the Voice of the Design Team (performance requirements). QFD helps to map the two voices to identify the few, critical requirements that will satisfy the most important customer needs. QFD is also called "the House of Quality."

QFD dates back to the late 1960s, when Professors Shigeru Mizuno and Yoji Akao leveraged the work of Dr. Juran, Dr. Ishikawa, and others to make quality assurance integral to manufacturing management. Earlier quality assurance programs were reactive in nature, fixing problems as they arose. Mizuno and Akao sought to design a quality assurance methodology in which customer satisfaction was designed into the product before production began.[58]

The first Japanese organization to use QFD was the Kobe Shipyards of Mitsubishi Heavy Industry. Kobe Shipyards used QFD to replace unwieldy fishbones. QFD evolved to become a comprehensive quality product and service design system. QFD came to America in 1983.[59] I was trained in the House of Quality a few years later when it was presented as a manual whiteboard brainstorming technique. There was nothing like it then, in a period before Just-in-Time, Lean, and Six Sigma took hold. But it was a time when American manufacturing was waking up to stiff Japanese competitors who were making products with quality levels superior to those of their American counterparts.

The wakeup call for the United States and Europe came in 1984 when the results of Toyota's vehicle design success became known. By using QFD over the previous seven years, vehicle development time was reduced by 33% and cost was reduced by 61%, while developing well-accepted cars.[60] The West quickly adopted a simplified version of QFD, which became the tool of choice in designing new products and services, and the backbone of most Design for Six Sigma projects. QFD's use moved beyond automotive manufacturing to all major industries.[61]

QFD has continued to evolve and generate global acceptance. In 2015, International Organization for Standardization (ISO) released ISO 16355, which will standardize and popularize QFDs usage.[62] QFDs value will continue to grow now that all marketplaces are global, and as customers demand ever greater quality and shorter product development cycles.

Today's successful QFD applications will accept that the Voice of the Customer (VOC) is not as straightforward to determine or set in concrete as in the past. Today's customer likes and dislikes will change as new product and service features are rolled out. This is the major difference between the traditional waterfall approach to new product introduction (NPI) and nimbler Agile NPI efforts that accept changing customer requirements. This can happen for a variety of reasons. Customers may try a new feature they requested and realize it is not really what they wanted. Or they may try a new feature and realize they want additional features.

Structure of the House of Quality Used in QFD

Why QFDs are commonly known as the House of Quality can be seen in Exhibit 3.54, with a series of boxes and a roof-shaped top representing the structure. There are seven rooms, but not all of them need to be constructed. The core rooms are:

- Voice of the Customer (VOC)
- Voice of Design (VOD)
- Correlations between VOC and VOD

Even at this basic level a QFD is valuable, but the next level of analysis sets it apart from other tools by adding how competitors meet those needs, weighing their importance, and benchmarking how competitors meet VOD:

- Analysis of competitors
- Importance of VOC
- Benchmark of competitors

One last comparison is saved for the roof of the house. This is the correlation between the VOD and VOC requirements. Customers and designers can ask for conflicting things. For example, the desire for a powerful car engine but one that gets 30 miles per gallon, or the desire for a low car price but a car with all leather seats, GPS, and so forth. Other requirements are complementary, such as a reliable car with a good warranty. Capturing conflicts and inconsistency in customer requirements is a unique advantage in using a QFD.

Exhibit 3.54 shows the seven rooms of the House of Quality.

EXHIBIT 3.54   The House of Quality

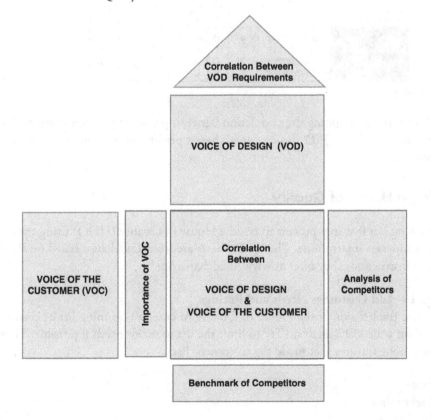

Exhibit 3.55 shows an example of a QFD for selecting a new computer system.[63] Notice the 9, 3, 1 scoring system to respectively indicate a strong, medium, or weak relationship. This is the typical scoring regime in use, but other QFD creators use other regimes.

EXHIBIT 3.55   A QFD example

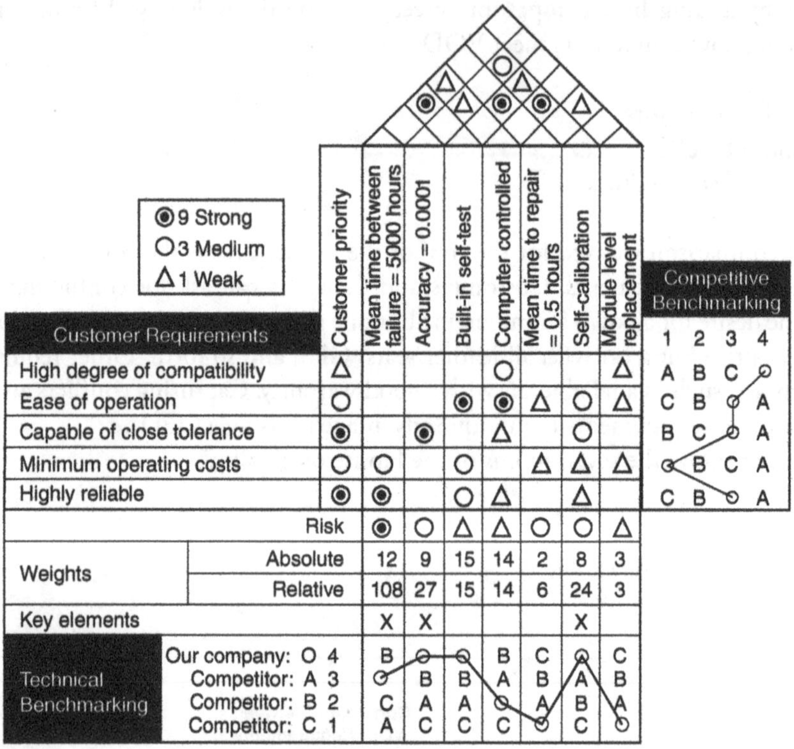

Source: From ASQ.

Besides its matrix mapping the correlation between performance requirements (VOD) and customer needs (VOC), QFD also captures where a performance meets one or more customer requirements.

Building a House of Quality

The following is a five-step process to build a House of Quality (HOQ) using the example of creating a superior smartphone. The QFD charts are from Lucidchart based on the free templates and instructions they offer at www.lucidchart.com.[64]

Step 1 – Add Customer Needs and Ratings.
Use the left side of the House of Quality to enter a beginning list of customer needs based on your VOC analysis. Try to limit the list to major needs if possible. These are the smartphone features that made the customers' list:

- Size
- Lightweight
- Easy to use

- Reliable
- Cheap
- Big screen
- Long-lasting battery
- High-quality camera

Next, each customer need is ranked and rated in terms of its importance. In this example, a scale of 1 to 5 is used, with 5 being the most important. There can be multiple items with the same score. In the column to the right of the rating, the percent of customer importance is calculated.

Exhibit 3.56 shows the importance rating and percent of importance for the eight customer wants.

EXHIBIT 3.56 Customer needs and ratings

	Customer Importance	% of Customers
Size	1	4%
Lightweight	2	8%
Easy to use	3	12%
Reliable	3	12%
Cheap	4	16%
Big screen	3	12%
Long-lasting battery	4	16%
High-quality camera	5	20%

Source: From Lucidchart.

Step 2 – List Design Requirements.

In this step the design product or service requirements are listed horizontally above the relationship matrix. Exhibit 3.57 shows the eight design requirements for the smart phone.

EXHIBIT 3.57 Smartphone design requirements

Weight	Cost of production	Expected life	Operating system	Camera piece	Speaker	Battery	Glass

Source: From Lucidchart.

Step 3 – Weigh the Relationship Between Customer Needs and Design Requirements.

In this step each design requirement is rated as to how it meets customer wants. Exhibit 3.58 shows the three symbols for strong, medium, and weak importance plus the scoring for the eight design requirements.

EXHIBIT 3.58   Importance symbols and requirement scoring

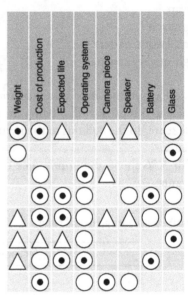

Source: From Lucidchart.

As with the customer needs, each design requirement is given an importance rating and percent of importance for each design requirement. The importance rating is calculated by multiplying the percent of importance rating by the relationship score for each of the customer needs.

Step 4 – Complete the Correlation Matrix.

In this step, the correlation matrix is calculated, helping to determine if design requirements complement or hinder each other. Exhibit 3.59 shows that the correlations are scored from strongly positive, which means they help each other, to strongly negative, which means they hinder or conflict with each other.

EXHIBIT 3.59   Correlation matrix

+ +	Strong positive
+	Positive
–	Negative
– –	Strong negative
	Not correlated

In this step each of the design requirements is marked with either an up arrow or a down arrow. This indicates whether the goal is to increase or decrease the requirement. For the smartphone weight, a down arrow indicates the desire to lower its weight. For the battery, an up arrow indicates the desire to increase battery life. The up and down arrows determine the correlation between different design requirements.

Exhibit 3.60 shows two down and six up arrows in the correlation matrix.

EXHIBIT 3.60 Arrows in the correlation matrix

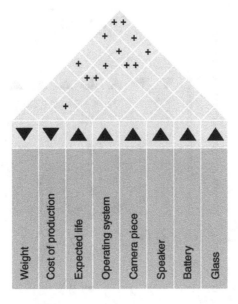

Source: From Lucidchart.

Step 5 – Add Competitor Research.

In the final step, a competitive analysis is made by ranking each of the companies as to how they meet customer needs. While correlation matrix and competitor research will not impact importance ratings, they do provide valuable insights to customer needs and design requirements.

Exhibit 3.61 shows the completed QFD House of Quality.

Benefits of Using QFD

Since its widespread adoption in the 1980s, QFD has provided many benefits to its users. The following summary is from Lucidchart.[65]

Understand Customers. Many times, the VOC is filled with emotion and contradictions. Customers may have a general idea of a potential solution, but they may be unable to provide any specifics. The goal of QFD is to understand customers better than they know themselves.

Predict How Customers Perceive a Product's Value. The success of a new product or service depends on understanding how customers will react to various options.

Obtain Buy-In from Stakeholders. QFD is an ideal tool to survey stakeholders in the design process to gain their acceptance of the selected product or service.

Develop Performance Goals from Customer Wants. Only by detailing performance goals can customer needs be fulfilled.

EXHIBIT 3.61　QFD House of Quality

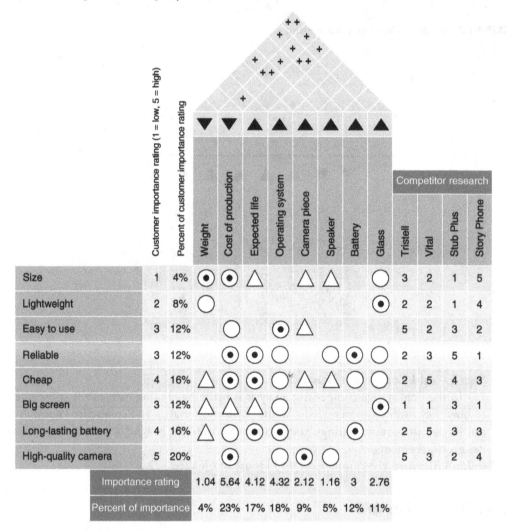

Document Requirements. Each process step in the QFD needs to be documented, especially since QFD is an iterative process that may be repeated over an extended period of time.

Provide Structure. The QFD framework of building the house, one step at a time, will provide the needed structure to run a complex QFD with a cross-functional team.

Prioritize Resources. QFD's scoring system will identify the significant few customer and design requirements to focus on.

Using QFD in Combination with the Pugh Matrix

QFD and the Pugh Matrix share a common goal of matching customer wants with product and service design offerings. Both tools use cross-functional teams to brainstorm customer needs and to match them with potential design options. The goal of both is to select the best product or service offering.

One approach is to use them serially in a two-step process. Start by using QFD to identify and brainstorm the merits of various solutions that satisfy the VOC. This should include applying competitor benchmarking to identify the datum as a baseline against which to compare the various design options. The second step is to use the Pugh Matrix to determine the final design selections. One reason to start with QFD and then finish with Pugh is that Pugh is faster and simpler to use.[66]

Smart Technologies to Automate QFD

While QFD and its House of Quality were envisioned as a manual brainstorming process, efforts to use Smart Technologies, such as machine learning to automate the QFD process, have been underway for many years. Kim E. Stansfield and Freeha Azmat describe the advancements: "Recent significant progress in development of cognitive artificial intelligence (AI) systems such as IBM's Watson system, provides opportunities to gather and analyze significant volumes of market and technological information to support the core objectives of QFD, i.e. aligning new product system design to customer and stakeholder priorities. This involves targeting information sources, refining analysis algorithms to ensure market priorities with associated underlying trends are identified reliably."[67]

Neural networks are modeled on the human brain and the human nervous system. While research in the use of machine learning and AI to automate QFD date back several years, neural networks offer a major advancement in introducing new products and services. Neural networks learn from examples and existing product solutions. They will help optimize and automate QFDs with a targeted analysis of historical customer behavior and in determining what design options will best meet customer requirements. [68]

Exhibit 3.62 shows an example of a neural network model that automates the creation of a QFD.

EXHIBIT 3.62 Neural network model

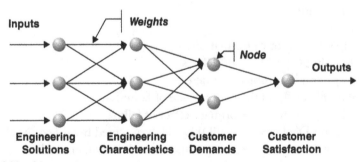

Source: From Stansfield and Azmat.

Summary

In this chapter we presented more than 20 popular process improvements tools and techniques that have proven their value over decades of use in a wide variety of industries and government agencies. Most of the tools were originally designed to be used manually or with simple electronic spreadsheets. As the use of smart technologies increases, each of these tools gains a new lease on life. While many of these tools are qualitative in nature, their value will grow as physical operations are digitized creating new sources of data. When combined, the new data sources will create a much more accurate version of the truth than the informal information sources they replace. The good news is that all these tools are being used by all sizes of organizations and in teams as small as a few people. None of them require advanced statistical or technical expertise.

Sample Questions

1. What are the major differences between a process map and value stream map?
   a. A process map captures work time and total cycle time.
   b. A value stream map captures work time and total cycle time.
   c. A process map always includes swim lanes.
   d. Value stream maps typically take less effort to create than process maps.
2. The growing popularity of Net Promoter Score (NPS) as a survey technique comes from its
   a. Use of a comprehensive list of questions
   b. Ease of use
   c. Use of simple-to-understand emoji
   d. Ability to be completed in a less than a minute
   e. b, c, and d
3. The power of the Delphi Technique comes from its
   a. Exclusive use of confidential surveys
   b. Combining confidential surveys with brainstorming meetings
   c. Exclusive use of brainstorming meetings
   d. Origins from Oracle Corporation
4. A brainstorming tool used to organize and summarize many ideas around common themes is a/an
   a. Kaizen
   b. Kano Model
   c. Affinity Diagram
   d. Critical to Quality
5. FMEA is superior to root cause analysis
   a. In tackling simple problems or defects
   b. In tackling complex problems or defects
   c. Because it only analyzes the major mode of failure
   d. Because it only analyzes the primary effect of failure
6. Quality Function Deployment (QFD) is widely accepted because of its ability to
   a. Create a matrix to compare how products match customer needs
   b. Compares product design options against competitors
   c. Rank customer wants
   d. Rank design options
   e. All of the above
7. A decision-making tool to compare new concepts based on customer needs
   a. Root cause analysis
   b. Pugh Matrix
   c. FMEA
   d. Gage R&R

8. PDCA is a continuous improvement program. PDCA stands for
   a. Prepare, Do, Check, Achieve
   b. Plan, Do, Check, Act
   c. Plan, Do, Check, Analyze
   d. Participate, Do, Concentrate, Achieve
9. An example of _____ is a form that cannot be submitted if a field is left blank.
   a. Kanban
   b. Heijunka
   c. Kaizen
   d. Poka-Yoke
10. Kanbans are effective in
    a. Supporting JIT/pull systems
    b. Reducing work-in-process inventory levels
    c. Replacing push systems and their large lots of production
    d. All of the above
    e. None of the above

Notes

1. Six Sigma. DMAIC Process - Define Phase - Capturing Voice of Customer. The Six Sigma Institute. https://www.sixsigma-institute.org/Six_Sigma_DMAIC_Process_Define_Phase_Capturing_Voice_Of_Customer_VOC.php (accessed June 18, 2021).
2. https://blogs.gartner.com/jake-sorofman/gartner-surveys-confirm-customer-experience-new-battlefield/.
3. https://www.bain.com/insights/are-you-experienced-infographic/.
4. https://medium.com/@infopulseglobal_9037/how-does-data-driven-voice-of-customer-reinvent-cx-and-ux-a5c129aa5d4e
5. Luck, I. (February 14, 2018). How the Internet of Things will revolutionize the world of customer experience. *Customer Gauge*. https://customergauge.com/blog/internet-of-things-revolution/
6. Holst, A. (January 20, 2021). Number of Internet of Things (IoT) connected devices worldwide from 2019 to 2030. Statistica. https://www.statista.com/statistics/1183457/iot-connected-devices-worldwide/.
7. What is NPS? Your ultimate guide to Net Promoter Score. *Qualtrics*, https://www.qualtrics.com/experience-management/customer/net-promoter-score/ (accessed 19 June 2021).
8. What is NPS? Your ultimate guide to Net Promoter Score. Qualtrics. https://www.qualtrics.com/experience-management/customer/net-promoter-score/.
9. Team Support. https://www.teamsupport.com/blog/customer-success-net-promoter-score-nps.
10. Mulder, P. (2017). Delphi Technique. Tools Hero. https://www.toolshero.com/decision-making/delphi-technique/.
11. The Delphi Technique/Method. Expert Program Management. https://expertprogrammanagement.com/2011/03/the-delphi-technique-method/.
12. The Kano Model. (2020). Expert Program Management. https://expertprogrammanagement.com/2020/11/the-kano-model/.
13. The Kano Model: Practical applications. What is Six Sigma.net. https://www.whatissixsigma.net/kano-model-practical-applications/.
14. Zhou, F., Ayoub, J., Xu, Q., and Yang, X. (January 2020). A machine learning approach to customer needs analysis for product ecosystems. *Journal of Mechanical Design* 142: 011101-2–011101-13. https://deepblue.lib.umich.edu/bitstream/handle/2027.42/153965/A%20Machine%20Learning%20Approach%20to%20Customer%20Needs%20Analysis%20for%20Product%20Ecosystems.pdf?sequence=1.

15. Affinity Diagram – Kawakita Jiro or KJ Method. (February 1, 2017). Project Management.com. https://project-management.com/affinity-diagram-kawakita-jiro-or-kj-method/.

16. Ahmed, A. (November 18, 2020). Affinity Diagram. ProjectManagement.com. https://www.projectmanagement.com/contentPages/wiki.cfm?ID=347898&thisPageURL=/wikis/347898/Affinity-Diagram#_=_.

17. 6Sigma. (March 22, 2017). Use of affinity diagram as a brainstorming tool. Lean Six Sigma Certification. https://www.6sigma.us/six-sigma-articles/affinity-diagram-as-brainstorming-tool/

18. The Affinity Diagram Tool. (December 25, 2012). Six Sigma *Daily*. https://www.sixsigmadaily.com/the-affinity-diagram-tool/.

19. Stroud, J. Defining CTQ outputs: A key step in the design process. iSixSigma. https://www.isixsigma.com/methodology/voc-customer-focus/defining-ctq-outputs-key-step-design-process/.

20. Open Source Six Sigma. https://www.opensourcesixsigma.com/.

21. Stroud, J. Understanding the purpose and use of benchmarking, iSixSigma. https://www.isixsigma.com/methodology/benchmarking/understanding-purpose-and-use-benchmarking/.

22. Hassan, M. (2013). SIPOC diagram for the welding wire manufacturing process. *American Journal of Industrial Engineering* 1 (2): 28–35. http://pubs.sciepub.com/ajie/1/2/4/Table/1.

23. Swimlane flowchart template. Cacoo.com. https://cacoo.com/templates/swimlane-flowchart-template.

24. Beels, M. (February 8, 2019). Is your work adding value? Michigan Manufacturing Technology Center. https://www.the-center.org/Blog/February-2019/What-is-Value-Added-vs-Non-Value-Added-Work.

25. Fishbone Diagram. Lean Six Sigma Definition. https://www.leansixsigmadefinition.com/glossary/fishbone-diagram/.

26. Boers, D. (September 13, 2017). Beyond the Risk Matrix. ARMS *Reliability*. https://www.thereliabilityblog.com/2017/09/13/beyond-the-risk-matrix/.

27. Chegg.com. https://www.chegg.com/homework-help/quality-6th-edition-chapter-4-problem-40p-solution-9780134413273.

28. Tarun Kanti Bose. (2012). *International Journal of Managing Value and Supply Chains (IJMVSC)* 3, (2): 17–24. http://www.airccse.org/journal/mvsc/papers/3212ijmvsc02.pdf.

29. JM. (February 5, 2015). Six Sigma and the Pareto Chart Six Sigma Daily. https://www.sixsigmadaily.com/six-sigma-pareto-chart/.

30. What is a pull system? Details and benefits. *Kanbanize*. https://kanbanize.com/lean-management/pull/what-is-pull-system.

31. Kanban software development. Kanban Tool. https://kanbantool.com/kanban-software-development.

32. 60 common examples of poka yoke. Lean Factories. https://leanfactories.com/poka-yoke-examples-error-proofing-in-manufacturing-daily-life/.

33. Schrage, M. (February 4, 2010). Poka-yoke is not a joke. *Harvard Business Review*. https://hbr.org/2010/02/my-favorite-anecdote-about-des.

34. Poka-yoke in manufacturing. RNA Automation. https://www.rnaautomation.com/blog/poka-yoke-in-manufacturing/.

35. History of 5S: An Overview of the 5S Methodology. (May 4, 2010). Bright Hub PM. https://www.brighthubpm.com/monitoring-projects/70488-history-of-the-5s-methodology/.

36. Akers, P. (September 7, 2011). Sustaining 5S. *Beyond Lean*. https://beyondlean.wordpress.com/2011/09/07/keys-to-sustaining-5s/.

37. Wrye, M. (September 7, 2011). Keys to sustaining 5S. *Beyond Lean*. https://beyondlean.wordpress.com/2011/09/07/keys-to-sustaining-5s/.

38. Heijunka – definition and meaning. *Market Business News*. https://marketbusinessnews.com/financial-glossary/heijunka/.

39. Shmula. (May 17, 2012). Why Heijunka is a block in the foundation of the Toyota House. Shumla. com. https://www.shmula.com/why-heijunka-is-a-block-in-the-foundation-of-the-toyota-house/ 10414/.

40. Kanban software development. Kanban Tool.

41. The PDCA Cycle: What is it and why you should use it. (April 14, 2021). Kanban Zone. https:// kanbanzone.com/2021/what-is-pdca-cycle/.

42. Ibid.

43. Exploring PDCA examples: Use the Plan-Do-Check-Act Process in Your Company. (July 12, 2010). Bright Hub PM. https://www.brighthubpm.com/methods-strategies/77327-pdca-examples-implementing-change-in-your-company/.

44. Kaizen. The Lean Six Sigma Company. https://www.theleansixsigmacompany.co.uk/kaizen/.

45. Combining Kaizen with Six Sigma ensures continuous improvement. iSixSigma. https://www. isixsigma.com/methodology/kaizen/kaizen-six-sigma-ensures-continuous-improvement/.

46. Roser, C. (March 2, 2014). The history of quick changeover (SMED). *All About Lean*. https:// www.allaboutlean.com/smed-history/.

47. How to do SMDE. Web Archive. https://web.archive.org/web/20060323033009/http://www. sevenrings.co.uk/SMED/HowtodoSMED.asp.

48. Muelaner, J. (November 10, 2016). Gage Repeatability and Reproducibility, Gage R&R in Excel. *Dr. Jody Muelaner: Simplifying Complexity*. https://www.muelaner.com/quality-assurance/ gage-r-and-r-excel/.

49. Gage R&R template in Excel. QI Macros. https://www.qimacros.com/gage-r-and-r-study/gage-r-and-r-excel-template/#anova (accessed 23 June 2021).

50. Schaper, D. (January 8, 2021). Boeing to pay $2.5 billion settlement over deadly 737 Max crashes. NPR. https://www.npr.org/2021/01/08/954782512/boeing-to-pay-2-5-billion-settlement-over-deadly-737-max-crashes.

51. Srinivasan, N. (August 29, 2020). FMEA example for manufacturing. YouTube. https://www. youtube.com/watch?v=Yp_6RVUWDGY.

52. Pugh, S. (1991). *Total Design: Integrated Methods for Successful Product Engineering*. Boston: Addison-Wesley.

53. What is Pugh Matrix and how to use it? MSG Management Study Guide. https://www. managementstudyguide.com/pugh-matrix.htm.

54. George, M., Maxey, J., Rowlands, D., and Upton, M. (2004). *The Lean Six Sigma Pocket Tool Book: A Quick Reference Guide to 70 Tools for Improving Quality and Speed*. New York: McGraw-Hill Education.

55. What is Pugh Matrix?

56. Pugh Matrix. iSixSigma. https://www.isixsigma.com/dictionary/pugh-matrix/.

57. Prasad, Y. Pugh matrix concept evaluation in Design. *SlideShare*. https://www.slideshare.net/ yogeshprasad5/pugh-matrix-concept-evaluation-in-design.

58. Mazur, G. History of QFD. QFD Institute. http://www.qfdi.org/what_is_qfd/history_of_qfd.html (accessed June 25, 2021).

59. Ibid.

60. Akao, Y. (1990). *Quality Function Deployment: Integrating Customer Requirements into Product Design*. Portland: Productivity Press.

61. Stansfield, K., and Azmat, F. (2017). Developing high value IoT solutions using AI enhanced ISO 16355 for QFD, integrating market drivers into the design of IoT Offerings. *IEEE International Conference on Communication, Computing and Digital Systems (C-CODE)*.

62. ISO 16355-1:2015. (2015). Application of statistical and related methods to new technology and product development process — Part 1: General principles and perspectives of Quality Function

Deployment (QFD). International Organization for Standardization (ISO). https://www.iso.org/standard/62626.html.

63. House of Quality. American Society for Quality. https://asq.org/quality-resources/house-of-quality (accessed June 10, 2021).
64. Lucid Content Team. How to build a House of Quality (QFD). Lucidchart. https://www.lucidchart.com/blog/qfd-house-of-quality.
65. Ibid.
66. Pugh Matrix: When to use this tool. MoreSteam. https://www.moresteam.com/help/engineroom/pugh-analysis.
67. Stansfield and Azmat, Developing high value IoT solutions.
68. Ibid.

CHAPTER 4

Improving Supply Chain Resiliency Using Smart Technologies

Anthony Tarantino, PhD

Introduction

Enterprises face market, credit, reputation, legal, liquidity, and operational risks. Supply chain risk can be seen as a subset of operational risk, and arguably it may be the most challenging type of risk to mitigate. To provide effective supply chain resiliency, risk managers must account for risks from a wide variety of internal and external sources that can include rare but drastic black swan events. The 2003 SARS epidemic, the 2011 tsunami in Japan, and the 2019 COVID-19 pandemic created major shocks to global supply chains that were very difficult to predict with even with the most innovative modeling tools and contingency planning.

While it is common to think of global supply chains and their management as creations of the late twentieth century, global supply chains date back over 2,000 years, when trading prospered between China and Rome (silk and paper) and India and Indonesia (spices). We have evidence of effective supply chain resiliency in 215 BC, when Roman legions outsourced their rescue from Spain to merchant ships that were not part of the empire. The process included formal contracts and insurance.[1]

Much of what we consider to be best practices in managing supply chain risks have their origins dating back hundreds of years. What has changed is the extent and velocity of change. According to the Brookings Institution, global trade increased from 39% of global GDP to 58% from 1990 to 2019, in less than 30 years.[2] Outsourcing and offshoring have become more common as enterprises chased the lowest labor costs in a process known as global labor arbitrage.

While most all of us who grew up in procurement were taught to balance price, quality, and reliability, the lowest price almost always trumped reliability and quality. Most procurement professionals are expected to continue to lower costs regardless of the commodity they are managing. In some cases, this was done at the expense of supply chain resiliency, resulting in an undermining of the financial stability of their suppliers.

Along with the pressure to lower prices, major supply chain organizations adopted Lean principles, first introduced by Taiichi Ohno as the foundation for the Toyota Production System in the 1950s and 1960s. Lean advocates for just-in-time (JIT) inventory over traditional practices, which viewed inventory as good insurance against supply chain disruptions. We make the argument later in the chapter that it was the misapplication of Lean practices that caused the COVID-19 supply chain disruptions.

The combination of global labor arbitrage and the misapplication of Lean inventory practices exposed supply chains to major disruptions in the past decades. Tsunamis, pandemics, trade wars, extreme weather events, national rivalries (such as the one between China and the United States), and civil wars are some of the more common causes that have forced a rethinking of globalization, outsourcing, and offshoring. As a result, over the past 12 years, global trade has shrunk from 61% to 58% of global GDP. The trend was significantly accelerated by the COVID-19 pandemic, precipitating a global economic crisis. The crisis further exposed structural problems in global supply chains.[3] With China representing almost one third of all global manufacturing, the United States and Europe have become overreliant on China for many commodities. The lack of critical medical supplies during the COVID-19 pandemic demonstrated this overreliance.

The combined effect of the COVID-19 pandemic, the trade wars that started under the Trump administration, and the growth of nationalism have exposed the fragile nature of today's supply chains and the need to redesign them with greater agility and resilience. National governments and corporations have come to realize that lowest costs and Lean inventories cannot be at the expense of resilience that ensures that supply can meet major and sudden fluctuations in demand.

Supply Chain Resilience

The Brookings Institution has a good definition of supply chain resilience: "Supply chain resilience refers to the ability of a given supply chain to prepare for and adapt to unexpected events; to quickly adjust to sudden disruptive changes that negatively affect supply chain performance; to continue functioning during a disruption (sometimes referred to as 'robustness'); and to recover quickly to its pre-disruption state or a more desirable state."[4]

Resiliency requires the ability to rapidly detect pending disruptions, respond to them effectively, and then recover from the changed status. To do all this, total supply chain visibility is essential. This is much more demanding than one might imagine because of the need for visibility into suppliers' status, possibly into third-tier suppliers.

Exhibit 4.1 is a graphic of the multiple tiers in supply chains.[5]

EXHIBIT 4.1   Tiers in supply chains

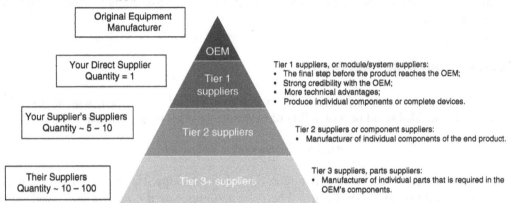

Source: Courtesy of Sophie Luo, Insight Solutions Global.

I have added some examples of how difficult it can be to achieve true multilevel or multi-tiered supply chain visibility. A direct supplier may be receiving raw materials, parts, and subassemblies from a wide variety of suppliers. The numbers and locations will vary widely, especially when so many materials and components are coming from offshore. It gets even more complex and opaque when going down to the third levels of suppliers. At the third level, there may 100 or more suppliers in the supply chain.

All it takes is one part from one supplier to shut down a production line. I learned this the hard way as a materials/supply chain manager earlier in my career, when we manufactured surgical operating tables for hospitals. There are several hundred individual components in a typical table, but a table costing several thousand dollars cannot be shipped lacking even one inexpensive part or label. In manufacturing, 99.99% availability is not good enough.

However, there are a variety of Smart Technology tools to improve resiliency for organizations of all sizes, and the costs of these technologies continues to decline. While much of this technology is not new, the costs have dropped (following Moore's Law), and it is now affordable to smaller manufacturers. By using Big Data, machine learning, computer vision, and Industrial Internet of Things (IIoT) devices, organizations of all sizes can enhance their supply chain resiliency. How many of these tools and how extensively they are applied will depend on the organization and the risks they face.

The number of organizations using technology to improve their supply chain management and resiliency is growing. The Business Continuity Institute (BCI) reports that more than half of surveyed organizations are using technology to help address supply chain disruptions. As expected, COVID-19 was the major reason cited for large increases in supply chain disruptions – the highest level ever reported in the BCI survey. The BCI report also found a 40% increase in disruptions beyond tier 1 suppliers during the pandemic, highlighting the need to greatly improve multitier supply chain visibility.[6] The BCI survey also found a growing number of manufacturers requesting their suppliers to share their business continuity plans – from two thirds to three quarters of respondents. Most respondents did not cite COVID-19 for their increased interest, rather citing ongoing reviews to improve supplier resiliency.[7]

Supply Chain Risk Heat Maps

At its most basic level, a heat map is a matrix that multiplies the consequence or severity of a given risk by the likelihood or probability of its occurring. This is the formula:

$$\text{Severity} \times \text{Likelihood} = \text{Risk score}$$

Applied to supply chain, a heat map rates the risk for its suppliers and/or for critical purchased commodities and components. The riskiest supplier or the riskiest purchased item would be painted red in the upper-right-hand corner. There are a variety of available heat map software tools, from simple Excel-based templates to sophisticated Cloud-based tools. Even a limited heat map to the most obvious high-risk items is a valuable exercise. It works best as a collaborative team effort in which all inputs are welcome.

Exhibit 4.2 shows a typical heat map, with the greatest risks in the upper right-hand corner.[8]

EXHIBIT 4.2   Typical heat map

LIKELIHOOD					
almost certain	Moderate	Major	Critical	Critical	Critical
likely	Moderate	Major	Major	Critical	Critical
possible	Moderate	Moderate	Major	Major	Critical
unlikely	Minor	Moderate	Moderate	Major	Critical
rare	Minor	Minor	Moderate	Moderate	Major
	insignificant	minor	moderate	major	critical
			CONSEQUENCE		

Source: Courtesy of Pencil Focus.

Exhibit 4.3 shows a construction project heat map. For this project, financing delays present the greatest risks in terms of both probability (likelihood) and severity (impact).[9]

EXHIBIT 4.3   Construction project heat map

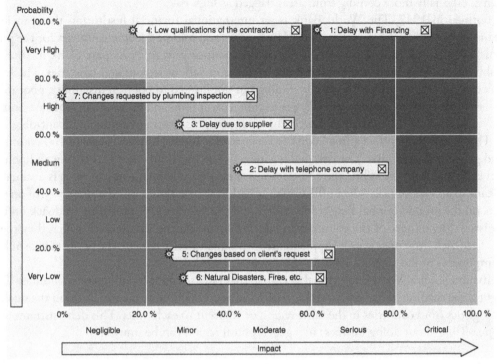

Source: Courtesy of Intaver Institute.

I learned over my decades of running supply chain departments and later running dozens of supply chain projects that visual graphics are powerful. Their power comes from their simplicity. This is especially true with higher-level executives who have little patience or time to digest long-winded reports. In our first example, it is obvious that the areas in the upper-right-hand corner present the most critical risks.

While supply chain mapping goes back centuries, accurate supply chain maps only became possible with online maps and social networks. The first online supply chain maps were developed in 2008 by the Massachusetts Institute of Technology (MIT). The work of MIT demonstrated that online supply chain mapping had a number of key advantages. Today's supply chain mapping tools may be too complex for a smaller organization to trace products from raw material to finished good. Online mapping promotes collaboration on a large scale in which teams work together from all companies in a supply chain to account for every item, process, and shipment.[10]

Supply Chain Mapping at a Macro and Micro Level

On a macro level, a supply chain map should graphically identify the risk levels for doing business in various regions and countries of the world. On a more granular level, a supply chain map is a graphical representation of an organization's material suppliers providing everything, from raw materials to component parts to assemblies to finished products. On its most micro level, the map should capture the most critical long-lead-time and sole-sourced purchased items, especially those coming from areas flagged as high risk.

Formed in 1947, The World Bank is an international financial institution that provides loans and grants to the governments of low income and middle-income countries for the purpose of pursuing capital projects. Its goal is to reduce poverty.[11] As part of its mission it publishes statistics about the risks of doing business in various regions and countries. The high risk can come from several sources, including but not limited to political instability, poor rule-of-law, little intellectual property (IP) protection, corruption, and governments that are hostile to your home country, unfair or dangerous labor practices, poor environmental controls.

The World Bank offers a free Logistics Performance Index (LPI) to compare 160 countries and geographical regions. The free tool is interactive so users can compare existing and potential suppler countries. While the LPI does not measure individual companies, it is a valuable means to access the risks when offshoring. "The LPI is based on a worldwide survey of operators on the ground (global freight forwarders and express carriers), providing feedback on the logistics 'friendliness' of the countries in which they operate and those with which they trade . . . Feedback from operators is supplemented with quantitative data on the performance of key components of the logistics chain in the country of work."[12]

Exhibit 4.4 is a World Bank global heat map that rates the LPIs for all major economies. The lighter the shading, the great the risk.[13] Africa is a good example of the variations in the quality of logistics from countries in the same region or adjacent to each other. This demonstrates that generalizing about doing business in a geographical region can be unwise.

EXHIBIT 4.4   Global heat map

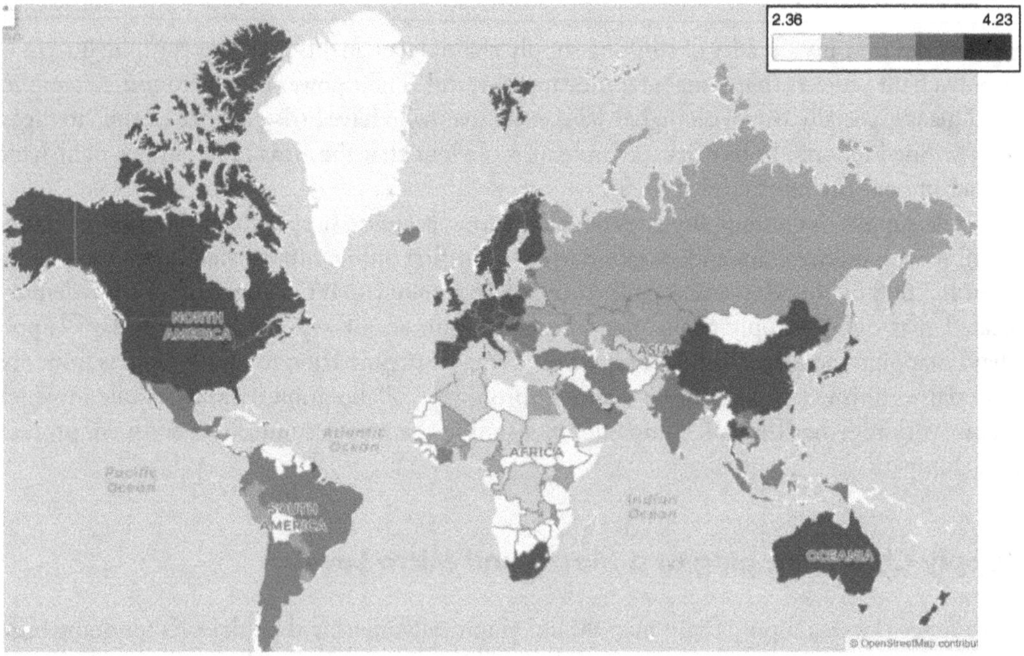

The World Bank logistics performance for the top, middle, and bottom rated countries is captured in Exhibit 4.5.[14]

EXHIBIT 4.5   World Bank logistics performance

Table 1.1   Top 10 LPI economies, 2018

Economy	2018 Rank	2018 Score	2016 Rank	2016 Score	2014 Rank	2014 Score	2012 Rank	2012 Score
Germany	1	4.20	1	4.23	1	4.12	4	4.03
Sweden	2	4.05	3	4.20	6	3.96	13	3.85
Belgium	3	4.04	6	4.11	3	4.04	7	3.98
Austria	4	4.03	7	4.10	22	3.65	11	3.89
Japan	5	4.03	12	3.97	10	3.91	8	3.93
Netherlands	6	4.02	4	4.19	2	4.05	5	4.02
Singapore	7	4.00	5	4.14	5	4.00	1	4.13
Denmark	8	3.99	17	3.82	17	3.78	6	4.02
United Kingdom	9	3.99	8	4.07	4	4.01	10	3.90
Finland	10	3.97	15	3.92	24	3.62	3	4.05

Source: Logistics Performance Index 2012, 2014, 2016, and 2018.

Table 1.3   Top-performing upper-middle-income economies, 2018

Economy	2018 Rank	2018 Score	2016 Rank	2016 Score	2014 Rank	2014 Score	2012 Rank	2012 Score
China	26	3.61	27	366	28	3.53	26	3.52
Thailand	32	3.41	45	3.26	35	3.43	38	3.18
South Africa	33	3.38	20	3.78	34	3.43	23	3.67
Panama	38	3.28	40	3.34	45	3.19	61	2.93
Malaysia	41	3.22	32	3.43	25	3.59	29	3.49
Turkey	47	3.15	34	3.42	30	3.50	27	3.51
Romania	48	3.12	60	2.99	40	3.26	54	3.00
Croatia	49	3.10	51	3.16	55	3.05	42	3.16
Mexico	51	3.05	54	3.11	50	3.13	47	3.06
Bulgaria	52	3.03	72	2.81	47	3.16	36	3.21

Source: Logistics Performance Index 2012, 2014, 2016, and 2018.

Table 1.2   Bottom 10 LPI economies, 2018

Economy	2018 Rank	2018 Score	2016 Rank	2016 Score	2014 Rank	2014 Score	2012 Rank	2012 Score
Afghanistan	160	1.95	150	2.14	158	2.07	135	2.30
Angola	159	2.05	139	2.24	112	2.54	138	2.28
Burundi	158	2.06	107	2.51	107	2.57	155	1.61
Niger	157	2.07	100	2.56	130	2.39	87	2.69

(Continued)

EXHIBIT 4.5 (*Continued*)

| | Table 1.2   Bottom 10 LPI economies, 2018 | | | | | | | |
	2018		2016		2014		2012	
Sierra Leone	156	2.08	155	2.03	na	na	150	2.08
Eritrea	155	2.09	144	2.17	156	2.08	147	2.11
Libya	154	2.11	137	2.26	118	2.50	137	2.28
Haiti	153	2.11	159	1.72	144	2.27	153	2.03
Zimbabwe	152	212	151	2.08	137	2.34	103	2.55
Central African Republic	151	2.15	na	na	134	2.36	98	2.57

na: is not available.
Source: Logistics Performance Index 2012, 2014, 2016, and 2018.

The World Bank also offers country-by-country comparisons for corruption, the rule of law, regulatory quality, voice, and accountability. Exhibit 4.6 compares the control of corruption for countries with high levels and low levels of corruption controls.[15]

EXHIBIT 4.6   Comparing the control of corruption

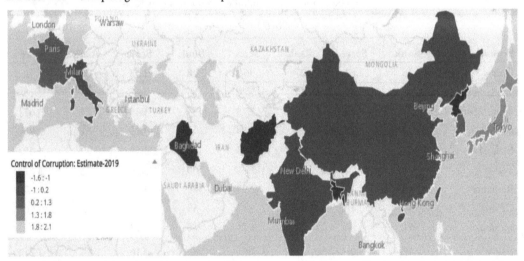

An illustrative example of high risk in a specific country and industry came with the 2013 Dhaka garment factory collapse in Bangladesh. It killed over 1,100 people and injured approximately 2,500. It is typically cited as the deadliest structural failure accident in modern history and the deadliest garment-factory disaster ever recorded.[16] Many high-fashion retailers in the United States, UK, and EU were forced to confront their poor social responsibility in sourcing items made in facilities that became death traps. One of the first lessons I learned in procurement is that there is no substitute for source inspections of your critical suppliers. It did not take a structural engineer to realize that the Bangladesh garment facilities did not meet any basic safety standards and the working conditions were very poor.

While the World Bank maps and databases are helpful in calculating the risks for a given country or region, the World Bank does not drill down to a micro level for individual supplier or critical commodity and components performance. This data comes from implementing and maintaining preferred supplier programs, bill of material (BOM) risk grading tools, and supplier mapping solutions that go beyond tier 1 suppliers.

Preferred Supplier Programs

It is not unusual for a supplier master file to contain thousands if not several thousand supplier names. Even when narrowed down to direct materials used in production, the supplier list can easily total several hundred, even for more simple consumer goods. A proactive means to improve supply chain resiliency is to create a preferred supplier program using a variety of software tools. These tools will help capture the performance history for suppliers around their on-time and quality performance. With preferred supplier programs it is possible to identify the best suppliers out of the thousands in an organization's supplier master list.

Creating a preferred supplier program can be a true win-win in which your best suppliers are rewarded with more business and long-term relationships. Preferred suppliers have greater incentives to invest capital to ensure a consistent source of supply and to maintain high quality levels.

Software tools can help automate the process by capturing on-time delivery rates, rejection rates, financial ratings (Dun & Bradstreet, Equifax, etc.), diversity programs, social responsibility, business resiliency planning, and so forth. These tools can also be effective in flagging bad suppliers that should be barred from further business, or at least put on probation.

Procurement and Supply Australasia (PASA) is a leading provider of information and education to procurement and supply professionals throughout Australia and New Zealand. PASA makes the following case for creating a preferred supplier program:

- **Reduced costs** – Preferred suppliers enjoy more business and longer-term relationships, so they are more open to reducing prices.
- **Improved quality of service** – Preferred suppliers are better educated in your business requirements because of the closer working relationship, so they are more capable in meeting and exceeding your expectations.
- **Communication is more straightforward** – Because of the closer working relationship, it is easier to resolve problems that arise.
- **Lower risk** – Risks are reduced when dealing with trusted suppliers who have demonstrated a history of meeting and exceeding purchase specifications and delivery commitments. They are also more likely to make long-term investments to increase capacity.[17]

Bill of Material Risk Grading Tools

The ability to grade the individual raw materials and purchased components that go into products is essential to good supply chain resiliency. Typically a bill of material (BOM) is graded on its multisourcing, inventory, lifecycle, environmental, and social responsibility risks. The following is a short description of each risk category.

- **Sole-Sourcing Risk.** Sole-sourced components always present risk and should be avoided unless the design is unique or if it is cost-prohibitive to develop additional suppliers.
- **Inventory Risk.** Inventory shortages of BOM items will vary based on the marketplace and on major swings in supply and demand.
- **Lifecycle Risk.** Product lifecycles continue to shrink. When a major component approaches its end of life, developing replacements may be challenging. This may require introducing new suppliers and can affect other components.
- **Environmental Risk.** Environmental regulations continue to change and grow more complex at a rapid pace and can vary widely from nation to nation. The drive for more sustainable supply chains will tend to increase environmental risks.[18]
- **Social Responsibility Risk.** The responsibility of your suppliers now includes actions impacting society and promotes sustainable development, including the health and the welfare of society. It considers the expectations of shareholders, employees, and the community.

Exhibit 4.7 shows a sample BOM report card.[19] In this demo report the BOM receives only a "C" grade because of poor lifecycle and multisourcing scores. Over 70 components are either sole-sourced or have no source identified. Lifecycle risks can be expected to grow as product life cycles continue to shrink.

EXHIBIT 4.7   Sample BOM report card

C	16%	11%	83%	10%
OVERALL GRADE 19 MPNs at risk	**LIFECYCLE** 38 MPNs at risk	**MULTISOURCING** 55 MPNs at risk	**COMPLIANCE** 0 MPNs at risk	**INVENTORY** 62 MPNs at risk

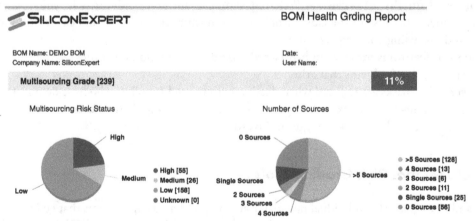

Source: Courtesy of SiliconExpert.

Environmental Risk Solutions

The CDP (formerly known as the Carbon Disclosure Project) is an international nonprofit organization based in the UK, Germany, and the United States that helps companies and local governments disclose their environmental impact.[20] CDP surveyed 8,000 companies in 2020 about the environmental risks they face. The respondents estimated a total of $120 billion from environmental risks in their supply chains over the next five years. The industries that reported the highest risks are "manufacturing ($64 billion), food, beverage and agriculture ($17 billion), and power generation ($11 billion)."[21] These costs can be expected to be passed down the supply chain to their tier 1 and tier 2 customers.

The rising costs from environmental risks come from deforestation, water-related impacts, and climate change. Some of the physical impacts include increased severity and frequency of cyclones/hurricanes and floods, increased raw material costs, increased regulations, and changes in the global marketplace. To address the growing risk, organizations are demanding greater transparency and direct action by their suppliers to address the environmental impact on their supply chains.

One key finding of the CDP report highlights the importance of reducing greenhouse gases (GHGs) among suppliers. The report found supply chain GHG emissions are on average 11.4 times higher than from their own operations. This is twice as high as earlier estimates due to improved emissions reporting. Unfortunately, the survey found that only about one-third of manufacturers are engaging their suppliers to reduce emissions. The good news is the growth in buyers requesting emission data from their suppliers – an increase of 19% from 2019 to 2020.[22]

The COVID-19 global pandemic dramatically showcased environmental risks unlike any event in modern history. Exhibit 4.8 is charts the major changes in supply chain disruptions from 2019 to 2020 with a 414% increase in security (border closures), 137% in manufacturing, and 163% in health.[23]

EXHIBIT 4.8   Changes in supply chain disruptions, 2019–2020

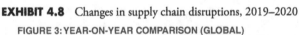

FIGURE 3: YEAR-ON-YEAR COMPARISON (GLOBAL)

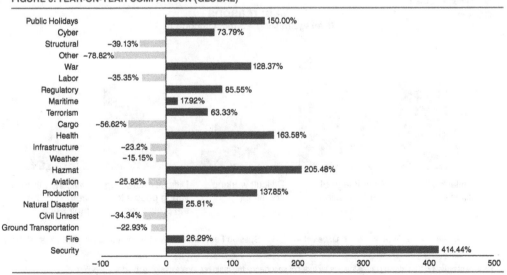

Source: Courtesy of CDP.

The Global Driver Shortage and Poor Utilization

A 2019 analysis by the American Trucking Associations (ATA) predicted a US shortage of over 60,000 truck drivers. This will mean a driver shortage of 160,000 by 2028 if the trend continues. With the average driver's age nearing 50 and only 6% of drivers being women, the demographics are easy to understand.[24] The ATA has been predicting a driver shortage for several years. For example, Brenny Transportation, based in Minnesota, increased driver pay 15% in an effort to attract additional drivers. Even with many drivers earning $80,000 per year, the company could not find enough drivers.[25] Exacerbating the driver shortage is the rapid growth of online retailing with fast delivery services. The American Trucking Association estimates that 50,000 more drivers will be needed to meet the growing demand.[26]

The problem is not unique to the United States. Research by the Geneva-based International Road Transport Union (IRU) shows that the problem is global. The IRU surveyed about 800 trucking firms in over 20 countries. The IRU reports: "Transport companies forecast driver shortages to intensify again in 2021 as economies recover and demand for transport services increases. This shortfall is expected to reach 18% in Mexico, 20% in Turkey, 24% in Russia and almost 33% in Uzbekistan." The IRU also found diversity efforts failing to bring in younger drivers and attract more female drivers. As in the United States, the aging workforce has an average age about 50 and remains a major issue. Globally, the lack of female drivers is worse than in the United States, with only 2% of drivers being females, and declining participation. The number of drivers under 25 also fell to 5% in Europe and Russia, 6% in Mexico, and 7% in Turkey.[27]

A common problem is the poor reputation of the industry. Earlier generations were attracted by the good pay and independent lifestyle. It is now difficult to attract new drivers, who want to spend more time closer to home. Exhibit 4.9 is a chart demonstrating the major reasons the industry is losing people.

EXHIBIT 4.9

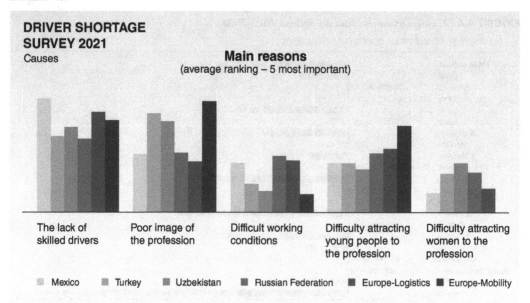

Results of question "Could you please indcate by order of importance the main reasons explaining the difficulties you find to hire drivers? (please rank the items by order of importance)" ; Europe = EU-27 + Norway, Switzerland, UK.

Source: IRU's 2021 Driver Shortage Survey.

Besides chronic driver shortages, poor driver utilization also plagues the industry. Research conducted by the MIT Freight Lab shows that the average driver is only driving about 60% of the time (about 6.5 hours per day out of 11 available hours).[28]

FreightWaves estimates the total cost of operating a truck at about $210,000 annually (or about $1.76 per mile) based on the average truck driver driving 120,000 miles per year (400 miles per day, 6 days a week, 50 weeks per year).[29] Increasing utilization by even 10% could save the average truck operator $20,000 per vehicle.

Vehicle Monitoring Tools

Smart Technology tools provide the means to help improve driver efficiency, vehicle utilization, and safety. More advanced technology is coming in the next few years to automate long-haul driving – the industry sector with the greatest shortages.

The most widely deployed solutions use the driver's smartphone and a microcontroller based on GPS/GSM/GPRS technology. GPS is great at capturing geographic coordinates while the GSM/GPRS updates the vehicle location to a database.[30] Once a vehicle's speed, direction, and location are determined from the GPS system, they can be transmitted via the Cloud to a fleet management software application. Terrestrial data transmission typically works well in populated areas, while satellite tracking may be needed in more remote areas.

There are limitations to GPS and related technologies. While good at locating a vehicle's road location, they lack the granularity to precisely track vehicles within a facility. Delays within truck terminals, shipping docks, and distribution centers can be costly, lower productivity, and hurt customer satisfaction levels.

Computer Vision Systems Using Smart Cameras

Computer vision systems use a combination of cameras and small computers to analyze all visible yard activity. This includes entry and exit times, loading and unloading times, weigh station times, and license plate and scale readings. These systems typically rely on a combination of Edge computing for real-time alerts and the Cloud for reporting and analysis.

Unlike passive security cameras that require humans to review hours of images, computer vision's smart cameras use AI, with deep-learning algorithms to signal events of interest or concern, foreign materials inside an empty van, debris in the bed of dump truck, people not wearing a hard hat in a designated area, and vehicles left unattended for an extended period.

Exhibit 4.10 is courtesy of Atollogy, a computer vision hardware and software provider that monitors yard activity.[31] It provides an example of real-time monitoring of a truck terminal. Each vehicle is logged in automatically by a reading of its license plate. Its total time in the yard and the number of daily trips it has taken/it should take are also displayed. Unlike GPS-based systems, computer vision tracks all vehicles in a facility, not just the organization's fleet vehicles. The need for guards to manually log trucks and drivers into and out of a facility is reduced with a digital guard, which never blinks or takes breaks, working around the clock.

EXHIBIT 4.10　Real-time monitoring of a truck terminal

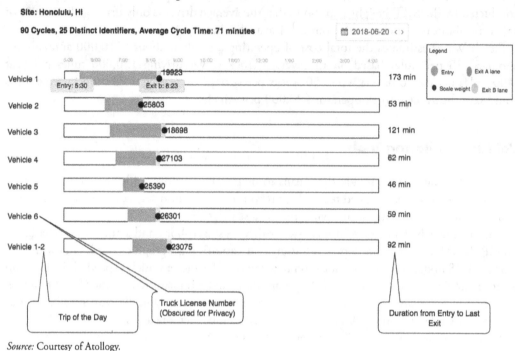

Source: Courtesy of Atollogy.

Computer vision is also an ideal way to address internal and external theft. In developing applications for computer vision over the past five years, yard managers have often complained about the difficulty in detecting theft. It may sound strange that anyone would steal a truck full of gravel or asphalt, but that can be worth several hundred dollars.

Finally, computer vision is a good tool to help identify drivers who need additional training or need to be removed from the job to prevent accidents. It is fairly easy to capture speeding vehicles or drivers going in the wrong direction or not wearing their safety equipment. For example, AI can help identify high-risk drivers and avoid assigning them to high-priority loads traveling through challenging weather or road conditions. [32]

Autonomous Trucks

Several US companies are developing self-driving trucks, which are essentially mobile robots. It can be argued that self-driving trucks are advancing faster than self-driving cars. While trucks are much larger than cars and require long distances to brake, they have the advantage of following fixed routes along major highways, and they enjoy far greater visibility than cars with their 13- to 15-foot heights.

Lidar is a detection system that uses laser light rather than radar. Lidar sensors are limited to about 200 yards. Because heavy trucks may require twice the distance to stop as passenger cars and lidar's limited range, trucks can mount high-definition cameras on top of their cabs to extend forward visibility to 1,000 yards. Due to their large size, trucks can install more powerful computers, cameras, and sensors. Therefore, trucks are a better platform for autonomy.

Exhibit 4.11 is from TuSimple, a San Diego–based startup using a combination of GPS, radar, lidar, and several HD cameras to navigate trucks.[33]

EXHIBIT 4.11 Using cameras to navigate trucks

Source: TuSimple.

The goal of the self-driving truck industry is to achieve level 4 autonomy in the next few years, which means that trucks will be able to operate without a human driver under normal driving conditions, those not impacted by traffic conditions, weather, and time of day.[34]

Supply Chain Resiliency in a Post-COVID World

The COVID-19 pandemic created shock waves to supply chains on a massive and global scale.

It has forced a rethinking of supply chain resiliency, challenging many long-accepted conventions. It has exposed the vulnerabilities and complexities in global supply chains that were hard to detect or predict. To gauge the impact of COVID-19 on global supply chains, Capgemini surveyed 1,000 organizations from consumer products, retail, discrete manufacturing, and life sciences sectors.

The survey found widespread agreement that supply chain planning will need to adapt post-COVID-19.[35]

Exhibit 4.12 demonstrates the need for change.

EXHIBIT 4.12 The need for change

"We believe our supply chain strategy will need to change significantly, in order to adapt to a new normal post-COVID-19"
% of respondents who agree

Overall	66%
China	74%
France	73%
Spain	73%
Sweden	69%
Norway	69%
US	65%
Netherlands	61%
Germany	60 %
Italy	60 %
India	60 %
UK	60 %

Source: Capgemini Research Institute, Supply Chain Survey, August–September 2020.

Criticism and Defense of Lean Inventory Management

The Capgemini survey also found organizations reducing their reliance on Lean's just-in-time sourcing philosophy by 10%, from 39% to 29%.[36] Modern Lean philosophy is based on Toyota's just-in-time (JIT) production system developed in the 1950s and 1960s that became known globally as Lean in the 1970s and 1980s. Its underlying principle is continuous improvement that strives to reduce waste of all types – waste in transportation, waste in overproduction, waste in rejected components, and waste in excess inventory. Unfortunately, most organizations applied their own interpretation of Lean, not realizing what was unique to Toyota operating in Japan and how global offshoring impacted sourcing and inventory planning.

Toyota was very successful in achieving dramatic inventory reductions, in part because its key suppliers were located close to Toyota's manufacturing plants and were conditioned to stock inventory in order to respond to any major shifts in demand. Once materials hit the production lines, Toyota was able greatly reduce work-in-progress (WIP) inventory by continuously striving to slash setup times. Unlike the US model, pioneered by Henry Ford, that sought to run very large lots of materials in a batch system, Toyota was able to run small lots that met immediate demand. Setup times of three to four hours were reduced to a handful of minutes in many cases.

This system worked well in Japan with its short geographic distances, homogeneous culture, and close supplier relationships. The problem was that trying to achieve just-in-time inventory levels is challenging when suppliers are a continent away and there is a lack of alternate suppliers close to the point of consumption.

It is instructive to examine how Toyota, the creator of Lean and just-in-time (JIT), fared during the COVID-19 pandemic and resulting supply chain shocks. It was widely reported that the Lean philosophy was in part responsible for widespread shortages. Interestingly, Toyota has fared far better than its competitors because it applied robust business continuity planning. For example, Toyota production levels were not crippled by the global shortage of semiconductors. Bloomberg's Anjani Trivedi argued in a February 2021 article that this was because Toyota broke its own JIT rules.[37] There is a good argument that the opposite is true. Toyota's Taiichi Ohno, the creator of JIT, was a realist and never argued in his pioneering work, *The Toyota Production System*, against developing alternative sources or business continuity plans (called disaster recovery planning in the past).[38]

Taiichi Ohno was a revolutionary who advocated empowering individual workers, drastically cutting setup times, making only what was needed, using simple visual controls over complex systems, and pulling parts through production with kanbans. He showed how plant safety stocks were counterproductive and how prevention was better than healing. It is doubtful that Ohno would have approved of the Lean label being used as a reason for the shortsighted outsourcing and offshoring practices that created so many supply chain shocks during the COVID-19 pandemic.

Good Sourcing Strategies

The traditional approach to reducing supply chain risk is to introduce alternate suppliers as a backup, even if the alternate suppliers have higher costs. Ideally, the alternate suppliers are not located in the same geographic region as the primary supplier. The dangers of sourcing within only one region became apparent during the 1997 Asian financial crisis and the 2004 tsunami in Japan.[39]

There is a good argument for sourcing materials in the same region where they are to be used in manufacturing. The obvious example for American producers is to shift their more labor-intensive work from distant regions such as China to Mexico and Central America. Western European manufacturers are now considering sourcing closer to home from Eastern Europe, Ukraine, or Turkey.[40]

The difficulty in finding alternative suppliers to the Chinese supply base will depend on the types of commodities. Simpler consumer goods may be more easily moved than complex manufactured items. It has taken China over 20 years to build a robust national supply and logistics base to support its dominance in manufacturing. Vietnam and other Southeast Asian countries cannot be expected to provide the same level of supply chain and transportation sophistication for some time.

Supply Chain Stress Testing

Manufacturing organizations can learn some valuable lessons from major financial institutions that are compelled by central bank regulators to conduct periodic stress tests of their ability to respond to significant and sudden changes in the economy. The capital reserves of major banks are determined by the results of these stress tests.

Major banks employ PhD-level economists using advanced modeling programs to run stress tests on a regular basis. Most global corporations have the financial resources to run supply chain resiliency modeling simulations, but few have demonstrated that they have the expertise needed to do so. According to the Capgemini survey, a large majority of organizations now accept the need to increase their stress-testing capabilities.[41] Smart Technologies follow Moore's Law, meaning that the cost of modeling software tools will continue to drop.

History has shown that corporate priorities will change the further we move away from the pandemic. The banks have no choice. They are statutorily required to run regular stress tests that are closely scrutinized by regulators. It may make sense for regulators to consider mandating supply chain stress tests for producers of critical commodities. Unfortunately, the pandemic taught us that even simple household items such as bleach and hand sanitizer can become critical commodities. Some commodities such as semiconductors and rare earth minerals have been and will continue to be critical to national defense, and therefore warrant periodic supply chain stress tests. In fairness, today's supply chain modeling is far more complex than models used by banking and beyond the capacity of most small to mid-size manufacturers. The costs of AI will continue to drop following Moore's Law, making supply chain modeling more affordable for smaller organizations in the future.

Summary

From 1975 to 2000, I wore a global sourcing and inventory management hat. I often wonder what it would be like to run sourcing and supply chains today with shorter and shorter product life cycles, the constant pressure to use the lowest cost labor sources, little support for the costs of developing alternate and higher-cost suppliers close to home, and trade wars spurred by growing nationalism and a rejection of globalization. On the other hand, I would now have an amazing array of Smart Technology tools today to assist in improving supply chain resiliency. Smart tools can digitize supply chains to support continuous improvement, to create comprehensive models, and to conduct stress tests. Best of all, you do not have to be a data scientist or software engineer to use these tools.

The COVID-19 pandemic shocked supply chains across the globe and helped alter corporate sourcing and inventory strategies to better reflect the need for more agile supply chains. As mentioned earlier, memories will likely fade quickly the farther we move away from the pandemic. It is difficult to predict if the post-COVID normal will endure. One thing, however, is easy to predict. Supply chains will continue to become more complex and dynamic, requiring strong professionals to manage them and the risks that accompany them.

Sample Questions

1. Supply chain resiliency has become more difficult in recent years because of
   a. Reductions in outsourcing
   b. Increases in offsourcing
   c. The reduction in the number of trade wars
   d. Lean inventory management principles

2. Lean's just-in-time system was created by
   a. Toyota
   b. Ford
   c. Apple
   d. Amazon
3. Global trade represents what percentage of global GDP?
   a. Under 10%
   b. Over 80%
   c. Over 50%
   d. 100%
4. Trade wars impact supply chain by
   a. Not having any impact
   b. Increasing the number of available suppliers
   c. Reducing risk
   d. Increasing risk
5. Self-driving trucks will
   a. Help reduce the shortage of drivers
   b. Use radar, lidar, and cameras to navigate
   c. Be tied to GPS systems
   d. All of the above
6. Supply chain stress tests
   a. Model worst-case scenarios
   b. Are simpler than financial stress tests
   c. Are used by all major global manufacturers
   d. Can be effectively done manually
7. Computer vision
   a. Can only analyze vehicles
   b. Does not typically require artificial intelligence (AI)
   c. Digitizes all physical operations to improve efficiencies
   d. Is too expensive for most organizations
8. Suppliers collectively emit how many times more greenhouse gases than the manufacturers they serve?
   a. 3 times
   b. 5 times
   c. 11 times
   d. 20 times
9. The advantages of a preferred supplier program include
   a. Suppliers having more incentives to improve quality and service
   b. Good suppliers being rewarded with more business
   c. Good suppliers being better at understanding your business
   d. All of the above
10. The World Bank helps to improve supply chain resiliency by
   a. Providing loans to US businesses
   b. Measuring the risk levels of 160 countries
   c. Measuring the risk levels of individual companies
   d. Measuring the risk levels for critical components and raw materials

Notes

1. Munn, N. A history of procurement: Supply chain through the ages. SourceSuite.com. https://www.sourcesuite.com/procurement-learning/purchasing-articles/A-History-of-Procurement-Supply-Chain-through-the-Ages.jsp.
2. Iakovou, E., and White, E. (December 3, 2020). How to build more secure, resilient, next-gen U.S. supply chains. The Brookings Institution. https://www.brookings.edu/techstream/how-to-build-more-secure-resilient-next-gen-u-s-supply-chains/.
3. Ibid.
4. Ibid.
5. Luo, S. (December 18, 2018). What is a tier 1 company or supplier? Insight Solutions *Global*. https://insightsolutionsglobal.com/what-is-a-tier-1-company-or-supplier/.
6. Business Continuity Institute (BCI). (2021). *Supply Chain Resilience Report 2021*. BCI Thought Leadership. www.thebci.org/uploads/assets/e02a3e5f-82e5-4ff1-b8bc61de9657e9c8/BCI-0007h-Supply-Chain-Resilience-ReportLow-Singles.pdf.
7. Ibid.
8. Ekram, Z. (August 4, 2016). Supply Chain Management-2. Pencil Focus. https://www.pencilfocus.com/2016/08/supply-chain-risk-management-2.html.
9. Risk Matrix and how to use it. Intaver Institute. https://intaver.com/blog-project-management-project-risk-analysis/risk-matrix-and-how-to-use-it/.
10. What is Supply Chain Mapping? SourceMap. (April 14, 2015). https://sourcemap.com/blog/what-is-supply-chain-mapping.
11. https://www.worldbank.org/en/home.
12. The World Bank. Aggregated LPI 2012–2018. LPI Home. https://lpi.worldbank.org/.
13. Arvis, J.-F., Ojala, L., Wiederer, C., et al. (2018). *Connecting to Compete 2018: Trade logistics in the global economy. The Logistics Performance Index and its indicators*. The International Bank for Reconstruction and Development/The World Bank. https://openknowledge.worldbank.org/bitstream/handle/10986/29971/LPI2018.pdf.
14. The World Bank. Aggregated LPI.
15. Control of corruption. Worldwide Governance Indicators. The World Bank Data Bank. https://databank.worldbank.org/databases/control-of-corruption.
16. Hopkins, T. (April 23, 2015). Reliving the Rana Plaza factory collapse: a history of cities in 50 buildings, day 22. *The Observer*. https://www.theguardian.com/cities/2015/apr/23/rana-plaza-factory-collapse-history-cities-50-buildings (accessed 18 May 2021).
17. PASA. (June 1, 2020). Why do you need a preferred supplier programme? PASA Thought Leadership. https://procurementandsupply.com/2020/06/why-do-you-need-a-preferred-supplier-programme/.
18. Arrow. (March 12, 2020). Understanding & analyzing BOM supply chain risk. *Research and Events*. https://www.arrow.com/en/research-and-events/articles/understanding-and-analyzing-bom-supply-chain-risk.
19. BOM Demo Report. (January 2021). SiliconExpert. https://www.siliconexpert.com/wp-content/uploads/2021/01/SiliconExpert_BOM-DEMO-Report.pdf.
20. CDP.Net. https://www.cdp.net/en/ (accessed May 19, 2021).
21. Bhonsle, S. (February 2021). Transparency to transformation: A chain reaction. CDP Global Supply Chain Report 2020. CDP. https://6fefcbb86e61af1b2fc4-c70d8ead6ced550b4d987d7c03fcdd1d.ssl.cf3.rackcdn.com/cms/reports/documents/000/005/554/original/CDP_SC_Report_2020.pdf?1614160765.
22. Ibid.
23. Ibid.

24. Costello, B., and Karickhoff, A. (July 2019). *Truck Driver Shortage Report 2019*. American Trucking Association. https://www.trucking.org/sites/default/files/2020-01/ATAs%20Driver%20Shortage%20Report%202019%20with%20cover.pdf.

25. Long, H. (May 28, 2018). America has a massive truck driver shortage. Here's why few want an $80,000 job. *Washington Post*. https://www.washingtonpost.com/news/wonk/wp/2018/05/28/america-has-a-massive-truck-driver-shortage-heres-why-few-want-an-80000-job/?noredirect=on&utm_term=.f3611620f7e4.

26. https://www.trucking.org/News_and_Information_Reports_Driver_Shortage.aspx.

27. International Road Transport Union (IRU). (2021). *IRU Annual Report* https://www.iru.org/system/files/IRU%20Annual%20Report%202020.pdf.

28. Cassidy, W. (January 20, 2021). Outlook 2021: Latest US driver shortage requires long-term solutions. *Journal of Commerce*. https://www.joc.com/trucking-logistics/labor/outlook-2021-latest-us-driver-shortage-requires-long-term-solutions_20210120.html.

29. Robinson, A. (September 11, 2020). Infographic: The cost of operating a truck. FreightWaves Sonar. https://sonar.freightwaves.com/freight-market-blog/operating-a-truck-infographic#:~:text=FreightWaves%20estimates%20the%20total%20cost,%2C%2050%20weeks%20per%20year).

30. Adam Robinson, Infographic: The Cost of Operating a Truck. FreightWaves Sonar, September 11, 2020 https://sonar.freightwaves.com/freight-market-blog/operating-a-truck-infographic#:~:text=Freight Waves%20estimates%20the%20total%20cost,%2C%2050%20weeks%20per%20year

31. www.Atollogy.com (accessed May 18, 2021).

32. Pyzyk, K. (March 1, 2019). AI in trucking. *Transport Topics*. https://www.ttnews.com/articles/ai-trucking.

33. TuSimple, www.tusimple.ai.

34. Ackerman, E. (January 4, 2021). This year, autonomous trucks will take to the road with no one on board. *IEEE Spectrum*. https://spectrum.ieee.org/transportation/self-driving/this-year-autonomous-trucks-will-take-to-the-road-with-no-one-on-board.

35. Capgemini Research Institute. (2020). Fast forward: Rethinking supply chain resilience for a post-pandemic world. Capgemini. https://www.capgemini.com/us-en/research/fast-forward/?utm_source=google&utm_medium=cpc&utm_campaign=growth_cprd_intelligent_supply_chain&utm_content=fast_forward&utm_source=adwords&utm_campaign=&utm_medium=ppc&utm_term=digital%20supply%20chain&hsa_ver=3&hsa_grp=110319934545&hsa_cam=11312715767&hsa_ad=510964291437&hsa_acc=2114937760&hsa_src=g&hsa_tgt=kwd-73287654682&hsa_kw=digital%20supply%20chain&hsa_mt=p&hsa_net=adwords&gclid=Cj0KCQjw4v2EBhCtARIsACan3nxk77Rvg60nGhnqUE_DPjj05bnRuJlBSFc_T9uK8dEXBgZ4DIk41xQaAnfsEALw_wcB#.

36. Ibid.

37. Trivedi, A. (February 16, 2021). Toyota broke its just-in-time rule just in time for the chip shortage. *Bloomberg Opinion*. https://www.bloombergquint.com/gadfly/toyota-broke-its-just-in-time-rule-just-in-time-for-the-chip-shortage.

38. Ohno, T. (1988). *Toyota Production System: Beyond Large-Scale Production*. Cambridge, MA: Productivity Press.

39. Shih, W. (September–October 2020). Global supply chains in a post-pandemic world. *Harvard Business Review*. https://hbr.org/2020/09/global-supply-chains-in-a-post-pandemic-world.

40. Ibid.

41. Capgemini Research Institute, Fast forward.

CHAPTER 5

Improving Cybersecurity Using Smart Technology

Craig Martin

Introduction

The process of sourcing, procuring, building, and delivering goods, the supply chain for short, is a key cog in the engine that makes many companies successful. An effective supply chain contributes lower costs, a faster production cycle, improved profitability, access to advanced technology and specialized expertise, and improved customer satisfaction. Each decision made in developing a comprehensive supply chain strategy, however, is a balance of risk versus reward.

Manufacturing and supply chain teams are continually searching for ways to improve their contribution to the company while having systems and processes in place to mitigate risk. Many of the risk tradeoffs are well chronicled and understood – quality versus cost, for example. The potential of sourcing a component from an unproven source in a remote location to save 20% presents a very tempting opportunity, and one many organizations face daily. Establishing a strategy that is the most cost effective but takes longer to get to market in a very competitive world can be a dilemma. Sourcing new exciting but immature technology offers the potential of creating a competitive advantage but can be disastrous if units begin to fail in the field.

Exhibit 5.1 shows the balance between established and cutting-edge cybersecurity technologies.[1]

EXHIBIT 5.1   Balance between technologies

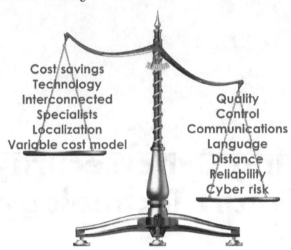

Source: From WiseGEEK.

In this chapter, we explore a relatively new and rapidly emerging challenge in pursuit of these benefits, that of cyber risk. Why is cyber risk a legitimate and growing threat to factories and the supply chain? How do hackers breach seemingly secure systems and for what purpose? Most importantly, we discuss how to take steps to reduce the probability of being breached, and to keep any breaches from being catastrophic. In this section, the supply chain includes all third-party vendors of component materials, manufacturing operations performed by contract manufacturers, and internal manufacturing operations.

Trends Increasing the Risk of Manufacturing and Supply Chain Cyberattacks

As noted, most of the tradeoffs mentioned have existed from the earliest days of manufacturing. However, not until the middle of the past century did the capability of being connected electronically with partners and customers even exist. Today's increasingly more sophisticated information technology has enabled a number of trends that conspire to create challenges in protecting a company's important assets.

Globalization and Specialization

Information technology enabling immediate access to suppliers through advanced search engines now makes it easy to identify sources in any corner of the world, as well as uncover companies providing very specific solutions. As a result, companies are more often engaged with a growing number of relationships. This creates a recipe for a scavenger hunt of sorts to find the optimal mix of suppliers to meet your sourcing needs – usually in search of lower costs or unique technology. This explosion of sourcing alternatives has led to an intriguing world of decentralized and specialized solutions, many of which are fledgling, small, distant, and unfamiliar. Compounding this phenomenon is the fact that many of these companies need to be connected to your infrastructure, and in some cases to each other.

Improved Security Within the Corporate Network

As organizations improve their perimeter defenses against attack, hackers are shifting their focus to third parties. Combined with the trend in specialization, outsourcing talent and expertise may help companies meet project timelines and goals, but inadequate vetting can increase the risk of being compromised. In one example, highly sensitive information from one company, including 13 million files, were stolen from an outsourced third-party legal counsel.

Artificial Intelligence and the Internet of Things

On the factory floor, the emergence of artificial intelligence (AI) and the explosion in new appliances and sensors under the category of Internet of Things (IoT) have greatly increased the attack surface that threat actors can take advantage of. Many of these IoT devices have inadequate security controls, making them an attractive target.

Software Supply Chain Compromises

One of the most concerning trends involves software-based cyberattacks, for a couple of reasons. Customers do not have the ability to perform rigorous tests on new releases from the myriad software vendors supporting their business and as such trust that the released code has been properly tested and is safe to install.

However, there have been an increasing number of compromises in which threat actors have found their way into the software development environment by using stolen certificates to bypass detection or otherwise subverting the upgrade process to introduce malicious payloads into what appears to be legitimate code. There were seven significant software supply chain events in 2018, and the number is growing.

In 2021, a very vivid and high-profile breach involved software supplier SolarWinds, in which attackers were able to insert malicious code into a software upgrade version without detection.[2] Despite all of the typical qualification testing done by software providers before releasing the upgrade to the public, the malicious content was not caught. Before the breach was discovered by a leading cybersecurity firm that had itself been victimized, 18,000 of SolarWinds' 30,000 clients, including a number of major corporations and government organizations, installed the upgrade. Until then, no one knew they had been breached. By attacking one organization – SolarWinds – the attackers were able to penetrate thousands of entities. This simple math makes companies that sell to many large organizations an attractive target.

The Emergence of Cloud Computing and the Public Cloud

Just as much of the manufacturing strategy has evolved over the past 50 years from companies building their own products to an outsourced model using contract manufacturers, managing data centers has taken the same path. Companies like Amazon Web Services (AWS) now manage more and more of the data, and ERP companies like Oracle and SAP now have Cloud versions of their software.

Just because the storage of data is moved to a third-party public Cloud provider does not absolve the customer of the need to maintain the same levels of hygiene and control over their data. Attackers are following the data as customers migrate to Cloud deployments. Hardware chip vulnerabilities have allowed intruders to access companies' protected memory spaces on Cloud services hosted on the same physical server. Successful exploitation provides access to memory locations that are normally forbidden. This is particularly problematic for Cloud services because while Cloud instances have their own virtual processors, they share pools of memory – meaning that a successful attack on a single physical system could result in data being leaked from several Cloud instances.

Exhibit 5.2 is a graphical representation of Cloud storage.[3]

EXHIBIT 5.2   Cloud storage

Source: From Shutterstock.

Targeting Small Companies

Another technique that even the most advanced threat actors use is to look for the easy path to target victims. If you are thinking, "I'm too small to be a target," think again. Strikingly, 92% of cybersecurity incidents occurred among small firms. The reason? Simply, the smaller vendors, having "less strategic" status with larger customers, are less likely to be audited for their security vulnerabilities as soft targets by threat actors to get to a larger fish. Also, smaller companies may not have the budget or security capabilities in staffing and infrastructure to protect against the sophisticated methods used by cyber terrorists. The term "island hopping" has been coined to refer to this method of attack.[4]

So Why Is Manufacturing and the Supply Chain an Attractive Target?

The traditional view of the supply chain involves those in the direct flow of goods – sourcing component materials, transforming them into finished goods, and distributing them to customers. However, the supply chain also includes suppliers to those partners, and a host of relationships hosted by other organizations within the company including Cloud and software vendors, business contractors, maintenance contractors, equipment manufacturers, logistics providers, outside legal counsel, and more.

A startling statistic cited by a Ponemon Institute survey in the fall of 2018 reported that the average number of vendors with access to a company's sensitive information has grown to 471. More concerning is that only 35% of the respondents even had a list of third parties with such access.[5]

In simple terms, the more connections in your supply chain, the higher the risk of compromise. By definition, the supply chain relies on a network of independent partners developing and sharing solutions through various forms of communication. The interconnected system of partners spread all over the globe creates an opportunity to infiltrate many links in a potential target's network. This broad, complex, and distributed attack surface offers access to multiple high-value targets. And the deeper you go, the harder an attack is to detect. Given the potential ramifications if reported publicly, many attacks go unreported.

Primary Motives Behind Manufacturing and Supply Chain Attacks

All of the motives criminals and political forces have to attack a target – financial gain, accessing personal information, political objectives, stealing intellectual property – have existed long before technology created the ability to get access to key information without physically breaking in. The next section discusses how cyberattacks change the landscape.

Stealing Proprietary Information and Intellectual Property

Much of the information exchange between a company and its supply chain partners involves design files and exchanging technical details including drawings, specifications, and the like. The contents of these files may represent a significant competitive advantage that could be lost if exposed.

The attackers may be aligned with a competitor seeking to gain access to your proprietary files or may be a cog in a more sophisticated nation-state initiative. In one example, state-sponsored attackers targeted multiple independent global aviation manufacturing companies providing components of the same project to gain a complete understanding of the supply chain and underlying technologies of the larger project. Data was stolen for more than three years before being detected, which supported the production of the Comac C919 aircraft to compete against Boeing and Airbus, bridging the technology gap needed to produce the same components locally by state-owned enterprises.[6] In another case, a high-end fashion designer had actual design drawings for purses stolen, answering the question of how knock-off purses seem to emulate the real thing so closely.[7]

Exhibit 5.3 shows the threats from the advanced persistent attack (APT) groups operating in China.[8]

EXHIBIT 5.3   Threats from APT groups in China

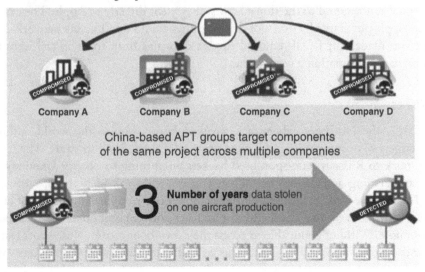

China-based APT groups target components of the same project across multiple companies

3 **Number of years** data stolen on one aircraft production

Source: FireEye.

Financial Gain from Ransomware

An increasingly more common threat is ransomware, in which cyber criminals penetrate a target network and hold its information hostage by encrypting it until a significant ransom is paid. In addition to the demand for payment, a victim's system can be paralyzed until the attacker unlocks the key to their network, resulting in a huge disruption to their operation.

The rapid rate of technology is developed largely for beneficial purposes, but those same tools can be used for criminal motives as well. Ransomware employs many of these techniques, taking advantage of our increasingly connected world. Many ransomware attacks are funded and managed by organized crime syndicates, emulating Silicon Valley's software as a service (SaaS) model, known as RaaS, or "ransomware as a service." These RaaS providers give criminals the software they need to attack and lock up target networks for a percentage of any ransom paid.

Ransomware attacks are very effective with organizations providing critical services, putting them in a very difficult situation. Health-care providers need to either cease treating patients or pay up. Companies with highly valuable corporate secrets risk having them released on publicly promoted websites. Paying ransom to regain control of their systems is a very difficult and expensive decision, all the while recognizing that paying up just adds to the hackers' coffers and encourages these threat groups further.[9]

In May 2021, Colonial Pipeline, an American oil pipeline system that carries fuel to much of the Southeastern United States, suffered a ransomware attack that affected the systems managing the pipeline. The attack caused gasoline shortages and panic buying throughout the region and resulted in a $4.4 million ransom payment, according to their CEO. Given Colonial's role within the framework of the nation's critical infrastructure, the government has increased efforts to protect these critical assets. Within two weeks, large meat producer JBS Foods was the victim of a similar ransomware attack, forcing them to close nine of its US beef processing plants until they made a payment of $11 million, but not before the nation's supply of beef was impacted with shortages and higher prices. The attack was credited to REvil, a Russian-based RaaS provider.[10]

This method of extortion can hit manufacturing organizations as well. Attackers also targeted one of Apple's key partners, Taiwanese laptop designer Quanta in early 2021.[11] Hackers using the REvil service originally demanded a $50 million ransom from Quanta, but since Quanta would not respond to the threat actors, they went directly to Apple. The extortionists demanded Apple purchase blueprints of drawings they found on Quanta's network. No public disclosure was made, but REvil removed Apple's drawings from their official website which leaks proprietary information a week later.

Political Motives

In June 2017, the NotPetya cyberattack wreaked havoc around the world, although the intended targets were companies in Ukraine using a local tax filing software. The CIA attributed the attack to Russian state-sponsored hackers attempting to disrupt Ukraine's financial system masquerading as ransomware. When customers downloaded the tax software, they installed NotPetya as well.[12] However, NotPetya took it one step further. Instead of encrypting a victim's records, it overwrote them with the attacker's own master boot record and displayed the ransom note. Without being able to access their own boot record, a user can't restore their system and recover their data. The virus wiped data from the computers of banks, governments, energy firms, and the like. This approach is taking ransomware to extract financial payments one step further into the category of extortion.

Denmark's Maersk, the world's largest container shipping firm, was critically impacted by NotPetya even though they were not an intended target, a form of collateral damage in this new form of warfare. Their operations were paralyzed for 10 days at an estimated cost of $300 million, and they had to replace 4,000 servers and 45,000 computers (and install 2,500 applications), which had their master boot records overwritten and could not be restored.[13]

Disruption of Operations

Cybersecurity experts are increasingly more concerned about the shift from data theft to disrupting critical services. Some cyberattacks are intended to disrupt the operations of a competitor or enemy, creating opportunities to level the playing field or to create chaos and uncertainty. The threats can be initiated by individual threat groups or sponsored by nation-states with different motives and methods.

As you can see, the reasons behind cyberattacks may have multiple dimensions to them, which can add to the impact created by them or blur the ability to identify the attackers. Bottom line, the supply chain is much broader than the traditional view of manufacturing partners. In a study by Symantec in 2019, supply chain cyberattacks were up 78% over 2018, with as many as 80% of breaches starting in the supply chain, according to a different study by Verizon Enterprise.[14]

Methods Used to Breach Target Systems

Given the potential gains as viewed by the perpetrators, the skills they use match the most advanced techniques in technology. They are continually morphing and growing in sophistication, partly to adopt the latest technologies and partly to avoid detection. There are numerous methods hackers use to accomplish their objectives. Let's look at a few of the most common methods that could target the supply chain or manufacturing floor.

Using a Third-Party Connection as a Means to Get to Your Network

Inadequate controls in the exchange of information between you and your suppliers can allow threat actors access to your internal systems, a side door when the target has established a strong perimeter defense around their own network. Once breached, an attacker can move laterally within the victim's network to get financial information, personnel records for key employees, intellectual property such as design files, confidential information regarding your customers, pricing and volumes, and more. Lax controls within your supplier's systems can also be a path to get access to your intellectual property and need to be audited regularly. We talk about this later in the section on prevention.

Using a Third-Party Connection as a Means to Get to Your Customers' Information

As discussed, you may not be the ultimate target when threat actors attack your information through a connection with one of your partners. It is critically important that you protect not only your sensitive information, but also that of your customers. The SolarWinds breach is a classic case of island hopping, leveraging a source with broad reach to get to many targets in a shotgun approach, even though there may only be a few specific targets.

Tampering with Components or Products in the Manufacturing Process

One key area product designers and manufacturing engineers must protect against is the potential for malicious code being inserted into custom parts or software used in the manufacturing process. It is particularly important that components with firmware or custom circuitry not be tampered with, either, in the production of the components themselves (similar to the SolarWinds breach, but in the hardware). Hidden code implanted in a finished unit shipped to a customer can eavesdrop or otherwise compromise the functionality of the unit. Without strong controls within the manufacturing environment, unauthorized production techniques or untrained personnel can also be a route to compromising your or your customers' environments.

Tampering with Manufacturing Process Equipment

IOT products are becoming rapidly adopted across multiple industries. The emergence of this new class of intelligent Edge product has caused the landscape of Cloud-connected devices to explode, with a seemingly endless set of applications, each with its own firmware layer and even its own operating system. Although there are many tools focused on evaluating vulnerabilities and threat detection, the number and variety of devices makes it a challenge to protect them at the firmware level.

These devices often integrate third-party software but may not always analyze the components they include, which could lead to devices shipping with unknown security vulnerabilities. It is critical that device builders assess device risk all the way from firmware to network levels, or chip-to-cloud. Chip-to-cloud is a technology to create secure-by-design devices using energy-efficient microchips that stay connected to the Cloud at all times.

Conversely, organizations are beginning to adopt AI as part of their security automation strategies, with more and more utilizing IoT devices. AI security platforms can reduce the cost of a compromised record and employ machine learning and analytics to help security analysts identify and contain breaches.

What Are the Potential Costs of a Cyberattack?

A recent poll among the National Association of Corporate Directors found that 42% of 500 leaders surveyed list cybersecurity risk as one of the top five issues they face. Board members look to managers of the supply chain to ensure that the company is protected from cyber threats emanating from third-party relationships. Managers are responsible for answering two key questions:

1. Do you know who your suppliers are and who their suppliers are?
2. Do your suppliers know how they are managing their product and service risks?

Lack of knowledge can create blind spots. Depending on the nature of the attacks, the motives behind them, and the effectiveness with which they are carried out, cyberattacks can have varied and devastating consequences to the victim.

- **Lost Revenue and Customer Loyalty.** Per a study done by Cisco, 29% of breached companies lost revenue as a direct result of a breach, in part due to 7% of customers ceasing to do business with the breached company.[15]
- **Reputation and Good Will.** Failure to protect customers' sensitive information can be considered a serious breach of the trust they place in you, exacerbated by the cost and pain they bear dealing with identity theft, and so forth.
- **Lawsuits.** Given the impact on customers whose personal identities were compromised, Target faced 92 lawsuits in response to their high-profile breach and paid out $39 million in lawsuit settlements to impacted banks.[16]
- **Fines.** Companies need to worry about the reaction from not only customers when a breach is disclosed, but government entities responsible for consumer protection as well. European Union privacy laws governed by the GDPR (General Data Protection Regulation) enacted in 2016 levies strong fines to companies found to be liable for breaches impacting their customers, with a maximum fine of 4% of annual turnover. For a company with $1 billion in annual revenues, the maximum would be $40 million.[17]
- **Loss of Shareholder Value.** Shareholders also have a strong interest in the companies they invest in. Ponemon cited an average 5% drop in shareholder value on the day a breach is disclosed publicly.[18]
- **Business Disruption.** Maersk was brought to its knees for 10 days while practically the entire IT infrastructure (servers and laptops) were replaced due to the NotPetya attack crippling access to their network. The entire company had to work manually for 10 days and reported a 20% drop in volume.[19]
- **Cost of Remediation.** Per Ponemon, the global average cost impact of a data breach reached $3.8 million in 2018, but in extreme cases can go much higher. The cost to Maersk for identification and remediation of the breach, replacing all their servers and laptops, and losing all productivity for 10 days was estimated to cost between $250 million and $300 million.[20]
- **Job Security.** Those held responsible for protecting the integrity of the company's sensitive information bear significant risk, even if they have solid controls in place but find themselves to be no match for the technical prowess of a threat group.

Protecting Against Cyberattacks

Given the high stakes associated with potential breaches and the degree of sophistication shown in many attacks, preventing such threats can be a daunting task. And while many security experts largely accept that breaches are inevitable due to many factors including human error, lax security procedures, and alert fatigue, to name a few, the impact of a breach does not have to be catastrophic.

In developing a strategy and actions to secure your sensitive information, these same security experts suggest assuming the worst and developing defenses in the event you are compromised.

Why? Putting a defensive strategy in place changes the decision matrix on what steps you take and how you implement them. A comprehensive set of tools can prevent, identify, contain, and remediate attacks within hours. "The most important takeaway from the recent spate of ransomware attacks on US, Irish, German and other organizations around the world is that companies that view ransomware as a threat to their core business operations rather than a simple risk of data theft will react and recover more effectively," according to Anne Neuberger, the top cyber official in the US National Security Council.[21] This clear escalation in high-profile attacks has become a major governmental policy challenge. Gaining agreement among competing nations that responsible states don't harbor and support ransomware criminals is the best approach to controlling this growing threat.

In parallel, organizations need to take these threats seriously and match their investment and prioritization of defenses with the potential threat. Putting controls in place for early detection is critical to minimizing the impact of a breach. The key is knowing how to react such that the impact is contained by reducing the response time to mitigate the damage and having a plan to recover quickly and completely.

Approaches to Mitigating Risk

As organizations take inventory of all the considerations that need to be included in a comprehensive cybersecurity strategy, it's important to start with a high-level view. This approach will be key to keeping executive and board-level visibility, which will drive key decisions including a cyber risk budget, associated headcount and equipment planning, and establishing internal controls.

The first question to ask is, "What is the probability of a breach?" given the nature of the business. The higher the perceived risk, the more of an investment is required. From a supply chain point of view, there are a few key considerations:

Does the product design absolutely require innovative and custom design content that could become a target for hackers to compromise?

If the answer is yes, intellectual property could be stolen and used for a number of purposes:

- Implementing your intellectual property and competitive advantage into competing products
- Inserting malicious content in the products you ship to your customers for eavesdropping purposes or as a route to their environment
- Impacting product quality to undermine your brand and reputation

Can you eliminate custom code within components that would allow a threat actor to target your design and your company far down in the supply chain where the levels of control may be lacking, or not audited fully?

If not, it's extremely important to develop and monitor a complete set of controls in the electronic communications, test validation, and qualification phases to ensure that the initial design is clean and to apply the same rigor to guarantee that the production process replicates the approved design and is free of any tampering or compromise.

All parts with firmware or code need to be vetted, preferably before being inserted in the manufacturing process but absolutely in final testing before the completed product is shipped. There are a number of tools to verify that actual parts are produced to the approved design, such as monitoring revision levels and checking digits against the latest published spec and adding test scripts to verify the parts are clean, visual imaging to identify unauthorized components, and so forth. It's also critical to verify that nothing was added or tampered with in the production process. Thorough verification in the test process, using visual inspection techniques, and so on can all perform important validation steps.

Protecting the flow of finished goods from the factory to the end user is also critical in preventing goods from being compromised in the distribution process. Auditing any distribution channel partners in much the same way as inbound component suppliers is important, as is the use of Customs Trade Partnership Against Terrorism (CTPAT)-certified carriers that provide the highest levels of cargo security through cooperative steps and policies among stakeholders in the international supply chain (importers, carriers, brokers, manufacturers, etc.). CTPAT is a voluntary public–private sector partnership working closely with the US government's Customs and Border Protection agency.[22]

Is the lowest-cost solution a requirement that will likely lead to outsourcing production offshore to a low-cost region?

This is one of the most common challenges facing an organization: the tradeoff between low costs and potential risks in quality, predictability, and protection of intellectual property. The risk of having intellectual property stolen grows as you add more elements of risk in the sourcing location strategy, countries known to have poor track records in protecting intellectual property, suppliers that are young, small, do not do a lot of business within your industry or set of competitors, and so forth. Are you willing to accept minor premiums by sourcing a solution with inherently less risk in these key areas (e.g., Mexico vs. China, or a well-known tier 1 provider vs. a somewhat unknown but aggressive emerging contender)?

To give one example, a leading security firm assessed their risk of compromise as being high given they would be a "trophy prize" to hackers due to the trove of security tools they possess and access to their customers' most sensitive information. Given their position in the industry, the cost of a breach would also be very damaging, so their risk tolerance was very low. To minimize the risk of a breach through their manufacturing supply chain, they made two strategic decisions.

First, the firm chose to adopt a hardware design philosophy of placing all the differentiation in the software sitting on top of the appliance. They have no custom electronic content in their products, sourcing industry standard "catalog" items. This approach eliminated the potential of a threat actor targeting them far down in the supply chain and allowed them to focus on the final manufacturing process itself, where peripheral products (disc drives, memory, etc.) were added.

Next, they selected an industry leading tier 1 contract manufacturer with rigorous hiring and training practices, building product in locations deemed less risky. All finished goods were shipped via CTPAT-certified carriers with no intermediate stocking locations of distribution centers, minimizing the chance of goods being overtaken in transit to the final customer.

Exhibit 5.4 shows the first two steps in determining a cybersecurity strategy.

EXHIBIT 5.4 Determining a cyber risk strategy

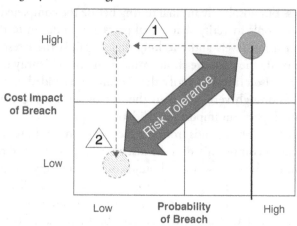

Finally, they minimized potential cost impact of a breach. There are two key ingredients to making sure a breach does not have a crippling effect – having the shortest response time to identify the breach and mitigate the damage and having a process to recover quickly and completely. As a cybersecurity vendor, the firm had a full suite of tools, from threat-monitoring intelligence to prevention to monitoring and remediating threats to reduce the impact of any breach that might slip through.

There is growing awareness in the industry of the need to reduce the time from breach to detection. In 2011, the average time from exploiting a network to being detected was a staggering 416 days, largely due to techniques where threat actors would penetrate a network and lay dormant for periods of time to avoid detection before moving laterally to their intended destination. That number was down to 78 days in 2019, partly due to attack vectors being more easily identified and better tools, but also partly due to the rise in ransomware attacks where criminals purposely make their presence known upon breach to extract financial bounties.[23]

Employing a full suite of tools starts with intelligence-led prevention, understanding what threat actors are doing in your industry segment, why they're doing it, and how. Vigilant monitoring of network traffic to identify legitimate threats is key (vs. approaches that identify and quarantine pedestrian threats that can number in the tens of thousands daily and are largely false positives). This often results in fatigue and complacency while missing the actual serious threats.

Developing Internal Processes and Controls

Before considering the process of vetting outside suppliers, it's important to establish strong internal policies and controls to reinforce a company's commitment to cybersecurity to its employees and partners. It's important to keep in mind that effective cybersecurity requires a multifaceted approach that is supported by the company's leadership – the tools used to protect company information, controls regarding employee policies, and comprehensive employee training.

The biggest cyber threat to any group is its employees. Most breaches are a result of human actions, whether unintentional or intentional. An IBM report found that insiders were behind

60% of cyber attacks, whether on purpose or by accident. Lots of people make simple mistakes: visiting malicious websites, using compromised USB drives or other personal devices at work, or sharing sensitive information and credentials with another person. Then there are malicious insiders who intend to do damage, at times planted or approached by organized crime syndicates to reach sensitive data through authorized access.[24]

Here are a few measures that you can implement to reduce the risk from insider threats:

- **Access policies.** Limit employees' access to only the resources they need, and limit accessing critical information from outside the network, using personal devices, and so on.
- **Employ strong authentication methods.** Require appropriate identity access and use tools like two-factor authentication before granting access to the network.
- **Unsecured device policy.** A mobile workforce is here to stay, but security cannot be compromised. Employees may not adhere to company guidelines or may use an unsecured Wi-Fi network, so a strong and well-communicated policy is necessary.
- **Providing cybersecurity risk training to employees.** Every employee plays a role in protecting and promoting digital safety. A comprehensive training program helps in reducing the threat from potential scams and phishing attacks and other unintentional actions that open doors for attackers.

With executive-level support, coordination across all functions in the supply chain is essential. This includes information technology, sourcing, supply chain, design engineering, legal, quality logistics, facilities, human resources, and so on. Many organizations have established a supply chain risk management council with an executive sponsor and representation across the enterprise to ensure that senior leaders are stakeholders in this critical aspect of protecting the company's resources.

It is also important to recognize there should be no gap between physical and information security. Threat actors can exploit physical security lapses to launch an attack, as in the case of Target's well-known breach when hackers used Target's air conditioning vendor's security access to break into their data center. Conversely, attackers can leverage cyber vulnerabilities to gain access to a physical location.

The foundation for establishing secure controls lies with the employees, given that most breaches are the result of human error, many times induced by threat actors through phishing campaigns to look like legitimate emails or the result of carelessness, fatigue, or complacency. It is essential that employees are aware of the risks they pose and are given the skills and techniques to protect their information. Establishing a model program internally can then become the benchmark for how third-party suppliers are measured.

Developing Secure Third-Party Relationships

So what are the best practices in reducing the risk of breaches within the supply chain? Here are some key principles to follow:

- **Know your suppliers and partners.** Identify and understand exactly who your suppliers and partners are and do basic research on them. Every company needs to understand their partners' cybersecurity risks, as well as their own.

- **Use approved suppliers.** Every company should have a supplier approval process, some type of methodology or protocols in place to vet each supplier that they are working with. This will also assist when determining if suppliers are breaking any social or ethical responsibility rules.
- **Know your contracts inside and out.** Make sure you have strong partnership agreements and contracts. This will make it harder for suppliers to do things under the cloak of darkness.
- **Constantly evaluate your suppliers.** On a regular basis, you should be formally evaluating your suppliers and their performance. This also means that your suppliers will have reason to maintain – or improve – the products and services that they deliver to you.

Establishing strong relationships with supply chain partners is a requirement in today's world. Before even selecting supply chain partners, it's important to be aware of these considerations and measure them against your risk tolerance profile relative to the potential benefits they bring.

Exhibit 5.5 shows the role of risk management in pursuing profits while avoiding losses.[25]

EXHIBIT 5.5   Risk management crossword

Source: From Shutterstock.

Avoid sourcing from low-cost suppliers that are:

- Not familiar to your organization and located in distant low-cost regions
- Young, small, or emerging companies with immature cybersecurity
- Making inadequate investments in process controls, quality management systems, and so on
- Weak in their vendor management processes to ensure that inbound parts meet specifications
- Lacking in their ethical standards and commitment to social responsibilities

Avoid sourcing technology from vendors that are:

- Lacking in internal controls when collaborating on proprietary development work
- Prone to reuse or sell your IP to competitors

Avoid selecting distribution partners that are:

- Smaller operators operating in remote markets
- Using lower internal control standards and operating in smaller markets
- Prone to violate agreements on gray-market policies

Once you are comfortable with a candidate's position on the risk/reward continuum, it's critical that you follow a rigorous auditing process to evaluate their security awareness and commitment, processes, and controls. Following are a few audit considerations before formally engaging with a new supplier.

Key categories in the qualification process and that should be included in contractual agreements are:

- Security governance
- Manufacturing/operational security
- Software engineering/architecture
- Incident management
- Transportation security
- Physical and environmental security
- Personnel security
- Information protection
- Vendor security (subtiers, service providers, Cloud providers, etc.)

Specifically, here are a few key questions to ask:

- Are the vendor's hardware and software design processes documented, repeatable, and measurable?
- Are known vulnerabilities mitigated in product design phase?
- How does the vendor stay current with emerging new "zero-day" threats?
- What controls are in place to monitor production output and quality?
- What steps are taken to eliminate risk of tampering? Risk of backdoors?
- How do they vet their personnel, particularly those with access to customer information and systems?
- How well do they vet their service providers? Even janitorial services can pose a risk.
- How well do they vet and control their own designs, particularly those with IP that will be integrated into customers' products, including your products?

Summary

The challenges of maintaining a secure environment to protect your company's and your suppliers' sensitive information is not a trivial task. That said, it is possible to establish an effective strategy that balances the rewards of leveraging the specific benefits supply chain partners bring while greatly reducing the likelihood of a breach and to contain any that do manage to slip through.

Here are 10 key takeaways:

1. Globalization, specialization, and IoT trends have increased cyber risk.
2. Your supply chain is much deeper and broader than you realize.
3. The supply chain is an attractive target for several reasons.
4. No supplier is too insignificant to be immune from risk.
5. Your security controls are only as strong as their weakest link.

6. Threat actors have a wide range of motives and methods.
7. Cyber risk can be mitigated by making business tradeoffs.
8. The impact of a breach can be contained and minimized with proper controls.
9. Cyber risks need to be considered in sourcing decisions.
10. The costs of a breach can be far-reaching and catastrophic, but they don't have to be.

Sample Questions

1. Avoid selecting distribution partners who are
   a. Smaller operators operating in remote markets
   b. Using lower internal control standards and operate in smaller markets
   c. Prone to violate agreements on gray market policies
   d. All of the above
   e. None of the above
2. What are the potential risks in using low-cost suppliers?
   a. They are familiar to your organization and located close to your operations.
   b. They are larger and well established.
   c. They make inadequate investments in process and quality controls.
   d. They may lack strong ethical standards.
3. The explosion in new appliances and IoT sensors on the factory floor will
   a. Decrease cyber threats
   b. Increase cybersecurity risks because many IoT devices have inadequate security controls
   c. Will have no impact on cybersecurity
   d. None of the above
4. The SolarWinds security breach was caused by
   a. Malicious code imbedded in IoT devices
   b. Poorly written software code
   c. Malicious code imbedded in SolarWinds' software
   d. The software providers selling their IP
5. Cyber criminals penetrating a network holding its information hostage by encrypting it is a description of
   a. Blackmail
   b. Hostage taking
   c. Ransomware
   d. Sabotage
6. What percentage of cybersecurity incidents occurred among small firms?
   a. Above 90%
   b. About 75%
   c. About 50%
   d. About 20%
7. Why do cyber hackers target small companies?
   a. They have more sensitive information than large companies.
   b. They are able to pay big ransoms.
   c. They typically have inadequate cybersecurity.
   d. All of the above

8. What is the advantage of using CTPAT-certified carriers?
   a. They cost less than other carriers.
   b. They offer more services than other carriers.
   c. They offer the highest levels of cargo security.
   d. They offer faster delivery times.
9. A 2018 survey reported that the average number of vendors with access to a company's sensitive information is
   a. Over 150
   b. Over 250
   c. Over 350
   d. Over 450
10. In a study by Symantec in 2019, what percentage of cyberattacks originated in supply chains?
   a. 15%
   b. 25%
   c. 50%
   d. 80%

Notes

1. WiseGEEK. https://www.wisegeek.com.
2. Schaller, J. (March 4, 2021). SolarWinds: Data security in 2021. exIT Technologies. https://www.exittechnologies.com/blog/tech-news/solarwinds-data-breach/.
3. Shutterstock. https://www.shutterstock.com/editor/image/file-storage-cloud-3d-computer-icon-109953752 (accessed August 14, 2021).
4. Verizon Enterprise Security Threat Report. https://www.verizon.com/business/solutions/secure-your-business/business-security-tips/?cmp=knc:ggl:ac:ent:security:8003162844&utm_term=verizon%20threat%20report&utm_medium=knc&utm_source=ggl&utm_campaign=security&utm_content=ac:ent:8003162844&utm_term=verizon%20threat%20report&gclid=CjwKCAjw092IBhAwEiwAxR1lRl-Z_opqpTag0l6tS5tQEUDC6a9zF-Ge-xDoUwLhqzW8Snhd_XAMcBoC8OYQAvD_BwE&gclsrc=aw.ds (accessed August 14, 2021).
5. Ponemon Institute. (2018). 2018 Cybersecurity Report. https://www.ponemon.org.
6. Constantin, L. (October 14, 2018). Report: China supported C919 airliner development through cyberespionage. CSO Online. https://www.csoonline.com/article/3445230/china-supported-c919-airliner-development-through-cyberespionage.html.
7. Chinese cyber espionage operation conducts supply-chain attack allowing access to multiple industry verticals. (August 17, 2017). FireEye. https://intelligence.fireeye.com/reports/17-00009106.
8. Advanced persistent threat groups. FireEye. https://www.fireeye.com/current-threats/apt-groups.html (accessed August 14, 2022).
9. Witkowski, W. (June 10, 2021). Ransomware boom comes from gangs that operate like cloud-software unicorns – "a truly incredible business model." *Money*. https://www.msn.com/en-us/money/technologyinvesting/ransomware-boom-comes-from-gangs-that-operate-like-cloud-software-unicorns-e2-80-94-e2-80-98a-truly-incredible-business-model-e2-80-99/ar-AAKPOpZ?ocid=uxbndlbing.
10. Uberti, D., and Stupp, C. (August 5, 2021). New hacking group shows similarities to gang that attacked Colonial Pipeline. *Wall Street Journal*. https://www.wsj.com/articles/new-hacking-group-shows-similarities-to-gang-that-attacked-colonial-pipeline-11628155802.

11. Lahiri, A. (April 21, 2021). Apple supplier Quanta hacked in $50M ransomware attack by Russian group: Bloomberg. Yahoo! Finance. https://finance.yahoo.com/news/apple-supplier-quanta-hacked-50m-113453234.html.

12. NotPetya attack. Cyber Search Tech. https://cyber-sectech.fandom.com/wiki/NotPetya_Attack.

13. Tung, L. (February 15, 2018). "Russian military behind NotPetya attacks": UK officially names and shames Kremlin. ZDNet. https://www.zdnet.com/article/russian-military-behind-notpetya-attacks-uk-officially-names-and-shames-kremlin/.

14. 2019 Internet security threat report now available. https://community.broadcom.com/symante centerprise/communities/community-home/digestviewer/viewthread?MessageKey=2b6b1950-9092-49af-ae9d-f2cf96d20bc7&CommunityKey=1ecf5f55-9545-44d6-b0f4-4e4a7f5f5e68&tab=digestviewer#bm2b6b1950-9092-49af-ae9d-f2cf96d20bc7 (accessed August 14, 2021.)

15. Cisco Cybersecurity Reports. Cisco.com. https://www.cisco.com/c/en/us/products/security/cyber security-reports.html.

16. Stempel, J., and Bose, N. (December 2, 2015). Target in $39.4 million settlement with banks over data breach. *Reuters*. https://www.reuters.com/article/us-target-breach-settlement/target-in-39-4-million-settlement-with-banks-over-data-breach-idUSKBN0TL20Y20151203.

17. Directive 95/46/EC of the European Parliament and of the Council. (October 14, 1995). https://eur-lex.europa.eu/legal-content/EN/TXT/PDF/?uri=CELEX:31995L0046&rid=5.

18. After a data breach is disclosed, stock prices fall an average of 5%. (May 16, 2017). Help Net Security. https://www.helpnetsecurity.com/2017/05/16/data-breach-stock-price/.

19. Lord, N. (August 7, 2020). The NotPetya fallout continues, with global transport and logistics conglomerate Maersk reporting up to $300 million in losses following a June cyberattack. *The Guardian*. https://digitalguardian.com/blog/cost-malware-infection-maersk-300-million.

20. Ibid.

21. Marquardt, A., and Sands, G. (June 3, 2021). First on CNN: White House pushes for companies to take ransomware more seriously after high-profile cyberattacks. *CNN*. https://edition.cnn.com/2021/06/03/politics/white-house-open-letter-ransomware-attacks-businesses/index.html?utm_campaign=General&utm_content=168928583&utm_medium=social&utm_source=linked in&hss_channel=lcp-19126566.

22. US Customs & Border Protection, US Department of Homeland Security. CTPAT. https://ctpat.cbp.dhs.gov.

23. GDPR more than halves the intrusion duration of cyber attacks. (February 21, 2020). The Defence Works. https://thedefenceworks.com/blog/gdpr-more-than-halves-the-intrusion-duration-of-cyber-attacks/.

24. Sengar, A. (May 25, 2021). Try these best practices to counter common cybersecurity risks. Security Intelligence. https://securityintelligence.com/posts/common-cybersecurity-risks/

25. Amalgami. Profit, loss and risk (buzzword crossword series). Image ID 6715479. 123RF. https://www.123rf.com/photo_6715479_profit-loss-and-risk-buzzword-crossword-series.html (accessed August 14, 2021).

CHAPTER 6

Improving Logistics Using Smart Technology

Frank Poon

Introduction: Why Logistics?

As with all physical goods in today's world, logistics and manufacturing go hand in hand, because all physical goods, be they raw materials or finished goods, have to be at the right place (e.g., factory, customer, etc.) at the right time to make a successful supply chain.

From the perspective of traditional manufacturers, logistics may not have been their primary focus for the past 30 years. This is because the traditional logistics needs of most manufacturers are primarily in two main areas: first, the logistics of getting raw materials to the factory, and second, the logistics of shipping finished goods to the retailer or business customers.

In the past 30 years, the globalization of manufacturing has made these two main areas more complex with cross-border logistics. But the majority of these manufacturers did not experience these complexities; they were shielded by companies specializing in manufacturing and logistics domains that helped lessen the brand owner's responsibilities to the complexities of globalized logistics.

A brand's logistics capability was not seen as a differentiator to the end customer. For this reason it was not a focus area of most manufacturers. In many cases, this was due to the homogeneous nature of the traditional logistics model, with primarily bulk cargo movements, relatively stable pricing, and a predictable service level. However, the attention to logistics has changed significantly in the recent years, especially with the COVID-19 pandemic, which created global supply chain shortages in items from semiconductors to cars to lumber.

In this chapter, we start by exploring the megatrends of logistics. Why should manufacturers pay more attention to the logistics part of the supply chain? What is the cost for not paying attention to logistics? Should manufacturers make logistics a competitive advantage and market differentiator? How does technology help make logistics even smarter, and what additional benefits can be gained?

Most importantly, this chapter presents high-level steps that will help fellow manufacturers contemplate designing and executing a supply chain strategy that incorporates logistics as a key component to improve supply chain competitiveness and align with the corporate strategy. In addition, this chapter attempts to highlight the key areas in logistics where Smart Technologies can be employed with use case examples. By the end of this chapter, readers will be able to decide whether logistics should be added to their strategy war chest as a competitive weapon and, if so, how to employ the corresponding technology to make it a reality.

Megatrends in Logistics That Impact Brands/Manufacturers

With the global e-commerce market expected to grow from $3.53 trillion in 2020 to $6.54 trillion in 2022, ignoring this megatrend is almost impossible.[1]

As demand-driven supply chain leveraging the use of Lean Six Sigma and just-in-time practices combined with the powerful forces of growing e-commerce dominate, logistics has become increasingly important as a key competitive capability for the growing number of brands and manufacturers. Customers are demanding faster and more reliable logistics capabilities from whoever they order from; this in turn has changed the "deliver" and "return" capability drastically in the past 10 years.

Specifically, 10 years ago, most brand owners/manufacturers sold through channels, including retailers. The primary focus of their logistics function was to deliver finished goods into retailer distribution centers and retail stores. However, this has now changed significantly with the growing dominance of e-commerce, with the following options:

1. Customers who order from platform/retailers such as Amazon, who in turn request their brand/manufacturer to deliver to their warehouse to mimic the traditional retailer's model.
2. Customers who order from platform/retailers, who in turn request their brand/manufacturer to deliver directly (drop ship) to the end consumer's hands.
3. Customers who order directly from the brand/manufacturer, who now owns the relationship with the end user, and now must perform the final mile of delivery – from the factory to the end customer's hands.

A growing number of brand/manufacturers are taking on the challenge of the third option because they see two major benefits:

1. Developing a more intimate relationship with customers
2. Enjoying higher margins through disintermediation (the reduction in the use of intermediaries between producers and consumers)

As a consequence, brand/manufacturer logistics teams are being asked to perform tasks such as last mile delivery planning and execution in areas they had no prior experience in. Furthermore, customers are taking the brands to task, as evidenced by the direct correlation between logistics performance and sales volume. In one recent study on logistics performance and its impact on customer purchasing behavior and sales in e-commerce platforms, it was estimated that reducing delivery times from three days to two days would translate to a 13.3% increase in average daily sales.[2] In a separate study, more than 55% of consumers prefer to order directly from a brand; that number is projected to go up by 14% in 2022.[3]

With changing expectations and the lucrative upside on revenue, brands/manufacturers are starting to take on the task of working directly with end customers instead of going through traditional retailer channels. Their entire customer-facing capability has to be upgraded, from customer service to finance to operations/logistics.

Making the challenges more daunting, there is a growing expectation of easy product returns in the e-commerce world. End customers are now demanding easy returns of their shipments with local and free drop-offs in convenient locations of their choice or pick-up directly from their homes. This latest trend is putting tremendous pressure on many brand/manufacturer logistics teams.

The move to e-commerce has been accelerated by the COVID-19 pandemic, and the direct correlation between logistics performance and a consumer's purchasing behavior makes it very unwise for any brand provider to ignore the importance of logistics.

The Different Expectation of Your Customer-by-Customer Type

As a brand/manufacturer who just now started to embed logistics as a key competitive strategic component, it will be prudent to understand the different customer segments of this value chain and what expectation they may have:

1. **Business-to-Business Model.** If you are in this model, your primary customers are other businesses, which can be divided primarily into two categories:
   a. **Direct materials customers.** Similar to individual consumers, end-user industrial/business customers are now expecting to get their shipments faster, with more transparency, and at a lower price. Here, the collaborations between customer and supplier become critical. End-to-end supply chain collaborations or JIT collaborations have been the theme for this, with reliable and predictable logistics service in between.
   b. **Indirect materials customers.** The proliferation of e-commerce in the individual consumer's space continues to build higher expectations for the same buyers in the indirect procurement space. Because of this, indirect procurement expectations have now become the same as those of individual customers. Customer experience becomes important and logistics performance become a key differentiator for these indirect materials customers as well.
2. **Business-to-Consumer Model.** If you are using this model, your primary customers are individual consumers. As consumer expectations continue to rise, brands/manufacturers have to keep up with those expectations. Here are some of the key expectations today's consumers have toward brands:
   a. **Personalization.** Personalization is how brands go beyond basic advertising and marketing and work to form a bond with each individual customer. Today, over 60% of individual consumers expect personalization when engaging a brand. It was only a

couple years ago that personalization was meant to be segmenting audiences based on behavior or other data to personalize messaging in email marketing. Today, however, new technologies allow brands to personalize individual marketing campaigns, to target website presentation, and to make product recommendations. In the future, consumers will expect the brand to be able to anticipate and personalize the product and/or service they receive. Every consumer will be expected to be treated like a VIP customer.

b. **Strong brand image.** Today's individual consumers favor brands that have personality. Brands need to have their own voice and style to appeal to individual consumers before they can sell something to these consumers.

c. **Quality customer experience.** Today's individual consumers are expecting a new level of customer experience, from the first time they were aware of the brand to the final delivery of the goods or services and the after-sale services. Every step in this customer journey is critical; the delivery and return of goods/services is a very important part of this entire experience. Think of an Apple store. Next-day free shipping is standard for the delivery of their product. On top of that, Apple seamlessly blends the after-sale repair service and makes the logistics part a competitive advantage. Apple has carefully curated the after-sale repair journey by providing the individual consumer with the right information at the right time, together with the right product at the right time, making their customer experience exemplary. It starts with the careful packaging of materials provided once the repair order is confirmed. Apple then provides the same level of tracking for the entire repair service cycle. Finally, Apple offers two-day shipping service of the fully repaired unit with tracking information.

The Cost of Not Paying Attention to Logistics

According to the 31st Annual Council of Supply Chain Management Professionals (CSCMP) State of Logistics Report, in 2019 US business logistics costs rose 0.6% to $1.63 trillion, or 7.6% of the nation's $21.43 trillion gross domestic product (GDP).[4] In fact, logistics cost is one of the major contributors to cost to many brands/manufacturers, putting significant pressure on their bottom lines.

Companies have spent a great deal of effort to lean out or optimize logistics since 1981 by adopting Lean Six Sigma (LSS) practices, reducing logistics costs from 16.2% of GDP in 1981 to around 8% in 2019.[5] The almost 40 years of practicing LSS with innovative business models has helped reduce excess inventory, movement, and packaging in filling truckloads going out and returning backhauls. Logistics assets are also optimized to be deployed more productively to offset fuel cost increases. All of these efforts have helped reduced the cost of logistics.

By adopting a logistics efficiency management approach, logistics-related costs can be significantly reduced, and that will translate directly into the bottom line.

The Benefits of Making Logistics a Strategic Competency

Given the new megatrend of e-commerce with rising customer expectations on brands/manufacturers in logistics and the cost of not paying attention to logistics for a company's bottom line, it is probably time for companies to elevate logistics into a strategic weapon. Here is a list of potential benefits of making logistics a strategic competency for the company:

- Lower overall cost
- More diversity in service capabilities (to support the omni-channel sales and fulfillment that today's customer demands)
- More flexible and reliable services (to provide a better customer experience)

Product, promotion, and price are the traditional competitive components, but more and more companies are realizing that time and place competencies will put the company into a superior position to compete in the marketplace and gain customer loyalty. While not every company can build their own supply chain/logistics capability in-house, this is leading to newer and more innovative business models to be formed in the supply chain space.

According to a research study into the practices of more than 1,000 manufacturers, retailers, and wholesalers by Donald J. Bowersox et al.,[6] some companies clearly stand head and shoulders above their competitors in logistics performance, and they use this superiority to gain and keep customer loyalty.

Steps to Make Logistics Your Competitive Advantage

Now that the overall benefits of making logistics a competitive advantage have been established, it is worthwhile to understand how doing so can be carried out. Following is a list of steps that help execute this concept.

1. Review your business/corporate strategy, map out your value chain, and identify where in the value chain logistics is a component.
2. For each area where logistics is a component of your value chain, perform an activity breakdown and a gap analysis to identify the gap between the future state and the current state.
3. Perform a cost–benefit analysis for each of the identified gap analysis areas.

Exhibit 6.1 illustrates a typical cost–benefit analysis for a logistics project.

EXHIBIT 6.1   Typical cost–benefit analysis

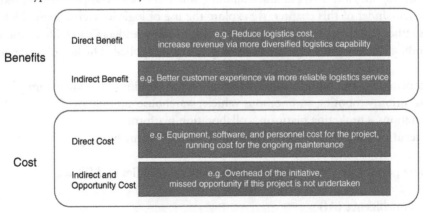

Decision : GO, IF Benefits > Cost

1. For those areas determined to have the benefits aligned with the business/corporate strategy with a good return on investment, take the gap analysis result of each identified area and perform a process mapping to identify the different processes involved.
2. With these processes mapped, identify the use cases and the technologies to be applied to achieve the future state.
3. Projects can now be initiated for each of these areas to achieve your future state.

Why Technology Is So Important to Logistics

Technology is changing everyone's life and almost every company is now on a path to digitize. This holds true for the logistics industry and applies to all logistics companies as well. Some may argue that the winners in the logistics industry of the future will be those who understand how to exploit the new waves of technologies available in the corresponding business scenarios, as identified in the aforementioned framework. Those who don't digitize their logistics risk obsolescence.

With so many technologies competing for management attention and investment, defining a clear digital strategy that is aligned and integrated into the overall corporate and business strategy is a must. According to MIT, there are five supply chain technologies that deliver a competitive advantage:[7]

1. The Internet of Things (IoT)
2. Blockchain
3. AI, machine learning, and analytics
4. Robots and automation
5. 3D printing

However, the industry has thus far been slow to seize the opportunity from new supply chain technology. In a recent industry study published by consulting firm PwC, the percentage of logistics companies that rated themselves as "advanced" on digitization was just 28%, well behind the 41% of automotive companies and 45% of electronics companies.[8] The lack of a "digital culture" and training is thus the biggest challenge for transportation and logistics companies. Thus, understanding the technologies themselves is probably not needed for any business leaders in the logistics industry, but understanding where and how to apply these technologies is.

For the remainder of this chapter, we explore the use of logistics technologies by their use-case areas rather than from a technology standpoint to allow readers a more direct and intuitive understanding on how some of these technologies can be applied. The key use-case areas are:

- Insight, planning, and monitoring – data analytics, AI-optimization algorithms
- Task execution – robotics, autonomous vehicle, warehouse automation
- Collaborations – brokering platform, collaboration platform
- Safety, security, and compliance – visual intelligence, data analytics

The more popular key technologies used in these case studies by MIT are:[9]

- Artificial intelligence (AI)
- Robotics

- Autonomous vehicles
- 3D printing
- Blockchain
- The Internet of Things (IoT)

For example, AI is enabled to improve the input parameters and data for planning, making it more accurate in predicting outcomes. AI is also used for enabling autonomous vehicles and robotics. Blockchain is enabled to decentralize the recording of every shipping container's movement while reducing errors and eliminating bottlenecks, and IoT sensors detect vehicle movements within facilities.

Let's explore each of these use-case areas and see how technologies can be applied to improve the specific logistics function.

Area 1: Insight/Planning/Monitoring

With the introduction of the advanced planning and scheduling(APS) system, warehouse management system (WMS), and transportation management system (TMS) over the past 20 years, experience-based decision making in supply chain organizations has been largely replaced by data-driven decision making through these different technologies. At the same time, these data-driven decision-making tools have always suffered from lack of real-time data and static input parameters, rendering some of the decision outcomes suboptimal. With newer and disruptive technologies such as data analytics, AI, IoT, and machine learning technologies, we are now finally at the cusp of solving these perennial issues.

Area 1, Use Case 1: Logistics Insight via Data Analytics

With vast amounts of data available to the logistics industry, 98% of third-party logistics (3PL) companies agreed that data analytics is critical to making intelligent decisions.[10] Data insights derived from this untapped wealth of data typically lead to actionable items such as optimization of resource consumption and improved delivery routings. This can also help to identify areas where automation or digitalization should be implemented.

Some of the questions in the logistics industry that can be answered via data analytics are:

- Which carrier/partner has the most compliance issues with wrongful filing and pre-alters (a freight forwarder's shipping notice to custom broker or forwarder to prepare for import clearance and delivery arrangements in advance)?
- Are on-time and loss issues tied to a particular provider or location?
- What will be the impact for the upcoming peak season pricing based on the current inflationary trend?
- Is there any second- or third-magnitude upstream logistics event that will impact the business?
- Are there better cost or service options available in the marketplace?

With better logistics insights, the following benefits can be achieved:

- Reduced costs to serve by optimizing resources and operations
- Improved customer satisfaction via improved service capability and quality

Area 1, Use Case 2: Advanced Forecasting via Machine Learning and AI-Based Prediction

By leveraging massive amounts of data from weather patterns, extreme weather events, earthquakes, or pandemics, and sales information, machine learning and AI-based prediction will keep improving forecast accuracy.

Exhibit 6.2 provides an overview of the machine learning pipeline to enable such predictions.[11]

EXHIBIT 6.2 Machine learning pipeline

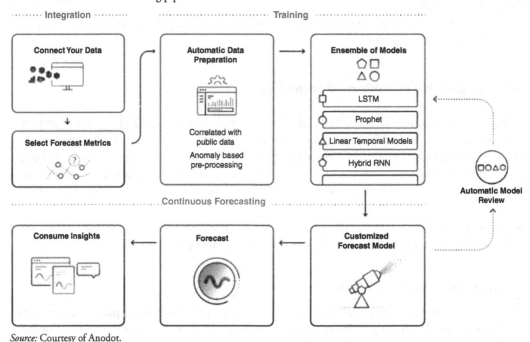

Source: Courtesy of Anodot.

There are plenty of examples of success. One such case is with consumer-packaged goods (CPG) manufacturer Danone. From a study by the Capgemini Research Institute by using machine learning, Danone can meet demand from product promotions and achieve its target service levels for channel- or store-level inventories. The system led to a 20% reduction in forecast error, a 30% reduction in lost sales, a 30% reduction in product obsolescence, and a 50% reduction in demand planners' workload.[12]

With better demand sensing/forecasting, customer satisfaction improves with prepositioned inventory, and inventory costs are reduced and are positioned correctly without any repositioning.

Area 1, Use Case 3: Dynamic Decision-Making via Machine Learning and AI-Based Prediction with Real-Time Data and Blockchain

With the growing availability of real-time data via the IoT transmission using 5G wireless networks coming soon, the future of dynamic logistics decision making has never been brighter.

Coupled with the use of Edge computing near the physical action, and with global parameters and considerations to be exchanged with the Cloud computing platform, AI-based dynamic decision making can create superior decision-making capability for areas such as ad-hoc delivery route optimization and automated dynamic job rerouting.

Additionally, new AI-based software solutions with real-time integration of all signals provides real-time supply chain visibility. This enables leveraging data analytics and crunching large data sets to optimize future performance and to make needed adjustments. Adding blockchain will allow immutability capability, substantially increasing trust in the process.

For example, one logistics provider leveraged real-time insights provided by multiparty integration, predictive alerts, and prescriptive recommendations generated by AI-based solutions to quickly avoid routing traffic through the Suez Canal within minutes of the March 2021 blockage that lasted six days. This prevented major customer order delays, winning the praise and loyalty of customers.

With better dynamic decision making, the following benefits can be achieved:

- Improved supply chain efficiency with a lower number of escalated firefighting events
- Lower costs to operate the new supply chain operations model, as automated decision making will take over some traditional manual tasks

Area 2: Task Execution

Logistics has been a labor-intensive industry and will continue to be so. However, recent innovations by several high-tech startups in key technology areas offers the promise of major labor savings. Specifically, the following three areas are most interesting.

Area 2, Use Case 1: Automation for Information Processing via Robotics Process Automation

This area applies to all the paperwork that needs to be processed in the logistics industry, from compliance to bills of lading and billings. Today, there are still too many processes and too much paperwork manually processed via email. Integration has been lagging in the entire industry, as the cost of integration is relatively high because logistics partners keep updating. The thin margin of the logistics player and the relatively low cost of entry has constantly played into this dynamic.

With the proliferation of robotics process automation (RPA) capabilities and the ease of adoption, there is now a good value proposition for even the smallest logistics players to adopt simple RPA technology. This is going to accelerate the digitization of logistics in much the same way that email replaced fax.

Here is a list of potential areas where RPA can help:

- **Data entry automation:** Any repetitive data entry tasks, such as downloading invoices and entering them into a different system, can be automated via RPA workflows.
- **Logistics alerting:** RPA bots can be used to trigger alerts if there is a delay in logistics shipments, reducing the manual monitoring and tracking required.
- **Automated reordering:** For logistics supplies, automated reordering can be triggered by RPA bots when the inventory level is below a certain threshold.

- **Load matching:** For manual tasks such as load matching with transport availability, RPA can help identify the different transport availability and perform matching scoring.
- **Vendor selection:** Vendor selections require a lot of manual due-diligence work such as checking credit ratings and other background information. RPA bots can help perform these repetitive tasks and streamline the process.

With less human touch, the following benefits can be achieved:

- Reduced human error
- Improved operation efficiency and cost

Area 2, Use Case 2: Automation for Physical Tasks via Robotics

With the advancement in AI-enabled robotics, the use of automating more complex and difficult physical activities has been accelerating over the past five years. With labor availability issues and the exponential growth of e-commerce, the robotics industry is also expected to grow exponentially. According to a study by Marketysers Global Consulting LLP, robotics technologies continue to grow at a compound annual growth rate (CAGR) of 21.3% to reach an estimated market size of $18.58 billion by 2026.[13]

The following is an overview of the different robotics applications being used in warehouses and distribution centers:

- **Packaging:** Robots can help in multiple packaging applications. Robots are becoming more versatile and affordable. Popular applications include robotic pick and place, robotic palletizers, and robotic case erecting.
- **Loading and Unloading:** Robots can also help with heavy-duty workloads such as loading and unloading. There are a variety of companies offering solutions such as automatic truck loading and unloading, plus exoskeleton robosuits to augment human strength for complex loading and unloading.
- **Piece Picking:** Piece picking using robotic arms has been advancing at an accelerated speed. Engineers are solving some of the toughest picking issues such as flimsy packaging, irregular shape packaging, and fragile contents.

With a collaborative approach between robotics that take on more mundane and physically demanding tasks while humans take on more complex tasks, the following benefits can be achieved:

- Reduced human error
- Improved operation efficiency
- Improved workplace safety

Area 2, Use Case 3: Autonomous Transportation via Autonomous Technologies

According to the American Trucking Associations (ATA), in 2018, the trucking industry was short roughly 60,800 drivers, which was up nearly 20% from 2017's figure of 50,700. The average age of over-the-road truck drivers in America is 46.[14] While every year there are more than 400,000 new drivers becoming heavy-duty truck drivers, the ATA also states that the real shortage is in retention and long-haul trucking. According to the ATA's own statistics, the average annual turnover rate for long-haul truckers at big trucking companies has been greater than 90% for decades.[15]

The key reasons for the driver shortage are long periods away from home, boredom, and job safety. With autonomous technology, there could be a reversal of this trend, as autonomous technology can assist in all these areas, as described next.

First, advances in technology can bring additional safety; with areas such as speed, acceleration, and breaking controlled by automation with zero reaction time, road safety can be improved.

Second, long-haul trucking requires long working hours; with autonomous technology advancing to the next level, one may imagine that the long-haul trucking autonomous capability could be similar to that of airplanes. While some of the simple cruising tasks are performed autonomously by the computers, the skills and expertise of the pilot are still crucial, especially during takeoff and landing and any other complex maneuvers. The technology could work alongside the human crew to reduce the stress level that is experienced today.

Third, with innovative operating models such as platooning (forming a so-called "truck-chain" by linking multiple trucks together), one long-haul driver could potentially platoon multiple trucks, increasing the total payload and thus increasing revenue for the truck service, translating to a higher take-home pay for the truck driver.

With a collaborative approach between autonomous technology and human crew, the following benefits can be achieved:

- Improved driver safety
- Improved operation efficiency, leading to improved cost efficiency

Area 3: Exchanges and Collaborations

With the proliferation of the Internet and Cloud technologies over the past 10 years, there have been many new logistics platforms introduced with the goal of creating value through one of the following means:

- Reducing the frictional cost of doing business
- Improving the visibility of disruption events and allowing the stakeholders to communicate more effectively

Area 3, Use Case 1: Digital Freight Brokerage via Cloud Technologies

A digital freight brokerage is an online "matchmaker" that connects shippers (individual companies or shipping carriers) with logistics service providers. It typically uses a mobile or online application with sophisticated algorithms to connect buyers (shippers) and sellers (logistics service providers) quickly and effectively. There are different types of digital freight brokerage nowadays. Many typically focus on serving a specific market, for example, US domestic trucking, international shipping, or domestic and international parcel. A majority of them tend to focus on ad-hoc shipment requirements of the shippers. The latest digital brokers use artificial intelligence and machine learning to optimize routes and predict supply and demand to help find the best match with the most optimal cost and service level for the shipper.

Here's how a digital freight brokerage process typically works:

1. A shipper has a load they need to send out. They have a customer account with this digital freight brokerage, but they don't have a contract with a logistics service provider directly.
2. The shipper goes on the website or app of the digital freight broker and inputs information about the delivery, such as the pickup location, weight, number of parcels/pallets, destination, timeline, special accommodations, and so forth.
3. The digital freight brokerage runs an algorithm to find the best-matched logistics service provider to accommodate those needs.
4. The digital freight brokerage either leverages a predefined service level contracted with the logistics service provider or alerts the logistics service provider of such a request and would request the logistics service provider to accept/deny the request, similar to the process of an Uber driver.
5. Once there is a match between the shipper and logistics service provider, the predefined terms between the shipper and digital freight brokerage will apply and the predefined terms between the digital freight brokerage will also apply. In some other cases, the terms between the shipper and the logistics service provider will be settled directly, without the digital freight brokerage being involved.
6. Once the transaction is confirmed, the logistics service provider will execute the request as directed and provide the corresponding logistics service.
7. The digital brokerage will bill the shipper and in turn settle the payment with the logistics service provider, receiving a commission for being the matchmaker.

With the digital freight brokerage, the following benefits can be achieved by the manufacturer:

- Providing greater availability and flexibility to shipper
- Reducing frictional cost of doing business with a large number of logistics provider
- Reducing human error, compared to the old days of manual phone-call matching

Area 3, Use Case 2: Supply Chain Collaboration via Cloud Technologies

With the complexity of today's global supply chains, any major enterprises that sources and manufactures in multiple countries will greatly benefit from the capabilities provided by supply chain visibility and collaborations solutions.

In supply chain visibility and collaborations solutions, the ideal scenario is for an AI-enabled platform to automatically detect any anomalies in the supply chain and to alert the corresponding stakeholders about the potential impacts, allowing stakeholders to work together to find a solution to mitigate those identified impacts. The ideal state is a supply chain that continuously runs smoothly without any backlog or issues.

Since supply chain collaboration involves sharing data with the different parties, the level of sharing must be considered properly by the sharing parties, as some information may be considered sensitive depending on the relationships of the sharing parties. According to Dominic Telaro,[16] there are three levels of supply chain collaboration:

Level 1: Transaction integration. This level of collaboration primarily focuses on transactional data such as sales order, work order, autonomous system number (ASN), and so on.

Level 2: Supply chain management information sharing. This level of sharing includes more forecasting and planning information, such as production or component forecast, production and transportation plan, bill of materials, promotions, inventory, and so on.

Level 3: Strategic collaboration. At this level the partners are taking part in joint planning and process redesign, as well as sharing some level of risk and reward. Typical collaborations include improving forecast accuracy, resolving critical supply chain, and joint production and fulfillment planning.

With supply chain visibility and collaborations, the following benefits can be achieved by the manufacturer:

- Reduced transportation and warehousing costs
- Reduced out-of-stock frequencies
- Reduced problem resolution cycle time in supply chain incidents
- Earlier and quicker decision making

Area 4: Safety, Security, and Compliance

Area 4, Use Case 1: Global Trade Compliance via Cloud Technologies

With today's global trade environment, cross-border logistics is commonplace in almost every industry. At the same time, actually shipping any item across borders is no small feat. Together with the heavily regulated nature of the transportation and logistics industry with myriad federal, state, local, and international laws to be followed, a considerable compliance burden is naturally borne by those who operate within the environment. This poses significant risks also for all the companies that are associated with them, from legal to financial to public relations risk. However, according to a study by PwC, only 15% of participants in the PwC webcast on global trade issues have reported that they deployed an integrated global trade solution across multiple locations.[17] Global trade compliance solutions can help to mitigate risk by actively monitoring all regulations relating to this topic for the countries monitored, as well as to alert the users.

Following are some key elements of trade compliance to be considered and the risks associated with them if not handled correctly.

1. **Goods classification.** The correct classification of goods using commodity and tariff codes is fundamental for customs compliance. If the classification is wrong and if Customs finds the error, it will raise a red flag and potentially lead to fines and penalties.

    Another part of product classification is to ensure that the product that is subject to export control has the correct license requirements established and licenses and approval obtained prior to any attempt to export it. It is important to understand that export control is critical, as noncompliance with export control will lead to substantial fines and even imprisonment.

2. **Country of origin.** Establishing the country of origin for the product to be imported will allow the determination of preferential or nonpreferential origin. Typically, if the country of origin is a preferential origin, it will lead to potential benefits of reducing tariffs if there is a trade agreement between the countries or block of countries. Such benefits should not be ignored, as the tax and duty benefit maybe significant.

3. **Incoterms.** Incoterms is an acronym standing for international commercial terms. It is used to define the responsibilities of the buyer and seller along the shipment lifecycle. It is integral to a contract between the buyer and seller so that both sides are clear on ownership, risk, and responsibilities. Getting incoterms wrong could lead to unclear ownership and responsibilities, leading to increased risks and exposure.

4. **Licenses and permits.** Licenses and permits are often required when importing and/or exporting certain products, such as those related to technology, medicine, food, and so on. It is important to get these right as not getting them correctly may be a criminal offense. Goods will also be detained and/or confiscated by the customs.

5. **Screening.** The screening of customers, vendors, vessels, and transaction data against global sanctions lists, sanctioned and risky entities, and other official lists flagged as relevant to trade counterparty risk will help ensure that you are not inadvertently doing business with undesirable party. At the same time, checking the destination such as cities, seaports, airports, and free trade zones in sanctioned countries is also a prudent step in mitigating risk. Screening should be done continuously to reduce potential breaches, as sanctions breaches are criminal offenses.

6. **Valuation.** All imported goods need to have a declared value in order to calculate customs duty, import value-added tax (VAT), and trade statistics. Thus, every shipment must have an appropriate value associated with it that should be defendable. Valuation errors can lead to fines and penalties, along with under- or overpayment of duties.

With global trade compliance solutions, the following benefits can be achieved by the manufacturer:

- Reduced fines and penalties due to noncompliance events
- Reduced chances of breaking the law
- Reduced incidents of detained shipments

Summary

In summary, logistics as an afterthought in supply chain management is outdated. Logistics is now front and center for every brand/manufacturer, and its strategic value has been elevated with the proliferation of e-commerce over the past 10 years. This megatrend will continue to push logistics to the forefront of every supply chain. The good news is, there is still time to catch up in your logistics capability by leveraging Smart Technologies in each of the logistics areas. Ideally, you may even be able to weaponize your newly found logistics capability to your competitive advantage.

Sample Questions

1. Logistics has become a global sensation recently for brand/manufacturer because of
   a. 3D printing
   b. e-commerce
   c. Globalization
   d. Product complexity
2. In 2019, logistics cost accounted for what percentage of US gross domestic product (GDP)?
   a. 6%
   b. 7.6%
   c. 8.1%
   d. 13%
3. By applying Lean Six Sigma practices, logistics costs were able to be reduced from 16.2% of GDP in 1981 to what percentage in 2019?
   a. 5%
   b. 6%
   c. 7%
   d. 8%
4. What is a potential benefit of making logistics a strategic competency?
   a. Increased cost
   b. Increased flexibility
   c. Increased product mix
   d. Increased risk
5. What is the main reason for a logistics company to be behind the digitization curve?
   a. Digitization culture
   b. Digitization training
   c. Digitization investment
   d. All of the above
6. What could be an example of AI-based dynamic decision making in logistics?
   a. Ad-hoc dynamic rerouting
   b. Predicting customers dropping their shopping cart
   c. Automatic matching of advertisement
   d. GPS repositioning
7. What is an AI-based robot in logistics?
   a. Robotic pick and place
   b. Robotic palletizer
   c. Robotic case erecting
   d. All of the above
8. Digital freight brokerage acts as
   a. Fighter
   b. Customs brokerage
   c. Matchmaker
   d. None of the above

9. A key assumption for successful supply chain collaboration between parties is:
   a. Transparency
   b. Training
   c. Trust
   d. Kindness
10. Technology-based global trade compliance allows:
    a. Reduced noncompliance incidents
    b. Reduced penalties and fines
    c. Reduced chances for criminal offenses
    d. All of the above

Notes

1. Chevalier, Stephanie. (July 7, 2021). Retail e-commerce sales worldwide from 2014 to 2024. Statista. https://www.statista.com/statistics/379046/worldwide-retail-e-commerce-sales/.
2. Deshpande, Vinayak, and Pendem, Pradeep. (September 2, 2020). Logistics performance, ratings, and its impact on customer purchasing behavior and sales in e-commerce platforms. Available at SSRN: https://papers.ssrn.com/sol3/papers.cfm?abstract_id=3696999 .
3. Ross, Lisa. The rise of direct to consumer (D2C) brands: Statistics and trends. Invesp. https://www.invespcro.com/blog/direct-to-consumer-brands/.
4. CSCMP. (2019). Executive summary: CSCMP's annual state of logistics report. Kearney. https://www.kearney.com/documents/20152/18946510/Executive+Summary-2020+State+of+Logistics.pdf/dc9ba2a6-5fbf-31ea-06a5-9198a6c0960c?t=1592501572361.
5. Topic 1.5: Logistics and the economy. MyEducator Online Course. https://go.myeducator.com/reader/web/869a/topic01/j53jl/.
6. Bowersox, D., et al. (1989). *Leading Edge Logistics: Competitive Positioning for the 1990s*. Oak Brook, IL: Council of Logistics Management.
7. Stackpole, Beth. (February 14, 2020). 5 supply chain technologies that deliver competitive advantage. MIT Management Sloan School. https://mitsloan.mit.edu/ideas-made-to-matter/5-supply-chain-technologies-deliver-competitive-advantage.
8. Tipping, Andrew, and Kauschke, Peter. (2016). Shifting patterns: The future of the logistics industry. PwC. https://www.pwc.com/sg/en/publications/assets/future-of-the-logistics-industry.pdf.
9. Stackpole, 5 supply chain technologies.
10. Council of Supply Chain Management Professionals (CSCMP). https://cscmp.org/CSCMP/Develop/Reports_and_Surveys/Research_Survey_Results/CSCMP/Develop/Reports_and_Surveys/Research_Survey_Results.aspx?hkey=76cbdda6-18ca-4945-b33a-540abada394c.
11. AI for Business Forecasting by Anodot. https://www.anodot.com/learning-center/business-forecasting/ (accessed August 14, 2021).
12. Capgemini. (2019). Scaling AI in manufacturing operations: A practitioners' perspective. https://www.capgemini.com/wp-content/uploads/2019/12/AI-in-manufacturing-operations.pdf.
13. Marketysers Global Consulting LLP. (July 2019). Logistics robots market size & share Global industry analysis. https://www.reportsanddata.com/report-detail/global-logistics-robots-market-2017-forecast-to-2022.
14. Costello, Bob, and Karickhoff, Alan. (July 2019). Truck driver shortage analysis 2019. American Trucking Associations. https://www.trucking.org/sites/default/files/2020-01/ATAs%20Driver%20Shortage%20Report%202019%20with%20cover.pdf.

15. Rosalsky, Greg. (May 25, 2021). Is there really a truck driver shortage? NPR. https://www.npr.org /sections/money/2021/05/25/999784202/is-there-really-a-truck-driver-shortage.

16. Telaro, Dominic. (February 21, 2017). 3 types of collaborations in supply chain management. I.B.I.S., Inc.https://ibisinc.com/blog/3-types-of-collaborations-in-supply-chain-management/.

17. Tennariello, Anthony, and Truchan, Mark. Why it's time to automate your global trade operations. PwC. https://www.pwc.com/us/en/services/consulting/risk-regulatory/global-trade-services/ trade-automation.html.

CHAPTER 7

Big Data for Small, Midsize, and Large Operations

Omar Abdon and Randy Shi

Introduction

In this chapter we describe the basics of data analytics and the migration to Big Data analytics when massive data volumes overwhelmed traditional methods of analysis, and detail the four process steps and the tools used in Big Data analytics. We demonstrate why Big Data is essential to manufacturing organizations of all sizes, and finally we demonstrate how Big Data analytics is helping small to midsize enterprises (SMEs), including best practices and affordable tools.

Data are facts and statistics collected together for reference or analysis. People have been analyzing data for thousands of years, but the Industrial Revolution (Industry 1.0) greatly increased available data for analysis, with the invention of the printing press making mass literacy possible.

The growth of data continued under Industry 2.0, with telegraph, telephone, and faster travel using trains and automobiles. The first uses of data analytics in business dates back to the turn to the early twentieth century, when Frederick Winslow Taylor initiated his time management exercises. Henry Ford's measuring the speed of his assembly lines is another early example.

Industry 3.0 accelerated the growth further with computers, software applications, the Internet, smart machines, barcoding, robots, and so on. Computers were key in the evolution of data analytics, as they were embraced as trusted decision-making support systems. Industry 3.0 created so much data that it outgrew traditional data analysis, fostering the introduction of Big Data analytics. Data warehouses, the Cloud, and a wide variety of software tools have accelerated the growth of Big Data over the past 20 years.

The first references to Big Data can be found in 2003 by The Data Center Program created by the Massachusetts Institute of Technology (MIT). Prior to this, the phrase "data analytics" was often employed as a crucial description in early research conducted in the late 1990s. It has become essential to clarify the words Big Data and predictive analytics to avoid confusion.[1]

To begin, it is important to understand the difference between analysis and analytics.

- **Data analysis** is a process to examine, transform, and arrange a data set to permit studying its individual components and then extracting useful information.
- **Data analytics** is an overarching discipline or science that incorporates the complete management of data. Data analytics includes not only analysis, but also data collection, data organization, data storage, and all the tools and techniques used for analysis.[2]

Next, it is important to understand the difference between structured data and unstructured data, especially given the rapid growth in unstructured data.

Structured Data and Relational Databases

Structured data usually resides in relational database management systems (RDBMSs). Examples include Social Security numbers, phone numbers, ZIP codes, sales orders, purchase orders, customer and supplier masters, item numbers in a bill of material, and so on. Relational databases are more than just structured data. The structure is supposed to reflect the relationships between the data as well. The columns, tables, rows, sheets, tabs, and so on of 2D, 3D, or 4D relational databases are supposed to permit the finding of subtle or hidden facts embedded in the data, as well as to make sorting, generating reports, and so on much easier. A program evaluation and review technique (PERT)-style schedule is an example of a relational database when in the format of a tool like Microsoft Project. Not only is data such as start and stop dates included, but also information such as which events are dependent on past events, what kicks off what, and so on. Then a number of other bits of data can be included, like costs, who is responsible, and so on.

Structured data may be generated by people or by machines, as long as it created within an RDBMS structure. This format makes it easy to search, either with human-generated queries or by searching types of data and field names.[3]

While structured data is easy to search, maintaining its accuracy can be a challenge to anyone using material planning systems (MRPs), enterprise planning systems (ERPs), and inventory/ logistics systems. In an age of continuing mergers and acquisitions there is an ongoing merging of item, customer, and supplier masters, resulting in the same items, customers, and suppliers being easily duplicated and not always easily discovered through queries or even using specially designed tools.

My favorite story about part or item number duplication comes from the time I was managing the supply chain for a facility making dental X-rays and operating tables. One day my production control manager advised that production had halted due to a shortage of simple twist ties used to secure wire harnesses. Walking the floor, one of the plant's more senior production workers told me that there were plenty of twist ties in-house but they were stocked under four different item numbers – one for each of the small companies they had acquired over the years. The engineering department had never bothered to update the item masters to show the equivalency or to merge numbers. Ironically, the largest inventory was stocked under an item number designated as excess inventory and being written off by our KPMG auditors.

Another problem in using structured data is in maintaining accurate promise or due dates for sales, purchase, and work orders. I found this to be a chronic issue in both large and smaller organizations over the years. Past-due orders make it very difficult to execute production and meet customer commitment dates.

Unstructured Data

Unstructured data is basically everything else. While unstructured data has an internal structure, it is not structured using predefined schema or data models. It may be textual or nontextual, machine- or human-generated. Typical sources of machine- and human-generated data include:

- **Human-generated unstructured data** comes from word processing, spreadsheets, presentations, emails, logs, social media such as, websites, mobile data, chat, instant messaging, phone recordings, Microsoft Office documents, productivity applications, and so on.
- **Machine-generated unstructured data** comes from satellite imagery, weather data, oil and gas exploration, space exploration, seismic imagery, atmospheric data, digital surveillance, traffic, oceanographic sensors, and so on.

Large amounts of unstructured data are challenging to define. Data sets are increasing exponentially in size, becoming too vast, too raw, or too unstructured for analysis using traditional relational database methods. This is typically what is considered Big Data.

According to some projections, the quantity of accessible data is expected to double every two years.[4] Data is coming from various sources, including conventional sources like industrial equipment, automobiles, electricity meters, and shipping containers, to name a few. Data is also coming from a variety of newer Smart Technology sources such as the Industrial Internet of Things (IIoT) and computer vision. Smart Manufacturing generates Big Data to measure location, movement, vibration, temperature, humidity, electric current, and vehicle identifications, to name a few.

Exhibit 7.1 shows the exponential growth in unstructured data from 2010 to 2025 in zettabytes.[5] (A zettabyte is a measure of storage capacity, equal to $1,000^7$ (1,000,000,000,000,000,000,000 bytes). One zettabyte is equal to a thousand exabytes, a billion terabytes, or a trillion gigabytes.)[6]

EXHIBIT 7.1   Growth in unstructured data, 2010–2025

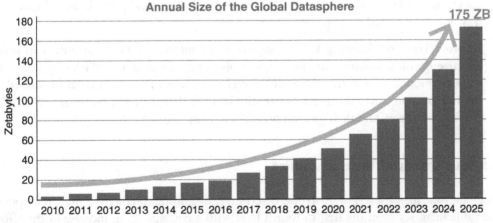

Source: Data Age 2025, sponsored by Seagate with data from IDC Global Datasphere, Nov 2018

Why Manufacturing Needs Big Data Analytics

Binny Vyas, writing in Softweb Solutions, details the significant data challenges manufacturers face.[7]

Distributed Data. There is a growing volume of distributed data collected from a variety of sources and often presented in an inconsistent manner. While organizations typically capture data correctly, they may fail to effectively analyze and utilize the data efficiently.

Integrating New Data. To stay competitive, manufacturing companies bring on new technologies and integrate them into legacy systems.

Growing Volumes and Complexity of Data. As the volume of data and complexity of data grow, they challenge existing visualization and interaction tools.

Connected Tools. Adding industrial control systems and connected tools may overload hardware gateways that connect all the IoT devices in a facility. The proliferation of connected tools also exposes manufacturers to external hackers and security breaches.

Vyas goes on to describe how Big Data analytics can meet these challenges and the many benefits it brings to manufacturers.

Lower Operating Costs. With search-driven data analytics, employees have the ability to create ad-hoc queries at any time. Sales dashboards monitor sales trends and spot problems quickly. Supply chain dashboards project pending material shortages and overages. Results are available in seconds and come in the form of visualization models and easy-to-read data, embedded in portals and shared workflows.

Select the Optimal Areas to Automate. Using workforce analytics helps manufacturers introduce workable staffing solutions and monitor their success over time. With this information they can select the optimal areas to automate.

Reduce Cyber Threats and Data Errors. Deploying a data security solution will help keep data safe from unintentional errors and external cyberattacks. Ironically, many manufacturers suffer greater data issues from their own employees' errors than from external threats. When combined, specific authorization permissions, security layers over every data level and data rows, and robust data governance policies and procedures will go a long way in reducing costly data problems.

Improve Decision Making. Using analytics in manufacturing does not only help make effective decision making but also helps resolve operational issues. With easy access to massive amounts of data that has gone through comprehensive data analysis (described in the next section), manufacturers can readily identify their best opportunities, their least efficient processes, and their greatest operational risks.

The Four Levels of Data Analytics

There are four major levels of analytics: descriptive, diagnostic, predictive, and prescriptive. The value increases from the first to fourth level with a corresponding increase in the effort required to use it.

Exhibit 7.2 shows the four levels of data analytics, the value, and the effort required.[8]

EXHIBIT 7.2   Four levels of data analytics

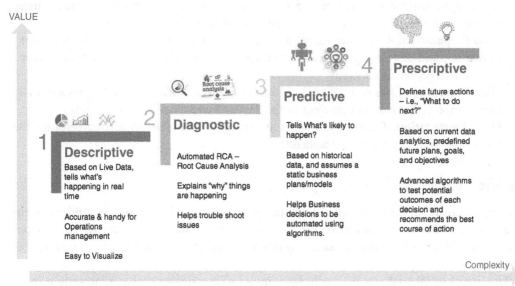

Source: From Vinit Kachchi and Yogesh Kothiya.

Descriptive Analytics – What Happened?

Descriptive analytics is a frequently used data analysis methodology that utilizes historical data collection, organization, and display to convey information clearly and concisely. In contrast to other analysis techniques, descriptive analytics focuses on what has already occurred in an organization and is not used to make conclusions or predictions from its results. In other words, descriptive analytics is a fundamental building block to inform or prepare data for further analysis.

The simplest way to employ data analytics is to use straightforward mathematics and statistical techniques, such as arithmetic, averages, and percent changes. Using visual tools like line graphs and pie and bar charts, it is possible to communicate results so that a large corporate audience can understand them quickly.

How does descriptive analytics work? The two most often used techniques in descriptive analytics are data aggregation and data mining (also known as data discovery). Data aggregation is the act of gathering and organizing data in order to generate manageable data sets. During the data mining phase, patterns, trends, and meaning are found in the data, and then this information is rendered in an intelligible manner.

The Five Phases of Descriptive Analytics

Descriptive analytics are broken down into five main phases.[9]

1. **Business Metrics Identification:** The first step is to build metrics that analyze how well a business's goals, such as increasing operational efficiency or boosting revenue, are achieved. Key performance indicator (KPI) governance is critical to the overall effectiveness of

descriptive analytics. Without governance, analytics may have a limited influence on decision-making.

2. **Data Identification:** Data is obtained from databases and reports. To correctly gauge KPIs, organizations must record and precompute data sources specific to the business's present condition to derive reliable data and calculate metrics.
3. **Data Preparation:** All the necessary data is gathered and arranged. Normalization, transformation, and cleaning are all important steps prior to analysis. This step can be very labor-intensive, but is necessary.
4. **Data Processing:** Data summarization, clustering, pattern tracking, and regression analysis are employed to discover patterns in the data and assess performance.
5. **Data Presentation:** Charts and graphs are typically employed to display findings that nonanalytic specialists can comprehend.

The Value of Descriptive Analytics

Descriptive analytics is widely utilized within organizations daily. Companies utilize descriptive analytics such as data on inventory, workflow, sales, and revenue to analyze their past operations. Reports of this type make it easy to get a comprehensive picture of an organization's activities.

The use of social analytics is usually an example of descriptive analytics, according to the DeZyre online learning platform.[10] One way of understanding descriptive analytics is by studying what people share on social media such as Facebook and Instagram. This study would typically include capturing the number of followers, likes, dislikes, posts, reply posts, and so on.

Surface Data

The primary function of descriptive analytics is to collect surface data and perform a limited analysis of it. Further research and insights drawn from the data are not applicable to prediction and inference. Therefore, descriptive analytics cannot utilize them for either prediction or inference. What this approach is capable of revealing, however, are patterns and significance when data from different time periods are compared.

The Advantages and Disadvantages of Descriptive Analytics

Since descriptive analysis depends solely on historical data and simple computations, it may be performed simply and on a daily basis. Its application does not necessarily require considerable analytical skills. This implies that companies may report on performance reasonably fast and simply and obtain insight into improvements. In general, descriptive analytics is restricted analytics that usually cannot go beyond the data surface.

Diagnostic Analytics – Why Did It Happen?

After learning what happened, the next stage is to understand the reason for what happened. Diagnostic analytics can be challenging because of the need for domain knowledge. To be successful, the analyst must understand a business at a detailed level, including its processes, regulations, policies, target markets, and so on.[11]

The analyst is like a sleuth. For example, a grocery store experiences a large drop in vanilla ice cream sales in June. This fact was discovered using descriptive analytics. The next stage is for the analyst to investigate why this happened. The drop may have been caused by supply shortages or could have been caused by recent news reports of vanilla ice cream recalls over product contamination. This example demonstrates the hypothesis testing and the domain expertise required to understand why the drop in sales happened.[12]

Performing diagnostic analytics is especially challenging in larger organizations because departments often function in silos where data sharing is not the norm. This is when good interview skills and techniques will be very helpful. For example, imagine that a doctor examines a patient and just makes an observation that the patient is sick and leaves the room. Of course, good doctors use diagnostic analysis to determine the cause of a sickness. Data analytics works much the same way: the analyst makes an observation, identifies the descriptive analysis, and then moves forward to the diagnosis.[13]

Using diagnostic tools permits an organization to get the most out of its data by translating complex data into visualizations and insights that anyone in an organization can benefit from. Diagnostic analytics helps gain value from data by asking the right questions and then doing a deep analysis to obtain the answers.

Predictive Analytics – What May Have Happened?

As the name indicates, predictive analytics looks into the future and attempts to anticipate and understand what could happen. Analyzing previous data patterns and trends using historical data and consumer insights may anticipate what will happen in the future and, therefore, inform many business areas, such as creating realistic objectives, effective planning, managing performance expectations, and avoiding risks.

Probability-based predictive analytics is part of the predictive analytics discipline. Predictive analytics seeks to predict future outcomes and the likelihood of those events by employing various techniques, including data mining, statistical modeling (mathematical relationships between variables to predict outcomes), and machine learning algorithms (classification, regression, and clustering techniques). Predictions are made by machine learning algorithms, for example, by attempting to anticipate missing data with the best feasible estimates based on known data. A new type of machine learning, deep learning, resembles the design of human brain networks. Applications for deep learning range from social and environmental analysis for credit scoring to automated processing of digital medical X-rays to anticipate medical diagnoses for clinicians.

Using predictive analytics, organizations are empowered to act more proactively by using data-driven decisions and strategies. Predictive analytics may be used by businesses to do forecasting and trend analysis for various purposes, including predicting consumer behavior and purchasing patterns and detecting sales trends. Just as forecasts help predict such things as supply chain, operations, and inventory demands, predictions may help anticipate those issues as well.

Although predictive analysis cannot achieve 100% accuracy, it can serve as a valuable forecasting and business strategy tool. Many aspects of a business can benefit from predictive analytics, including:

- Forecasting for efficiency
- Customer service and support

- Fraud detection and prevention
- Risk analysis and reduction

Prescriptive Analytics – What Is the Best Next Step?

Descriptive analytics explains what has happened, diagnostic analytics explains why it happened, predictive analytics explains what may happen, prescriptive analytics explains what should be done in a given circumstance. With prescriptive analytics an organization has the information it needs to take action.

Prescriptive analytics uses what was learned in descriptive, diagnostic, and predictive analytics to recommend the best potential courses of action. A high level of specialized analytics expertise is required to be successful. Prescriptive analytics uses artificial intelligence (AI), and specifically machine learning, which incorporates models and algorithms that allow computers to make decisions based on statistical data relationships and patterns.[14]

Prescriptive analytics systems are powerful and complex, requiring close monitoring and maintenance. There are especially sensitive to data quality issues such as incorrect or missing data, which can lead to false predictions, or inflexible predictions that are poor at handling data changes.[15] You should implement data quality standards and keep an eye on the models' predictions. Because of its complexity, it is not used by many manufacturing organizations, especially SMEs. It definitely requires the help of data scientists.

Examples of popular uses for prescriptive analytics include the following:

- Navigate using GPS to recommend the optimal routes for drivers to reach their destination.
- Track the fluctuations in the price of major commodities like oil and coffee.
- Improve equipment maintenance for capital equipment.
- Advise the advantages of various end-of-life product strategies.
- Optimize product design, pricing, and marketing strategies.
- Test various marketing strategies to determine which ones meet sales targets.
- Estimate how changing various factors impacts product demand.

Future of Big Data Analytics

Increasingly, businesses are turning to data to unearth insights that may help them develop business strategies, make choices, and provide customers with better goods, services, and personalized online experiences. However, when considering the four different descriptive, diagnostic, predictive, and prescriptive analytics techniques, these methodologies' potential utility is enormous, even though business analytics is a broad field. When utilized in conjunction with one another, these various ways of analysis are very complimentary and vital to the success and survival of any business.

A new phase in the Industrial Revolution, referred to as Industry 4.0, is characterized by a strong emphasis on interconnection, automatization, machine learning, and real-time data. In the context of Industry 4.0, which includes the Internet of Things (IoT), and Smart Manufacturing, physical production and operations are combined with smart digital technology, machine learning, and Big Data to create a more holistic and better-connected ecosystem for companies that focus on manufacturing and supply chain management, among other things. Every organization in the modern world faces a unique set of challenges. Still, they all share a fundamental requirement: the need for connectivity and access to real-time information across processes, partners, products, and people across all industries.

The following areas are some of the critical factors that will dictate the success of Big Data analytics in supporting Industry 4.0 and Smart Manufacturing.

- **The Need for Data Governance.** The risk of corporations misusing Big Data grows as they acquire vast amounts of information. As a result, many experts believe that data governance will receive increased attention. Peter Ballis, CEO of Sisu Data, shared his beliefs about the role of data governance: "As platforms for analysis and diagnosis expand, derived facts from data will be shared more seamlessly within a business as data governance tools will help ensure the confidentiality, proper use and integrity of data improve to the point they fade into the background again."[16]
- **The Rise of Augmented Analytics.** Gartner analysts believe that augmented analytics will shape Big Data's future trends. Using augmented analytics entails putting AI, machine learning, and natural language processing (NLP) to work on large data platforms. This enables businesses to make better judgments and detect trends more quickly.
- **Big Data and Researchers' Work Coexistence.** Many of today's big data systems are so powerful that it's natural for people to anticipate that they will be replaced in the not-too-distant future. Dr. Aidan Slingsby, a senior lecturer and head of a data science degree program at the City University of London, does not believe such an outcome is likely, particularly in applications such as market research employing big data. "Data science helps identify correlations. So data scientists can provide patterns, networks, dependencies that may not have been otherwise known. But, for data science to really add an extra layer of value, it needs a market researcher who understands the context of the information to interpret the 'what' from the 'why.'"[17]
- **Data Will Power Customer Experiences.** When it comes to Big Data developments, Cloud computing is a key topic of debate. Following are a few predictions from experts in the know about what's occurring now and what could happen shortly when consumers mix big data with Cloud computing.

  According to Nick Piette, director of product marketing and API services at Talend, one of the upcoming trends in Big Data analytics is leveraging the data to improve customer experiences. He also believes that adopting a Cloud-first attitude will be beneficial. Piette says, "More and more brand interactions are happening through digital services, so it's paramount that companies find ways to improve updates and deliver new products and services faster than they ever have before."[18]
- **Big Data Will Be More Accessible.** One of the most significant advantages of Cloud computing is that it allows users to access programs from any location. Andy Monfried, CEO and creator of Lotame, envisions a time when the majority of employees will be able to use self-service Big Data apps.

  Monfried explains, "In 20 years, big data analytics will likely be so pervasive throughout business that it will no longer be the domain of specialists. Every manager, and many non-managerial employees, will be assumed to be competent in working with big data, just as most knowledge workers today are assumed to know spreadsheets and PowerPoint. Analysis of large data sets will be a prerequisite to almost every business decision, much as a simple cost/benefit analysis is today."[19]

He then ties that prediction to Big Data technologies that work in the Cloud. "This doesn't mean, however, that everyone will have to become a data scientist. Self-service tools will make big data analysis broadly accessible. Managers will use simplified, spreadsheet-like interfaces to tap into the computing power of the Cloud and run advanced analytics from any device."[20]

Data Processing and Manipulation. Proper data analytics should avoid "garbage in, garbage out": inaccurate or incorrect data getting fed into analytical models, and analysts receiving deceiving results. To get the accurate, necessary, and robust data set, data handling will need to start in the beginning. Different methods and tools are required to support the process. Before having walked through the process step by step, one must be equipped with knowledge on what tools are out there to use.

Data Science Tools

Great cooks always have handy knives. Letting different tools handle different tasks is extremely important in the data science pipeline as well. The data pipeline procedures include data collection, data process, and data analytics. This section walks you through the dominant data tools in the market and illustrates how each tool is used in the pipeline. In addition, this section also contains sample codes to facilitate your reading.

Structured Query Language (SQL). In the data science world, SQL, and its different variations, is the most common language that any data science will use. Its primary function is storing and querying extensive volume data in SQL databases like MySQL, PostgreSQL, and others. In Industry 4.0, a firm could accumulate millions of data records within weeks or days, and it is nearly impossible for Excel to handle that much of data. Having an efficient database to store, retrieve, and update the data is the cornerstone of successful decision-driven data analytics. Using SQL databases, data across multiple departments can be easily aggregated, compared, and used for further analytics purposes. While you are waiting to open an Excel file with 100 megabytes, a SQL database user could have already gained the fundamental insights for that data.

Python. Python is a potent tool beyond data science. Such incredible power enables Python to be used in every step in the pipeline: from scraping third-party data to the integration of all different data sources or from featuring the existing data to create machine learning analytical reports. While Python can reach the furthest corners of data science and analytics, it is designed for computer science and data analytics beginners.

Business Intelligence (BI) Tools. Business intelligence tools such as Tableau and PowerBI are meant to transform the most complicated data set into easy-to-read visualization reports. The advantage of BI tools is that they are client-facing products. The designers of these tools are committed to making them easy to use and easy to maintain. Having the ability to connect and consolidate different data sources into one workbook, a user of most BI tools only needs to drag and drop to get the desirable dashboard.

Data Analytics Pipeline

Data pipelines can be described as moving data from one source to another so it can be stored, used for analytics, or combined with other data. Specifically, it is the process to consume, prepare, transform, and, finally, enrich unstructured, structured, and semistructured data in a controlled manner. This section discusses what the pipeline consists of and some details that one should be watching for during the process.

Collection

Collection starts with using different methodologies to grab data and ends with storing them in a data warehouse cleverly. Most analysts don't need to spend too much energy in the data collection process. In most cases, the analysts are the users of the collected data. However, in the managerial level of analytics, data collection/warehousing is a must-know procedure.

A proper data collection process requires a clear business goal. Is our future data for internal use or external use? Which parts of the data are we collecting in a business process? Are we observing the operational data or the production data? What are we trying to achieve with the data we will collect? Are there other data we should acquire to supplement the collected data? Many examples tell the stories of failure when these questions are not properly addressed:

> One of my clients uses cameras to gather their business customers' data. Despite the innovative method that brings significant monitor values to their business customers' operation, the data itself cannot capture the production results. In another words, when the analytics tell a story about improved operation, their business customers cannot easily find its effect on their production. Now, my client will need to find additional means to collect the production data.

On the other hand, there's always a tradeoff in precision and generics in the design of data warehouses, where collected data is stored. What are the collection sources of your data? What transformation do you need to do before putting those data into a datamart? Do you expect a large volume of data with different build-ups in the future? Without considering these questions, your data management will be chaotic. Data warehouse breakdowns are more likely to happen. Communication of data problems will be more confusing. Data insights will be more difficult to discover. Finally, when you want to renovate your data warehouse to improve efficiency, it is already too big, too complicated, and too costly for you to change.

Data Service System. The system/method for collecting data also matters. Using data management tools such as CRM tools, vehicle ticketing systems, and so on is usually an easy way to start digitizing your operation and analytics. Often, service providers offer matured data warehouse solutions and maintenance. As soon as the tools are installed in your system, you could immediately understand the basic internal operation of your firm or facility. Without a powerful and highly structured system, integrating these systems with external data will be difficult.

IoT. With proper design and configuration, IoT devices such as sensors and cameras generate valuable data. IOT configurations may require more technically expert employees to install the hardware, create a data feeding system, and construct a datamart. With technicians familiar with the system, adjustment and optimization of the data collection process is easier to achieve.

Cleaning and Standardization. In the initial step of a data pipeline, data cleaning and standardization create the foundation for analysis. The first step is to standardize data during the collection and storage process, including designing proper data structures. When dealing with unistructural data, one must not only think about the key components requiring extraction, but also consider initial raw data processing. Is the data stored in the same format, so as to reduce the process time and storage cost, or is the data in a tabular format so that it's easier for the analysts to query? Do the observed components have a constant number of attributes so that the data is in a column-wise format, or is there uncertainty on the collected attributes so a row-wise format table is more flexible? In addition, labeling of variable names requires robust and rigorous standardization. Countless workplace examples have shown that inaccurate labeling jeopardizes analytics efficiency. For example, the same attributes have completely different names when the attributes belong to two different observants. Without a proper dictionary, analysts waste time in communicating to find the definition for each attribute.

The phrase "garbage in, garbage out" describes how flawed data produces flawed results. Unfortunately, even with a rigorous processing of data collection, the data is never clean. Thus, data cleaning is inevitable in order to convert flawed data into usable data. Common areas needing cleaning include dealing with null values, merging different data sets, and unifying timestamps.

Analytical output is the final product presented to the users, who can be managers, stakeholders, customers, and others. In terms of dashboard design, maximizing the users' reading efficiency, the simplicity of the content is crucial. Removing distractions allows viewers to focus on the essentials more clearly. Standardization means unifying title names, font styles, font sizes, and so on. The analytical products should look professional and clear. Cleaning and standardization not only play a significant role in the design phase but also help analysts across different departments to communicate and collaborate. The color palette is usually a neglected area for standardization. Assigning different colors to the elements in charts is one of the most effective methods to improve usability.

The Benefits of Big Data for SMEs

There are significant benefits for SMEs to adopt Big Data analytics. Some of the good reasons to make the investment are:

Improved Customer Loyalty from Big Data. Big Data analytics is helping SMEs evaluate customer retention and customer loyalty to uncover changes in buying behaviors. These insights help SMEs maintain their current customers and expand their markets. Big Data can also be used to identify the hidden costs to acquire new customers and clients. This is critical, as the costs of bringing on some new customers may not be worth the travel, entertainment, and many hours to land that first order.[21]

Critical Customer Insights from Big Data. Big Data helps SMEs to focus on customer preferences, especially those of their highest-volume customers. Armed with these insights, organizations can personalize product offers, sales, and marketing campaigns, providing a competitive advantage.

Optimized Decision-Making from Big Data. The growing influence of social media warrants using a Big Data tool called social media data mining to gauge customer interests, assess reactions to quality issues, and evaluate responses to marketing campaigns and promotions. Social media data mining can look at these same factors for major competitors and peer organizations. This is a good way to discover best practices and to avoid landmines – major mistakes your competitors have made. Many organizations, big and small, fail because they miss major moves in the market. Traditional data analysis is reactive. Big Data analytics is proactive and predictive in nature, which helps to get in front of market changes.[22]

Big Data Tools for SMEs

Large manufacturing and distribution organizations are very adept at using enterprise-level analytics tools. These tools are essential for them to remain viable in today's global marketplace. While SMEs are not candidates for the highest-end platforms, there are good options available to them as well. SMEs will also need to use more affordable and easy-to-use Big Data tools for them to remain viable in today's marketplace. A few examples of suitable tools include:

Data Integration Tools. For most SMEs, like most larger organizations, data resides in a variety of disparate silos and is not easily extracted, scrubbed, and normalized for analysis. Data integration tools are able to connect siloed data in various locations and in different formats and then transform that data into one data set in one easily accessed location. Data sources may include web traffic, CRM systems, material requirements planning/enterprise resource planning (MRP/ERP) systems, marketing operations software, sales systems, and a variety of other sources.[23]

Data Preparation Tools. Because of the disparate nature of data regardless of the number of silos they reside in, many SMEs spend much of their time cleaning data, with few resources available to analyze it. Data preparation tools can help solve this problem by automating the data preparation process, reformatting data, correcting it, and combining data sets.[24]

Data Quality Tools. Because data comes in various levels of quality and reliability, data quality tools are essential to bring data up to the level where it valuable for analysis. Valuable insights emerge by eliminating duplicates and normalizing data.[25]

Data Governance Tools. Data governance is a collection of processes, standards, policies, procedures, and metrics for the efficient and effective use of information so that an organization can meet its goals and objectives. Data governance helps outline who takes action upon what data, in what situations, and using what methods. It helps improve data quality and the understanding of data. Data governance tools allow SMEs to gain value in their data.

Problems SMEs Face in Adopting Big Data Analytics

There are many data sources and data types available for SMEs. They come in a wide variety of conditions, from poor to high quality levels. For these reasons, data can be misused and misunderstood, which can lead to major problems if not rectified. Michael Keenan, writing in *Cyfe*, describes some major issues SMEs face in gaining value from Big Data analytics.[26]

Hesitancy to Use Data Analytics. Many SMEs fear using data analytics because it is new to them and to their peers. It is doubtful that many of them were exposed to the value of data analytics in their education, training, and conferences they attended. This can also be a generational issue, as the average manufacturing executive is over 45 years old, going to school before the notion of Big Data existed.[27] Even today, degrees in data analysis are not offered in many two-year and four-year schools. Another reason for the hesitancy is the hype surrounding Big Data analytics. Manufacturing folks tend to be fairly conservative and have been bombarded for years about the latest and greatest technology that is hyped to solve all their problems.

Using the Wrong Data Analytics Tools. There are dozens of widely available free reporting tools, but free is a relative term. One of the reasons they are free is that no in-house developers are managing it for you, which means the burden falls on your IT team. If support is offered, it may come at additional costs. Free reporting tools typically lack an intuitive user interface (UI) and are difficult to navigate.[28]

Looking at the Wrong Metrics. There is a tendency for large and small organizations alike to measure the wrong metrics. Traditionally, very large and complex organizations measure everything rather than focusing on the critical few metrics key to their success. A good test of whether the correct metrics are being measured is determining whether, when combined, the metrics tell an organization if they are getting better or worse.

A problem prevalent in small organizations is the tendency to focus on vanity metrics such as Instagram followers and Facebook Likes. The best advice in measuring metrics is to focus on the very few key performance metrics, especially forward-looking metrics that help predict future events. For example, the size of the sales pipeline when compared over time is more predictive than measuring bookings or shipments over time.[29]

No Longer Asking Why. It is not unusual for long-time business owners to become complacent in their knowledge of market conditions. This is a problem in all sizes and types of organizations. A great example is the complacency of the Big Three American automobile manufacturers in the 1970s and 1980s, which become so complacent that they failed to measure the success of Japanese imports that offered more reliable vehicles at lower costs. Big Data analytics can help even the best-managed organizations to gain valuable new insights.

Best Practices in Data Analytics for SMEs

There are some commonsense best practices in data analytics that have proved their value to SMEs over the years to expose critical business trends. Trend analysis quantifies and explains trends and patterns in a data set over time. A trend is an upward or downward shift in a data set over time. A trend is valuable, as it quantifies and explains data patterns over time. Armed with trend analysis, an organization can take actions to support good trends and address bad ones.

Michael Keenen provides a good list of proven best practices.[30]

Measuring Web Traffic Source. A web traffic source tool can show where visitors came from before landing on your website. A direct source is someone visiting your website by typing in its URL. An organic source is through a search engine such as Google. By tracking the sources of traffic, an organization can optimize the effective channels.

Determining Keyword Rankings. After determining the source of web traffic, it is important to track website keyword rankings. Determining keyword rankings identifies the most popular content.

Measure Duration of Website Visits. High volumes of visits are of limited value if they only last a few seconds.

Calculate Percentage of Visitors Converted to Subscriptions. This will help determine the effectiveness of your messaging. Many organizations use website content to move site visitors to subscribers.

Present Your Data Analytics on Mobile. This is especially important with diverse teams and road warriors who travel frequently.

Keep Your Data in One Location. This will make it easy to share with your employees, your customers, and your suppliers.

Create a Data Dashboard. A good data dashboard, accessible to all stakeholders, will provide real-time monitoring of key performance indicators and will give viewers live updates of your most important metrics, be they website traffic, online sales, or active sales funnels.

Summary

Data analytics is a critical component of Smart Manufacturing and Industry 4.0 for a very simple reason. All the new technology creates massive amounts of structured and unstructured data from many different sources. The large volumes, inconsistent quality, and disparate nature of all this new data is far beyond the scope of traditional data analytics. Big Data analytics has proven its utility in analyzing data for large and small manufacturing organizations alike. While most of the powerful data analytics tool require the help of talented data scientists, SMEs can enjoy the benefits of Big Data by availing themselves of a variety of affordable and easy-to-use analytic tools to lower operating costs, increase market share, and reduce quality issues.

Sample Questions

1. Tableau and PowerBI are examples of
   a. Python providers
   b. Business intelligence tools
   c. Machine learning tools
   d. Prescriptive analytics tools

2. The first references to Big Data were in 2003 from
   a. Stanford
   b. Harvard
   c. UC Berkeley
   d. MIT

3. "A process to examine, transform, and arrange a data set to permit studying its individual components and then extracting useful information" is the definition of
   a. Data Analytics
   b. Big Data
   c. Data processing and manipulation
   d. Data analysis

4. Why is structured data easier to analyze than unstructured data?
   a. It resides in a datamart.
   b. It resides in a database.
   c. It resides in a relational database.
   d. It resides anywhere.

5. An example of untrusted data is
   a. ZIP codes
   b. Instant messages
   c. Item masters
   d. Social Security numbers

6. Descriptive analytics explains
   a. Why it happened
   b. How it happened
   c. What could have happened
   d. None of the above

7. Why is data analytics important to manufacturing?
   a. The volume of distributed data collected from a variety of sources is growing.
   b. The existing visualization and interaction tools cannot handle the growing volumes of data.
   c. Gateways are overloaded by new industrial control systems and connected tools.
   d. All of the above

8. Of the four levels of data analytics, which is the simplest?
   a. Prescriptive
   b. Predictive
   c. Diagnostic
   d. Descriptive

9. The quantity of data accessible is expected to double how often?
   a. Every four years
   b. Every six months
   c. Every two years
   d. Every month

10. Moving data from one source to another so it can be stored, used for analytics, or combined with other data is one definition of
    a. Data pipelines
    b. Data migration
    c. Data normalization
    d. Data scrubbing

Notes

1. Foote, K. (December 14, 2017). A brief history of Big Data. *DataVersity*. https://medium.com/callforcode/the-amount-of-data-in-the-world-doubles-every-two-years-3c0be9263eb1.
2. What's the difference between data analytics and data analysis? (January 23, 2020). *GetSmarter*. https://www.getsmarter.com/blog/career-advice/difference-data-analytics-data-analysis/.
3. Taylor, C. (May 21, 2021). Structured vs. unstructured data. Datamation. https://www.datamation.com/big-data/structured-vs-unstructured-data/.
4. Gallenger, B. (October 7, 2020). The amount of data in the world doubles every two years. *Call for Code Digest*. https://medium.com/callforcode/the-amount-of-data-in-the-world-doubles-every-two-years-3c0be9263eb1.
5. Coughlin, T. (November 27, 2018). 175 zettabytes by 2025. *Forbes*. https://www.forbes.com/sites/tomcoughlin/2018/11/27/175-zettabytes-by-2025/?sh=4cdf41c05459.
6. Chojecki, P. (January 31, 2019). How big is Big Data? *Towards Data Science*. https://towardsdatascience.com/how-big-is-big-data-3fb14d5351ba.
7. Vyas, B. (February 24, 2021). Why the manufacturing industry should embrace data analytics. Soft Web Solutions. https://www.softwebsolutions.com/resources/data-analytics-in-manufacturing.html.
8. Kachchi, V., and Kothiya, Y. (May 8, 2021). 4 types of data analytics every analyst should know – descriptive, diagnostic, predictive, prescriptive. *Medium*. https://medium.com/co-learning-lounge/types-of-data-analytics-descriptive-diagnostic-predictive-prescriptive-922654ce8f8f.
9. Shung, K. (May 20, 2020). Four levels of analytics/data science. *Building Intelligence Together*. https://koopingshung.com/blog/four-levels-of-analytics-data-science-descriptive-diagnostic/.
10. DeZyre. Data science course. ProjectPro. https://www.dezyre.com/data-science-training-course-online/36.
11. Shung, Four levels.
12. Ibid.
13. Ibid.
14. Prescriptive analytics guide: Use cases & examples. Stitch. https://www.stitchdata.com/resources/prescriptive-analytics/ (accessed September 2, 2021).
15. Ibid.
16. Matthews, K. (June 9, 2020). 6 important big data future trends, according to experts. Smart Data Collective. https://www.smartdatacollective.com/6-important-big-data-future-trends/.
17. Ibid.
18. Ibid.
19. Expert predictions: The future of Big Data and business 20 years from now. The Future of Everything. https://www.futureofeverything.io/expert-predictions-the-future-of-big-data-and-business-20-years-from-now/.
20. Ibid.
21. How big data benefits small business. Talend. https://www.talend.com/resources/big-data-small-business/.

22. Ibid.
23. Morris, A. (April 16, 2021). 23 case studies and real-world examples of how business intelligence keeps top companies competitive. NetSuite. https://www.netsuite.com/portal/resource/articles /business-strategy/business-intelligence-examples.shtml.
24. King, T. (March 16, 2021). The 10 best data preparation tools and software for 2021. Solutions Review. https://solutionsreview.com/data-integration/the-best-data-preparation-tools-and-software/.
25. How big data benefits small business.
26. Keenan, M. (May 13, 2020). Small business analytics: 4 ugly truths and 6 best practices. *cyfe*. https://www.cyfe.com/blog/5-ugly-truths-small-business-analytics/.
27. Vice president of manufacturing: Demographics and statistics in the US. Zippia. https://www .zippia.com/vice-president-of-manufacturing-jobs/demographics/.
28. Adair, B. Open source reporting tools: Pros and cons. SelectHub. https://www.selecthub.com /business-intelligence/pros-cons-open-source-reporting-tools/.
29. Keenan, Small business analytics.
30. Ibid.

CHAPTER 8

Industrial Internet of Things (IIoT) Sensors

Deb Walkup and Jeff Little

Introduction

Human history is full of attempts to better understand and monitor our environment, often involving the development of objects and technologies to observe, measure, monitor, and predict our environment. Early civilizations developed a number of ways to monitor time and the seasons; Stonehenge is an example. Taking advantage of a naturally occurring glacial scar in the landscape of the Salisbury Plain, the ancient Britons used this landmark to measure time and predict the seasons, in particular, the Winter Solstice. They built the monument we see today over a period of centuries to enhance this prediction capability. Today, while most people use the monument to note the Summer Solstice, it was actually the Winter Solstice that the ancients were concerned about, as it marks the end of shortened of days and the beginning of longer days, leading to summer and the times for planting and harvest.

Throughout history, other measurement and monitoring devices have been created to monitor data both seen and unseen. Some common examples are:

- Telescopes to see things that are far away
- Compasses to measure directions
- Thermometers to measure temperature
- Barometers to measure air pressure to monitor and predict the weather
- Microscopes to see things invisible to the naked eye

In the age of electronics, these devices, along with thousands of other kinds of sensors, have all been designed to communicate electronically by a variety of means. When combining these electronically connected sensors with networking, computers, and application software, we have a technology that is now called the Internet of Things, or IoT, meaning that networking no longer involves just humans and computers. It now includes anything that can provide data to a network/computer complex for monitoring and control through electronic communications.

The origin of the term *Internet of Things* or *IoT* has been attributed to several sources. The two most likely are:

- The concept of and term Internet of Things first appears in a speech given by Peter T. Lewis in Washington, DC, in September 1985.
- The actual term "Internet of Things" was coined by Kevin Ashton in 1999 during his work at Procter & Gamble. Ashton, who was working in supply chain optimization, wanted to attract senior management's attention to an exciting new technology called radio-frequency identification (RFID).

Industrial Internet of Things (IIoT) refers to the use of IoT in industrial contexts. It is also called "ubiquitous computing," a term first coined in an article by Mark Weiser in 1991 and described in detail in a 1994 IEEE article by Reza Raji.[1] The Industrial Internet Consortium (IIC) was founded in March of 2014. This is not a standards organization. It brings together a number of major industry players such as AT&T, Cisco Systems, General Electric, IBM, and Intel as well as innovators from academia and governments to accelerate the development, adoption, and use of IIoT technologies. The IIC created the Industrial Internet Reference Architecture (IIRA) in 2015. The current version can be found at https://www.iiconsortium.org/IIRA.htm.[2]

The IIRA is an extensive discussion of all aspects of IIoT, including technical, business, and best practices.

PLCs

The beginning of industrial automation and the use of sensors to monitor and control processes and equipment in an industrial setting is generally thought to be the invention of a programmable logic controller (PLC), the Modicon 084, invented by Dick Morley for General Motors in 1968. The PLC is a ruggedized device with inputs and outputs that can be programmed to control machine operations. An input would run through a program stored in nonvolatile memory and an output would be sent back to a machine.

Exhibit 8.1 shows an early PLC.

Relay logic systems made up of hard-wired relays, cam timers, and mechanical drum sequencers were difficult to set up and maintain and were often unreliable. The Modicon used software programming to replace the hardwiring and difficult "programming" of mechanical devices like cams and drums. Over the years, PLCs improved in reliability, ease of programming, and capability. They are still widely used in industry today and range from large modular systems to single-board computers such as the Raspberry Pi, and even single-chip controllers such as the Nano ACE PLC from Data Device Corporation (DDC).[3]

EXHIBIT 8.1 An early PLC

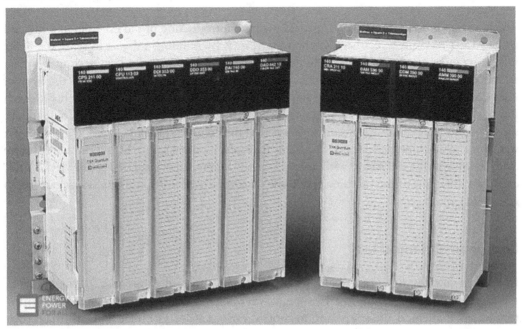

A great example of sensors revolutionizing an industry is the introduction of dive computers in 1983. The first dive computer was the Orca Edge. Prior to the use of dive computers, a diver would plan a dive that would follow a profile. A typical recreational dive lasts about 50 minutes, and a diver would plan on going to the deepest depth first (about 60 feet) and then working to more shallow depths with an estimate of how long they would be at each depth, with a 3-minute safety stop at 10 feet. The reason for this is that nitrogen builds up in a diver's body tissue the deeper they go and the longer they stay at depth. This can cause the bends, which can lead to death.

Divers would then use a table with depths and times to calculate the amount of nitrogen they would absorb. The use of the tables and simple profiles was a very conservative way of calculating the nitrogen level. With a dive computer you measure the exact depth and time for each second of the dive; with this more granular level of data, divers had a much better idea of their actual nitrogen absorption and could do more dives per day safely. Dive computers are now integrated in websites and logbooks to also record all dives with GPS tracking. An added level of safety locks a dive computer when a diver exceeds recreational limits or surfaces too fast, or the dive profile doesn't include the safety stop for outgassing. Operators will not allow a diver back into the water if their computer is maxed out. Even diving without an operator, a diver should cease diving until the computer clears. The upside is that it is possible to take a one- to two-week-long vacation on a live-aboard dive boat and safely dive up to five times a day. An entire industry was born.

Carnegie Mellon

The next leap in IoT technology happened with the introduction of the Ethernet in 1980 and a Coca-Cola machine at Carnegie Mellon University. By building a device to sense the lights indicating inventory levels for the columns of Coke in the machine, and connecting that data to the Ethernet, students David Nichols, Mike Kazar, and Ivor Durham, with the help of research engineer John Zsarnay, were able to build a device that monitored the lights for each row of the vending machine. A light going on and off was just someone buying a Coke. A light staying on for more than five seconds meant that the column was empty, and when the light went off the machine had been refilled and it would be three hours before the bottles were cold. Anyone at the university with access to the Ethernet, later extended to the ARPANET, could see the status of the machine and know whether it was worth walking across campus to get a cold Coke.[4]

Consumer-Oriented IoT

In 1985, a new company called Cloud Nine (CL 9) was founded by Steve Wozniak, one of the founders of Apple Computer. His dream was to have a universal control and monitoring technology that could allow a person to control various aspects of their home through their personal computer or a simple universal remote. This would include things like radio, stereo, TV, and video players, and eventually also lights, garage doors, security systems, and thermostats. Even kitchen appliances were included in this vision. While CL 9's intellectual property was eventually sold in 1988 and its other assets sold at auction, the idea had caught on.

In 1988, another new company called Echelon was founded to produce a technology called LonWorks. Lon stands for local operating network. This was a technology created for control applications using network technologies such as twisted pair cabling, high frequency over AC power lines, fiber optics, and radio. It was introduced as a technology for home and office building automation to control things like lighting, HVAC, security, and so on, but it is sometimes reported that its biggest successes were the control technology for lighting and special effects used in live stage events such as rock concerts and used by AC power utilities to manage their smart meters in certain countries in the EU. In 2010 it was reported that there were 90 million Lon-enabled devices installed. But this installed base has declined in recent years and in 2018 Echelon was acquired by Adesto Technologies.

Webcams

The first webcam was created at the University of Cambridge in the early 1990s to monitor a coffeepot. The coffee station serviced several labs in the building on multiple floors and no one enjoyed making the trip for a cup and finding the pot empty.

Exhibit 8.2 shows a sample of the images taken by this webcam.

Exhibit 8.3 shows Dr. Paul Jarderzky's setup of a Philips camera, to capture an image of the coffeepot three times per minute.

EXHIBIT 8.2   Sample webcam images

EXHIBIT 8.3   Dr. Paul Jarderzky's Philips camera

Dr. Quentin Stafford-Fraser, along with Dr. Jardetzky, created the software to serve the images to the Ethernet at Cambridge. In November of 1993, Dr. Martyn Johnson, who was not connected to the internal network, wrote the code that served the images to the World Wide Web (WWW). The program started with about 12 lines of code and anyone in the world could hit the website and see if there was coffee for the lab staff. This is most likely the first viral web sensation. Tech enthusiasts from around the world were checking it out and sharing it with their friends. There were requests to add a light to the area so that the status was visible at night when the other side of the world was awake.

Exhibit 8.4 shows the final image broadcast of a scientist's fingers pressing the "off" button.[5]

EXHIBIT 8.4   The final coffeepot image

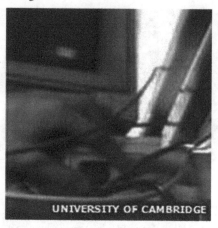

UNIVERSITY OF CAMBRIDGE

IIoT-Enabling Technologies

It was a natural evolution for PLCs and similar monitor/control systems to tie into communications technologies and eventually the Internet. With the emergence of Cloud technology around 2002, the ability to gather and process large amounts of data permitted a number of new capabilities to be applied to the industrial context. These included things like monitoring historical trends, predicting when maintenance or replacement might be needed on machines and tooling, efficiency studies, and safety alerts.

The number of technologies that enable and can be used in an IIoT environment are many and varied. Following is a listing of just a few. The key aspect is that the element being used to support IIoT generally has some form of electronic or RF communication capability so that it can report data without human intervention and may also receive commands to establish thresholds, perform actions, or adjust settings without human intervention. Some examples of these enabling technologies are:

Smart Sensors. Both analog and digital sensors with embedded controls are available for hundreds of applications. These use transmission technologies such as TCP/IP or other network technology, Wi-Fi, cellular, and Bluetooth. The most versatile and prolific sensors are cameras that can record still images or video.

High-Speed Secure Communications. When sending data to the Cloud, it is important to ensure the security of that data; Chapter 5 presents the important aspects of data and network security. Based on the amount of data and the time frame for transmission, the network should have the speed and bandwidth to accommodate the data in a timely manner.

Edge Computing. Performing analysis of the data at the source can reduce the amount of data sent to the Cloud as well as provide near-real-time alerts and exceptions to make corrections as soon as possible. For safety and/or security applications this is vital. This is covered in Chapter 11, "Edge Computing," by Vatsal Shah and Allison Yrungaray.

Cloud Computing. This is connecting to the Cloud through the Internet to deliver data for later analysis. This analysis will provide feedback to the IIoT implementation that allows the IIoT devices to take appropriate actions without direct human intervention.

Big Data Analytics. The real power in more and better data is being able to analyze historical data and understand trends. Analytics can be performed in the Cloud or at the Edge. This is covered in Chapter 7, "Big Data for Small, Midsize, and Large Operations," by Omar Abdon and Randy Shi.

Artificial Intelligence and Machine Learning. Machine learning is used to build models that can be used with artificial intelligence (AI) to determine whether processes are going smoothly, not quite right, or horrendously out of control. This can be supported at the Edge or in the Cloud. Edge analysis provides much faster feedback at the site. This is covered in Chapter 9, "Artificial Intelligence, Machine Learning, and Computer Vision," by Steven Herman.

Cyber-Physical Systems (CPSs). CPS is an intelligent and/or computer system that monitors and/or controls a mechanism and/or process through computer-based algorithms. CPS involves the merger of cybernetics, mechatronics, and design and process science through a multidisciplinary approach. This is the integration of physical functions and processes with communications, software, sensors, and so on. Examples include autonomous automobiles, industrial control systems, robotics, and auto pilots in aviation. This is covered in Chapter 5, "Improving Cybersecurity Using Smart Technology," by Craig Martin.

A Layered IIoT Architecture. As mentioned previously, the Industrial Internet Consortium (IIC) has published the Industrial Internet Reference Architecture (IIRA), which is an extensive discussion of all aspects of IIOT, including technical, business, and best practices. But similar to the kinds of layered architectural models found in networking, IIoT can also be described by a simple model with four layers.[6]

Exhibit 8.5 shows the four layers used in IIoT.

EXHIBIT 8.5   Four layers of IIoT

Layer Number	Layer Name	Function
1	Content	User interface devices: computer screens, point of sale stations, tablets, smart glasses, 3D goggles, smart surfaces, etc.
2	Service	Applications: software to analyze data and transform it into actionable information
3	Network	Communication protocols: Wi-Fi, Bluetooth, Ethernet, cellular, etc.
4	Device	Hardware: computers, network gear, cyber-physical systems, machines, sensors, etc.

A wide variety of IIoT implementations can be modeled using this simple layered model structure.

IIoT Platform Building Blocks

From a platform perspective the services needed for IIoT are:

Device Connectivity. How are devices connected to the network and how is the data being captured consistently?

Edge Analytics. Performing analytics at the Edge to provide immediate feedback makes every IIoT scenario more valuable and effective.

Data Integration. Is there a normalized set of data elements so that different data elements can be combined efficiently? An example is taking data from three different sensors and understanding the state based on that data by timestamp.

Application Deployment. Applications are the secret sauce that differentiate competitors in the IIoT market. The applications can be a combination of native code and third-party plug-ins. A typical example is capturing license plates using a third-party app and then using that data to understand the movement of vehicles through a yard and the duration at each point of interest. An example is a parking lot that provides parking for a fee. With a timestamp of each vehicle entering and leaving the lot, you have accurate data for how long the vehicle was there and there is no ticket to lose. In the event the driver leaves without paying you also have the license plate to tie the charge back to a vehicle and an owner.

Inputs to the platform might include data from:

- Sensors
- PLCs
- Manufacturing execution systems (MESs)
- Enterprise resource planning systems (ERPs)
- Materials requirement planning systems (MRPs)
- Historical databases

Services the platform should provide include:

- Cloud services
- Big Data
- Machine learning
- Applications

IIoT Sensors

A wide range of IIOT sensors are available today. Sensors have evolved from manual devices to analog components to digital semiconductor devices. The addition of cameras, as sensors, allows visual capture of manual devices and for the collection of data from existing components of your manufacturing environment. This allows a company to embark on an IIoT path without a complete revolution in the sensors and devices in current use.

Virtually any phenomena that can be measured now has a sensor that can monitor, record, and report that measurement. The following list is just a representative example of the phenomena that can be sensed by modern sensors today.

Light. Different sensors can measure different kinds of light, including infrared, visible, and ultraviolet, as well as distinct colors or light frequencies only. This was a component of the Carnegie Mellon Coke machine monitoring in the 1980s.

Current Sensor. Current sensors can be used to capture the power utilization of machines. Even a machine that is plugged in and unused draws power and uses energy. By capturing the power usage profile of machines and having a better understanding of how they are utilized and combining that information with smart switches, a machine shop could reduce their power consumption without a decrease in service levels.

Smoke and Gas Detectors. Common examples are the smoke detectors and carbon monoxide detectors found in homes. But there are many different kinds that can also measure many different gases and types of smoke and fumes. Gas detectors can be for specific gases and provide a level of safety in the event of a leak of toxic gas. A detector can also enable a cost savings if an expensive gas is leaking.

In industrial environments a thermal or infrared sensor could provide better and faster discovery of combustion. These sensors can identify the location of a hot spot before it ignites and before there is enough smoke to trigger a detector on the ceiling. An example is a paper recycling center. If a fire is started in this environment, by the time a smoke detector is triggered the building is lost; the goal is to get people out to safety.

Alcohol Sensor. This sensor measures the amount of alcohol in a fluid, which is very useful in brewing and winemaking.

Strain. This is a sensor that can measure the mechanical strain or force exerted on the sensor. This can be used to measure the mechanical strain in structures or machines.

Weight. Weight is the basic element of a scale; there are many variations of scales that cover weights from micro-ounces to tons. In yard applications this sensor can be used to measure tare (unladen) weights of vehicles and loaded weights and then an application can compare the difference and match it with the invoiced quantity to ensure that accurate amounts are provided to customers and all amounts are paid for.

Pressure. A pressure sensor can be used to measure the pressure in different enclosed spaces for air or fluids like oil and water. For presses or injection molding machines it can measure the pressure being applied mechanically. If the pressure is outside target limits, it could mean that maintenance is required. A common vehicle example is the air pressure sensor that measures the pressure in the tires on a continuous basis and engages a dashboard light when the pressure is too high or low.

Proximity. These sophisticated measurement sensors often use lasers or ultrasonic sound waves to measure the distance between objects. An example is enforcing social distancing requirements of six-foot spacing between people on the factory or warehouse floor. Another common example is the sensors in the rear of vehicles that warn of close objects in the path of the vehicle when backing up, especially when they are in a blind spot.

Position. Position sensors measure position by mechanical, optical, or other means. They are used to determine if something is in a particular position. An example could be an asphalt machine working on road resurfacing. Having an indicator that the machine is aligned with the road and moving in the right direction would be helpful.

Humidity. A humidity sensor measures the humidity in the air or in any gaseous environment.

Sound. Typically a sound sensor is a microphone using acoustic pressure sensors, pressure microphones, high-amplitude pressure microphone, prepolarized condenser microphones, high-amplitude pressure microphones, probe microphones, condenser microphones, and prepolarized free-field condenser microphones. These devices can measure sounds that are audible to humans, subsonic sounds (below the audible range of humans), and/or ultrasonic sound (above the audible range of humans). You could use this type of sensor to answer the age-old question, "If a tree falls in the woods and no one is there, does it make a sound?"

Lidar. By measuring distances using the time it takes for a laser to bounce off a surface, this technology can create a map of the ocean floor or an accurate representation of material in a closed bin or silo so that a volume can be calculated for material that isn't visible. An example is a cement plant that has different silos for cement, fly ash, and aggregate. By using lidar, the amount of material in the silos is a known entity and not a guess and replenishment is much more accurate. This technology is also used in self-driving vehicles to help in navigation.

Passive Infrared Sensor (PIR). A PIR is an electronic sensor that can measure the infrared light being radiated from an object within its field of view.

Flow and Level. A flow and level sensor can measure the flow and/or the level of gas or liquid in a pipe or container. A couple of common examples are the sensor that measures the amount of gas in a gas tank, and the gas meter on a house that measures the amount of gas consumed.

Accelerometer. This sensor measures the change in velocity of an object to detect abnormal vibrations in rotating machines, which can indicate that maintenance is required. They are also used in tablets and video games to detect a change in velocity. For a tablet, this means always showing the display correctly and in games it provides feedback to the game of the user's motion.

Radio-Frequency Identification (RFID). An RFID sensor uses electromagnetic fields to automatically identify and track tags attached to objects. An RFID system consists of a tiny radio transponder, a radio receiver, and a transmitter. When triggered by an electromagnetic interrogation pulse from a nearby RFID reader device, the tag transmits digital data, usually an identifying inventory number, back to the reader. This number can be used to track inventory goods. Passive tags are powered by energy from the RFID reader's interrogating radio waves. Active tags are powered by a battery and thus can be read at a greater range from the RFID reader, up to hundreds of meters.[7]

Barcode Reader. This is a specialized reader for barcodes, optically readable labels that can carry an ID or tracking number. The most common example can be found in the supermarket checkout stand. A common use is to identify retail goods at the checkout stand of a store so that the checkout person does not have to key in the identity of the goods being purchased. Barcodes are also used in shipping and warehousing to identify cartons, pallets, and containers. Barcodes are also readable by cameras; they don't always need a reader to work.

Quick Response (QR) Codes. QR codes are a type of matrix barcode that can be read by a camera attached to a sensor that can do the appropriate decoding. They were invented to support high-speed component scanning. Today they are widely used with smartphones. QR codes can carry more information than ordinary barcodes, as well as hyperlinks to more information.

Temperature Sensors. Digital temperature sensors are common, and most off-the-shelf sensors will use a device mounted on a circuit board inside a mechanical assembly. Digital devices typically measure temperature between –55° C and 150° C. Exhibit 8.6, an excerpt from an Analog Devices datasheet,[8] shows an example of one kind of digital temperature sensor.

Analog temperature sensors include thermistors, thermocouples, resistance thermometers, and silicon bandgap temperature sensors. A thermistor is a type of resistor whose resistance is strongly dependent on temperature, more so than in standard resistors. The word is a combination of thermal and resistor. Other types of analog temperature sensors vary a voltage according to the temperature. This can be a voltage either created by the sensor or measured across the sensor.

A typical temperature sensor application is in cold chain logistics. Some products, such as produce, require cooler temperatures to maintain freshness in transport. It is now possible to track the temperature of a refrigerated container as it moves across distances and different modes of transportation. The data is uploaded to the Cloud and an audit trail of the entire trip is available with alerts if the temperature falls outside the specific range – too cold and vegetables freeze; too hot and they go bad.

Exhibit 8.7 shows the specifications for five types of temperature sensors.

Cameras (Vision Sensor). The most versatile sensors used to capture IIoT data are cameras. With machine learning, cameras can become almost any type of sensing device. They can be used to capture still images or video. A camera can capture an image to show if a light is on or what color it is. A video can capture whether it is solid or flashing. Cameras can capture barcodes, QR codes, or text, and algorithms can be used to translate those images into values and data strings. Cameras can capture temperature readings, scale readings, dials, and gauges, and time-stamp those images. A group of images can be aggregated to understand processes. An example is a machine in idle mode. Whether an operator is present, what status the stack light is displaying, whether there is material to feed the machine, are all elements that can be combined to determine if the machine is in setup, being ignored, or planned to be idle.

Pairing these images with AI algorithms and computer vision you can monitor older manual sensors, safety issues, work in progress, quality control, inventory, and shipping. There are a lot of cameras on the market and there are more every day.

Three different types of cameras are available. An area scan camera takes an image of an area using a rectangular sensor. Line scan cameras take an image of one row of pixels and then stitch multiple images together to make videos (used for continuous process operations). The third is embedded smart cameras that can be either area or line scan and they are integrated into an Edge computing device.

EXHIBIT 8.6 Specifications for a low-cost temperature sensor switch

ANALOG DEVICES

Low Cost, 2.7 V to 5.5 V, Pin-Selectable Temperature Switches in SOT-23

ADT6401/AST6402

Data Sheet

FEATURES

±0.5°C (typical) threshold accuracy
Pin-selectable trip points from
 −45°C to +5°C in 10°C increments (undertemperature)
 45°C to 115°C in 10°C increments (overtemperature)
Maximum operating temperature of 125°C
Open-drain output (ADT6401)
Push-pull output (ADT6402)
Pin-selectable hysteresis of 2°C and 10°C
Supply current of 30 µA (typical)
Space-saving, 6-lead SOT-23 package

APPLICATIONS

Medical equipment
Automotive
Cell phones
Hard disk drives
Personal computers
Electronic test equipment
Domestic appliances
Process control

FUNCTIONAL BLOCK DIAGRAM

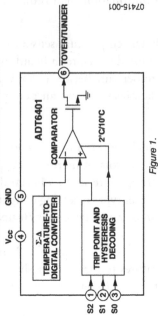

Figure 1.

EXHIBIT 8.7 Temperature sensor specifications

Temperature Sensor	Temperature Range	Output	Comments
Digital IC device	–55 to 150° C[9]	Voltage	Most common
Thermistor	–100 to 300° C[10]	Resistance	Ruggedized and immersible
Thermocouples	–200 to 2,500° C[11] in continuous operation	Voltage	Well adapted to high temperatures
Resistance thermometers	–200 to 500° C[12]	Resistance	Also called RTDs (resistance temperature detectors)
Silicon bandgap temperature sensors	Up to 250° C[13]	Voltage	Often used to measure the temperature of a silicon chip that the sensor is part of.

There are two types of sensors used, CCD and CMOS. CCD sensors capture an array of data and process it serially, and CMOS captures pixels and processes them in parallel. CCDs tend to be used in cameras that focus on high-quality images with lots of pixels and excellent light sensitivity. CMOS sensors traditionally have lower quality, lower resolution, and lower sensitivity. CMOS sensors are just now improving to the point where they reach near parity with CCD devices in some applications. CMOS cameras are usually less expensive and have great battery life.[14] Most phone cameras today have a CMOS because of the speed and battery life.

Sensor size matters. It's not just a matter of megapixels; the size of the sensor determines how much light is captured, which is a big factor in the quality of the image. Larger sensors are more expensive and require more space in the camera. Smaller sensors are typically paired with wide-angle lenses, which can introduce distortion to the image.

How the camera is mounted is very important. Most applications require a fixed camera with a study mount that will not be in the way of workers, processes, or hazardous environments (some cameras can be mounted in housings for placement in hazardous environments). Cleaning is also important, so you will need to access the camera or housing for scheduled cleaning.

Lighting the work area is critical if you are comparing images over time. An example is an outdoor application like an airport, where the sky may be overcast, it could be raining or snowing, or the sun could just be moving across the sky during the day. Even indoors the lights surrounding your area of interest could affect what is captured by the camera. Lights may go on and off, or flicker. Lighting can be provided to compensate for variations. An example is at the toll booth in a parking lot. A bright light comes on and a photo is taken of the license plate for reference, and this works in all conditions.

The field of view is dependent on the lens used with the camera and represents the angular size of the view cone. A large angle or wide-angle lens will introduce distortion of the image. Integrating data from a camera requires an understanding of the transmission interface.

Exhibit 8.8 shows a list of some of the common transmission interfaces for cameras.

EXHIBIT 8.8   Common transmission interfaces

Transmission Interface	Year Created	License Required	Underlying Protocols	Maintenance Association	Connection	Portable Computing Language
GigE Vision	2006	Yes	TCPIP, UDP	AIA	Ethernet	No
USB 3.0	2013	No		USB Implementation Forum	USB	Yes/No
CameraLink	2000	No	Channel Link	AIA	26 pin MDR 26 pin SDR	Yes
USB 2.0	2000	No		USB Implementation Forum	USB	Yes/No
CoaxPress	2013	No		JIIA	Coax Cable	No
IP Camera	1996	No	TCP/IP	ONVIF PSIA	USB Wi-Fi Ethernet	Yes/No

Application Areas for IIoT

IIOT can be utilized in support of a number of applications in the industrial space. Some of these are detailed next.

Quality Control. Some companies offer automated detailed part inspection to ensure quality. This has been an area that has needed automation for years. Typically inspection is done by humans, and they can be very subjective, they can be less than vigilant, and they can make mistakes. Quality control can also be expanded to automatically ensure that each step of a process is completed accurately so you can catch issues before parts have completed a line and the error has been propagated to many more parts. An example could be the requirement of an O-ring in an assembly. If the O-ring installation step becomes misaligned and the O-ring isn't installed correctly, the defect may not be discovered until the part is physically tested hours or days later, or when the customer receives the faulty part and it fails. Then there would be corrective action to try to find all the finished parts that could be affected and possibly a recall.

Inventory Monitoring/Asset Management. By attaching RFID tags or Bluetooth tags to items and positioning readers at different points of interest like doorways, inventory and assets can be tracked effortlessly. In the past, inventory technology has been centered around scanning barcodes. Each item is scanned individually and because it's a repetitive task, mistakes are made. The main barrier to wireless tag technology has been price. Over time, RFID tags have decreased in price to about 10% of the original cost ($0.10 instead of $1.00). Integration of this data with warehouse management software and ERP systems makes it faster and easier to receive goods, provides higher quality on fill rates by ensuring that the right parts are selected, and reduces leakage or theft by triggering an alert when items go out a door. It is now common for the RFID tag to be added to the item at the site of manufacturing and travel with the item for its lifespan, and because it's wireless technology it works for items in cartons, boxes, and on pallets.

Overall equipment effectiveness (OEE) is the ability to measure how well equipment is utilized. Many times the first response to a capacity issue is buying more machinery and hiring more people, but by increasing OEE all machines are more fully utilized during work shifts and capacity is increased without a large capital expenditure. Monitoring the actual uptime against the planned uptime is the first step. Going deeper, you can monitor whether an operator is present, whether there is material to feed the machine, and whether the machine requires maintenance (stack lights typically indicate this). In general, machines are utilized about 60% of the time and increasing this by even 10% can make a big impact to the bottom line.

Predictive maintenance has been accomplished by using a time interval or a usage target, similar to changing the oil in your car every 2,000 miles or two months. This is generally an inaccurate science. A better way to look at this issue is to monitor the machine for deviations from normal that can be defined through AI learning, so if a new vibration is detected it could mean repairs are needed. Most catastrophic failures occur randomly, and routine maintenance could be swapping good parts too soon. A great example of this is data centers full of server racks. A company may buy 10,000 servers a year and swap them out on a scheduled basis. Detecting temperature rises and power consumption could be a better way to understand deterioration of that hardware, and lower maintenance costs. Going forward, historical data and AI can now predict when repairs or replacement of equipment might be needed without monitoring.

Safety and Security. In large chemical or gas plants, monitoring people in restricted areas could prevent accidents. Also, knowing whether people are in a hazardous area and how many are in the area could be very helpful if a rescue operation were to be needed.

Security cameras do a good job of monitoring for theft, for example, an industrial truck being driven off the property by someone. But you really need to combine that monitoring with AI and machine learning to have a system that can also prevent theft or property damage. Most security cameras record hours of nothing and have to be reviewed in their entirety many hours after an incident occurs to try to piece together what happened, and the video quality is usually terrible. A smart system could determine whether the person was an employee or a stranger, exactly when the theft occurred, and it could also be connected to security devices that close and lock the exit gates or sound an alarm if the person is unauthorized and the truck is moving. This system would only record the time frames of interest.

By monitoring power consumption in conjunction with machine utilization and smart switches, a solution could turn off machines that are not in use and help reduce the carbon footprint for the company. Many homes today also have smart meters that help homeowners understand their power usage and gives them the tools to reduce their carbon footprint as well.

Industries Where IIoT Can and Does Play a Role

Where can IIOT be utilized? It can be used to improve the efficiency, yield, safety, profitability, and quality of almost any industrial activity. Following are a few examples with some of the benefits highlighted.

Manufacturing. This would include not only automation of formerly manual processes, but also inventory management and control, factory monitoring for predictive maintenance and repair, inspection and quality control, and many other aspects of a manufacturing facility that were formerly done manually or in a disconnected fashion.

Automotive Industry. IIOT enables new ways to move data from sales, engineering, and customers directly to the factory floor. This permits a greater degree of customization than any achieved before. It also supports new tools and processes to be included in the manufacturing process, such as 3D printing. It makes possible moving to 24-hour production with higher security, safety, and efficiency. Data analytics makes monitoring and trend prediction possible on a grand scale.

Cars themselves contain many sensors that provide feedback to the owner, the mechanic, and any Cloud-connected service like OnStar.[15] A list of typical sensors found in cars follows.

- Airbag
- Brake fluid pressure
- Cylinder head temperature gauge
- Exhaust temperatures
- Fuel level
- Oil level
- Oil pressure
- Oxygen
- Transmission fluid temperatures
- Water in fuel
- Curb feelers
- Parking
- Tire pressure
- ABS status

Oil, Gas, and Chemical. IIoT will not only improve the efficiency of complex chemical and energy-generation activities, but will also greatly improve safety through more complete and targeted monitoring. This would include widespread use of smart sensors to monitor for leaks, overheating, dangerous pressure, and safety violations. Due to the size of, or widespread distribution of, these kinds of facilities, personnel, public, and asset safety is greatly enhanced. Environmental impact can also be better mitigated. Companies will also be better able to adjust for fluctuations in demand, price, availability, and risk management.

Agriculture. With IIOT, farmers will be able to better monitor all kinds of aspects of their operations, they will be able to choose the best times to plant or harvest, apply fertilizer and irrigation for optimum results, monitor livestock more effectively, and monitor weather and soil conditions on a micro-basis. Currently there are drone offerings that will monitor a field for weeds, insects, and lack of water; theoretically you can use fewer people to cover more land.

Food Processing. From harvesting at the peak of ripeness, maintaining a cold chain from harvest to consumer, checking for foreign objects, and smart packaging and labeling, IIoT can make the food chain safer and increase the quality of the food.

Construction. Safety on a construction site is very important; accidents in this environment are usually fatal or life-changing. Being able to monitor worker actions with IIoT could prevent many accidents from happening. It typically isn't one thing that goes wrong, it's a chain of events where several things go wrong; catching even one of these events could make a difference.

In addition, by counting people in different work areas, an accurate model of costs can be developed so that the following proposals can be more accurate and less of a mystery.

Another area where mistakes can be avoided is in tracking the contractors delivering materials. When a skyscraper goes up there are different grades of cement for each section, with the strongest and most expensive being in the foundation. Verification that the right materials were used in the right place can help avoid costly rework and repairs or a collapse after construction is complete.

Retail Distribution and Wholesale Inventory Management. As mentioned earlier, the use of RFID tags or Bluetooth devices can give an exact picture of inventory positions across stores, warehouses, and distribution centers. With the blending of in-store and online shopping, being able to fulfill an order with inventory from any location within the network becomes a differentiator. The goods on the shelves can be monitored for replenishment and distribution centers have a better idea of what to stock to avoid excess or shortages. An example is a clothing manufacturer that thinks they have excess inventory in the stores but no way to aggregate the numbers, so they run a campaign for a sale on those items. If half of the stores have no inventory, they run the risk of disappointing customers who have been enticed to shop there and they have wasted their advertising budget.

Shipping and Transport (Ocean, Rail, Truck, Air). Using AI and IIoT in a transportation chain, a reliable model for predicted arrival times can be learned and it will have all the nuance and details that can tell you which ports are slow, which carriers bump containers, and where leakage occurs and why, providing a complete picture across ocean, truck, air, and rail. Once the transportation chain is understood with this type of model, you can start counting inventory in transit as available at a predicted date and lower safety stock levels.

Benefits may include better estimated times of arrival, asset tracking, loss reduction or prevention, security, and so on. With better data the next round of transportation contract negotiations will be more accurate and fairer.

Public Spaces. Wherever people gather – airports, restaurants, parks, or highways – there are applications for making the experience better. The security lines at airports can be ridiculous. If there is more than one checkpoint, knowing whether all of them or just one particular one are bad would be helpful. Passengers could even be directed to the shortest line through signage. Also, monitoring lines and crowding can be helpful in retail stores or airport gate check-ins. Staffing can be modified so that there are enough staff to accommodate the traffic.

Some parking garages now have sensors for each space to detect when a car is parked in it. Driving into the garage there are signs showing exactly how many spaces are available on each level to take the guesswork out of where to park. Previously there might have been a sign on each level that said full, and it might be accurate.

Future Trends in IIoT

John Burton, writing in *IIoT World*, has made some exciting predictions as to what we can expect from IIoT technology in the coming years.[16]

5G 3D Video Streaming. Models run locally on computer devices today, much like how we used to download movies for viewing. 5G will enable streaming of such information, similar to how we watch movies today. You can also expect to see full 3D video streaming for industrial uses.

Augmented Reality (AR). Virtual reality (VR) has found a niche in industry, typically for training. AR use cases will grow dramatically with sophisticated head-mounted gear that combines a display, a camera, and a microphone. This will include viewing live video, digital twins, and data sheets using voice commands.

3D Modeling. With 3D scans, 3D CAD, and 3D digital twins, the Smart Manufacturing world will model its environment in 3D.

Self-Training Machine Learning. As discussed in Steven Herman's Chapter 9, "Artificial Intelligence, Machine Learning, and Computer Vision," unsupervised learning for AI will greatly simplify machine learning by detecting normal usage patterns quickly and then monitoring for unusual patterns that exceed a preset threshold.[17]

Summary

The Industrial Internet of Things (IIoT) is now impacting more areas of Smart Manufacturing and Industry 4.0 than any other technology. IIoT is also the simplest and least expensive of all Smart Technologies to implement, which is especially important for smaller manufacturers and distributors. Proof of the IIoT's popularity is the dramatic growth of devices deployed. Juniper Research predicts an increase from 17 billion in 2020 to 37 billion in 2025.[18] There will also be a dramatic growth in revenue spent on the IIoT over the next four years. According to Market Research Report Reprint, IIoT revenue will double from 2021 to 2025.[19]

In this chapter we have showcased over 20 types of IIoT devices and how they can transform quality assurance, inventory control, and asset management safety, security, and operational efficiency. We can expect the power of these devices and their use in industry to continue to grow at an accelerated rate.

Sample Questions

1. What is the difference between IoT and IIoT?
   a. There is no difference.
   b. IoT refers to devices like sensors and IIoT to the intelligent systems employing the sensors.
   c. IoT is generally thought to be consumer-oriented and IIoT is IoT technologies in an industrial setting.
   d. IIoT must utilize connections to the Cloud and Big Data for analysis.
2. When was the first webcam introduced?
   a. 1980
   b. 1993
   c. 1968
   d. 2500 BC
3. Which of these are important application areas for IIoT?
   a. Asset monitoring and control
   b. Safety
   c. Automation
   d. All of the above

4. What industrial activity would not get much benefit from IIoT?
   a. Consumer retail
   b. Transportation
   c. Food preparation
   d. There are no industrial activities that would not be able to benefit from IIOT.
5. Security is an important subject for IIoT. What security applications would not gain much benefit from the implementation of IIoT?
   a. Personnel security in a factory or office
   b. Cybersecurity due to the vulnerability of IIoT
   c. Asset security such as controlling who can adjust a machine or process
   d. None of the above
6. What is the definition of a Smart sensor? How would we identify one?
   a. A Smart sensor has an embedded controller that provides some level of control or formatting of the data reported by the sensor.
   b. A Smart sensor has the capability to communicate directly with the Internet, which is what makes it Smart.
   c. A Smart sensor can perform its function autonomously, with no connection to anything else.
   d. All of the above
7. How many layers are there in the IIoT Layered Architectural Model?
   a. Seven, just like in the OSI Model
   b. The number of layers is variable depending on the IIoT technology involved
   c. Four
   d. Three: sensors, controllers, and high-speed communications
8. Which of these is not a real-world current example of IIoT?
   a. Self-driving cars and trucks
   b. Factory automation
   c. Video conferencing
   d. Facial recognition used in security systems for access control
9. The Industrial Internet Consortium (IIC) is a Standards organization made up of major industry players such as IBM, Cisco Systems, General Electric, and AT&T. True or false?
   a. True
   b. False
10. Which of the following are currently available sensors?
    a. Alcohol
    b. Proximity
    c. Cameras
    d. All of the above

Notes

1. Raji, R. S. (June 1994). Smart networks for control. *IEEE Spectrum* 31 (6): 49–55. https://doi.org/10.1109/6.28479.
2. The Industrial Internet Reference Architecture v.1.9. Industry IoT Consortium (IIC). https://www.iiconsortium.org/IIRA.htm.

3. Programmable logic controller (PLC). Wikipedia. https://en.wikipedia.org/wiki/Programmable_logic_controller (accessed August 23, 2021).

4. Teicher, J. (February 7, 2018). The little-known story of the first IoT device. IBM. https://www.ibm.com/blogs/industries/little-known-story-first-iot-device/.

5. Kesby, R. (November 22, 2012). How the world's first webcam made a coffeepot famous. BBC News. https://www.bbc.com/news/technology-20439301.

6. Hylving, L., and Schultze, U. (January 1, 2013). Evolving the modular layered architecture in digital innovation: The case of the car's instrument cluster. International Conference on Information Systems (ICIS 2013): Reshaping Society Through Information Systems Design, Milan.

7. Radio-frequency identification. Wikipedia. https://en.wikipedia.org/wiki/Radio-frequency_identification (accessed August 23, 2021).

8. Low cost, 2.7 V to 5.5 V, pin-selectable temperature switches in SOT-23: ADT6401/ADT6402. (2008–2013). Analog Devices. https://www.analog.com/media/en/technical-documentation/datasheets/ADT6401_6402.pdf.

9. List of temperature sensors. Wikipedia. https://en.wikipedia.org/wiki/List_of_temperature_sensors (accessed August 23, 2021).

10. Thermistor. Wikipedia. https://en.wikipedia.org/wiki/Thermistor (accessed August 23, 2021).

11. Thermocouple. Wikipedia. https://en.wikipedia.org/wiki/Thermocouple (accessed August 23, 2021).

12. Resistance thermometer. Wikipedia. https://en.wikipedia.org/wiki/Resistance_thermometer (accessed August 23, 2021).

13. Silicon bandgap temperature sensor. Wikipedia. https://en.wikipedia.org/wiki/Silicon_bandgap_temperature_sensor (accessed August 23, 2021).

14. What is the difference between CCD and CMOS image sensors in a digital camera? How Stuff Works. https://electronics.howstuffworks.com/cameras-photography/digital/question362.htm.

15. OnStar Corporation is a subsidiary of General Motors that provides subscription-based communications, in-vehicle security, emergency services, hands-free calling, turn-by-turn navigation, and remote diagnostics systems throughout the United States, Canada, China, Mexico, Europe, Brazil, and Argentina.

16. Burton, J. (January 5, 2012). IoT trends for industry 4.0 in 2021. *IIoT World*. https://iiot-world.com/industrial-iot/connected-industry/iot-trends-for-industry-4-0-in-2021 (accessed August 23, 2021).

17. Ibid.

18. O'Halloran, J. (October 3, 2020). Industrial IoT connections to reach 37 billion by 2025. *Computer Weekly*. https://www.computerweekly.com/news/252491495/Industrial-IoT-connections-to-reach-37-billion-by-2025 (accessed August 23, 2021).

19. Kawasaki, L., Huges, I., Zwakman, and Renaud, C. (April 29, 2021). IoT revenue expected to nearly double through 2025; data generation to triple. *Market Research Report Reprint*. https://f.hubspotusercontent10.net/hubfs/5413615/451_Reprint_Industrial-IoT_29APR2021.pdf?utm_campaign=451%20Reports&utm_source=resources%20pge (accessed August 23, 2021).

Artificial Intelligence, Machine Learning, and Computer Vision

Steven Herman

Introduction

Living in the twenty-first century, readers of this book almost inevitably have heard about artificial intelligence (AI), machine learning (ML), and deep learning (DL). Advances in AI and computer vision in many ways accelerated Smart Manufacturing. Rapid growth of computing power has enabled completely new forms of software development, driven by data. This shift in paradigm, sometimes referred to as Software 2.0,[1] can drive rapid innovation in the manufacturing sector and make development accessible to nonengineers.

With the deluge of articles mentioning artificial intelligence, machine learning, and deep learning, it is important to clarify the difference. Artificial intelligence refers to software performing tasks traditionally requiring human intelligence to complete. Machine learning is a subset of artificial intelligence wherein software "learns" or improves through data and/or experience. Deep learning is a subset of machine learning, usually distinguished by two characteristics: (1) the presence of three or more layers, and (2) automatic derivation of features.[2] For the remainder of the chapter, I use the term machine learning unless the subject necessitates differentiation.

Exhibit 9.1 shows the relationship among AI, ML, and computer vision.[3]

While machine learning is often discussed as black magic, the underlying concepts are relatively simple mathematics. At its core, the algorithm checks whether a point is above or below a line; points above the line are shaded darker while points below the line are shaded lighter. The line itself is learned from training data by optimizing the number of points correctly classified by the algorithm. Most machine learning operates on a similar principle with a massively scaled up number of dimensions.

EXHIBIT 9.1   The relationship among artificial intelligence, machine learning, and computer vision

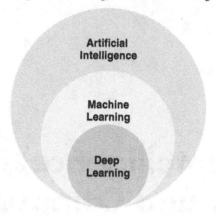

Exhibit 9.2 shows how a simple machine learning algorithm learns to differentiate between two clusters of points.

EXHIBIT 9.2   Learning to differentiate

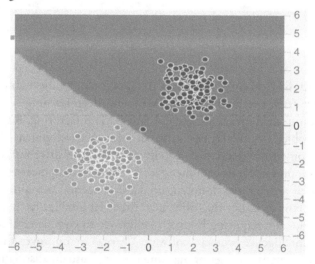

Imagine instead that the inputs were sensor data from a machine and the outputs were whether the machine would fail within the next day, month, or year. The algorithm then learns to pick out changes and relationships in the data that might not be obvious. This is not hypothetical because many companies are already using machine learning to drive predictive maintenance, reducing unnecessary downtime and avoiding catastrophic failures. Problems, such as predictive maintenance, rarely arise in academia, being driven instead by the needs of businesses and processes. This opens vast opportunities for advances in machine learning by applying it to the problems encountered in manufacturing.

Beyond sensors, computer vision offers huge opportunities to collect additional data about processes and systems. Cameras can detect operators, machine state, defects, barcodes, and other vital information, providing greater insight into the manufacturing process. This information can drive process improvement and optimization. Modern computer vision is intertwined with machine learning because most state-of-the-art computer vision is built with machine learning models. In many cases, it is much easier, cheaper, and more accurate for a computer than a human to monitor these processes.

While machine learning was once the purview of large and exceedingly well-funded organizations, it is now increasingly accessible to every business. Many free or paid tools allow custom models to be built with minimal technical experience. These resources lower the barrier to entry, allowing low-cost data collection. The goal of this chapter is to demystify machine learning and computer vision while providing a jumping-off point for what is currently possible with artificial intelligence so you can implement it in your company.

History of AI and Computer Vision

Since the early days of computing, humans have obsessed over teaching computers to think. The term *artificial intelligence* was first coined for the Dartmouth Summer Research Project on Artificial Intelligence in 1956.[4] From there, AI initially took off, spurred by cheaper computers, better algorithms, and funding. The perceptron, a precursor to modern neural networks, was invented only two years later in 1958. As the history of artificial intelligence progressed, it would follow a boom-bust cycle with AI "winters" exhibiting reduced research and funding, often followed by AI "springs" featuring renewed interest.

In the early years, success seemed imminent. Early achievements and funding created a generally optimistic view of artificial intelligence. Algorithms were developed to prove geometric and mathematical theorems, perform basic natural language tasks, and automate simple robotic tasks. In 1963, MIT received a $2.2 million grant from the Defense Advanced Research Projects Agency (DARPA), signaling intense interest in artificial intelligence. This optimism wouldn't last.

In the early 1970s, artificial intelligence saw its first winter, brought about by underperformance and a lack of progress. Limitations of computing power were one of the greatest driving factors. As funding collapsed, researchers turned away from neural networks toward logic and reasoning systems. These too proved difficult, requiring massive amounts of programmed knowledge about the world. In many ways, the first winter emphasized Moravec's Paradox: some tasks that are hard for humans are easy for computers while other tasks that are easy for humans are hard for computers.

Starting in the 1980s, artificial intelligence made a resurgence. The development of expert systems, or systems programmed using logical rules derived from domain experts, proved hugely successful and even commercially viable. With expert systems demonstrating commercial success, private funding poured into artificial intelligence research once again. Even neural networks saw a resurgence, driven by a focus on a new training mode, backpropagation, which drives almost all deep learning today.

Despite the advances of the 1980s, artificial intelligence proved fickle. Expert systems were brittle and difficult to maintain. Neural networks were still predominantly limited by computing power. Once again this led many to become disaffected with artificial intelligence. The second AI winter saw a round of funding cuts lasting through the early 1990s.

Starting in the early 1990s, advances in computing power started to drive artificial intelligence once again. Once artificial intelligence techniques achieved broad success in a particular field, they were often no longer labeled as artificial intelligence. Under the guise of data mining, robotics, optimization, and other new names, AI techniques were applied with great success. In 1997, Deep Blue beat Garry Kasparov, a reigning world champion, at chess. This signaled the success of artificial intelligence systems far outside their original domain. As AI progressed, many other games would fall to artificial intelligence mastery.

Heading into the 2000s, artificial intelligence continued to grow rapidly. The continuation of Moore's law and use of graphics processing units (GPUs) drove rapid increases in computing. By 2009, Google had started an autonomous car project, while 2010 saw the release of Siri on the iPhone 4s, with over 60 million units sold. Arguably, this was the moment artificial intelligence became mainstream.

As computers advanced, the success of computer vision would fall to machine learning. In 2021, AlexNet proved that deep convolutional neural networks (CNN) could beat the existing state of the art by a massive proportion in the ImageNet Large Scale Visual Recognition Challenge. With this new success, development surged, leading to multiple iterations of models and the dominance of CNNs in computer vision.

Computer vision saw rapid advancement with the application of deep neural networks. Object detection saw improvements with the development of Region-Based Convolutional Neural Network (R-CNN) features driving forward performance metrics in 2013. Iterative improvements would come with Fast R-CNN and Faster R-CNN, both improving inference time by 2015. While R-CNN derivatives would drive performance, lower-latency models based on single-shot detectors (SSD) would drive real-time object detection in 2015. While much of the progress was iterative, computer vision started to see machine–human parity on a number of tasks.

More recently, artificial intelligence has diversified its focus, taking on broad problems such as natural language processing, machine translation, and recommendation. In 2020, new techniques use more data and computation than ever before, while continuing to break existing records for accuracy. The gains in other areas of machine learning started to trickle back to computer vision with transformers, a model architecture originally developed for natural language processing, being applied to computer vision problems in 2020.[5]

From here, machine learning seems likely to grow well into the future. While concerns about the end of Moore's law have driven some uncertainty, specialized hardware has provided massive increases in the computation behind artificial intelligence. Further proliferation of computing power makes Edge machine learning and inference easier, while advances in computer vision create actionable data from images.

Understanding Machine Learning and Computer Vision

Types of Machine Learning

If the distinction between artificial intelligence, machine learning, and deep learning wasn't already confusing enough, machine learning algorithms can be broken up into three types, which can also be combined to form hybrid learning algorithms:

1. Supervised learning
2. Unsupervised learning
3. Reinforcement learning

Supervised Learning. Supervised learning is the most straightforward learning problem to understand. It aims to mimic a teacher–student relationship through labeled data. In supervised learning, you provide one or more examples of the output expected for given inputs. Through the process of training, the model learns to generate results that are most similar to the training data. Supervised learning further breaks down into regression (predicting a quantity) and classification (predicting a label). While supervised learning is likely the most common form of machine learning, it isn't the best to utilize the data present with Big Data because the data often is unlabeled. In order to know the expected output, data must be labeled, which can be a time-consuming and expensive process. Advances in few-shot learning help alleviate this problem and despite the high initial cost of supervised learning, it can be hugely beneficial. Detecting product defects or safety violations using models can easily save thousands of dollars. Once developed, these models can often perform at a similar level to human evaluation, but scale much more easily.

Unsupervised Learning. On the opposite side, unsupervised learning requires no labeled data, instead learning patterns directly from the data. While unsupervised learning is excellent at pattern recognition and anomaly detection, it can be more difficult to apply to problems. Still, unsupervised learning can utilize Big Data more efficiently without requiring human interaction. This means that unsupervised learning can offer a lower upfront cost, although it is not universally applicable. Commonly, unsupervised learning is used for clustering and dimensionality reduction.

Reinforcement Learning. Reinforcement learning relies on an interactive environment that an agent can explore. The algorithm tends to learn by trial and feedback. Most commonly this is used for robotics, although recently it has been used for scheduling and chip design as well. While developments in reinforcement learning may well drive Smart manufacturing in the future, reinforcement learning is less commonly applied to problems and may require more specialized knowledge to produce.

Different combinations of these techniques can form hybrid learning algorithms. Semi-supervised learning combines supervised and unsupervised learning to utilize unlabeled data to improve results from a small labeled data set. Recently, semi-supervised learning has driven improvements in computer vision by allowing a better internal representation of images created over a larger unlabeled data set. Self-supervised learning utilizes an unlabeled data set for supervised learning. Usually, the raw data set is the expected result, while some modified data serves as the input. One example of this is natural language processing, where the model is often taught to predict a word removed from a sentence.

Common Computer Vision Tasks

In computer vision, there are a number of commonly researched tasks that machine learning currently dominates. Each task focuses on producing different information given an image. While some tasks, such as barcode scanning or motion detection, don't require machine learning, many more complicated tasks are solely viable with machine learning. Simple tasks may still provide large productivity gains; for example, color detection is an easy computer vision task, but may be utilized to generate machine status information without interfacing with proprietary protocols. For many tasks, computer vision is not only cheaper than other ways of solving the issue, but is also more generalizable. Image classification is the task of assigning a label to an image. This may be used to identify a state; for example, a model may classify machines as in-use, waiting for repair, or available. This could be used to identify defective products where the location of the defect isn't important to the process. Alternatively, image classification may look to recognize one of several pre-known objects in an image. In this case, it is called object classification.

Exhibit 9.3 shows an example of image classification at work.[6]

EXHIBIT 9.3   Image classification

Cat   **Dog**

If the location of an object is important, object classification can be combined with object localization to perform the object detection task. The goal of object detection is to determine both class and location of an object in an image, if one or more exists. This may be a simple problem like locating all the people in an image, or more complex like identifying different parts for counting as they pass on a conveyor belt. Object detection is often used alongside other computer vision techniques to determine the state of a given scene. For example, instead of building an image classifier to determine whether an employee is wearing the appropriate personal protective equipment, building an object detector to recognize hard hats, safety vests, and safety glasses would allow greater information about why the behavior is unsafe. In many cases, the last level of logic is based on either regulation or general knowledge, which should be directly coded rather than learned.

Exhibit 9.4 shows an example of object detection at work.[7]

EXHIBIT 9.4   Object detection

If object detection can be seen as an advanced form of object classification, image segmentation is a more advanced form of object detection. Image segmentation aims to divide an image into multiple parts based on some attribute. It can be broken into two main forms: instance segmentation and object segmentation. Object segmentation involves segmentation based on classes. It might provide segmented areas for the conveyor belt, metallic objects, and plastic objects in a given image. Instance segmentation generates separate segments for each instance of an object. Rather than three segments in the previous example, each nut, bolt, and tray would be identified as a separate segmentation mask.

Exhibit 9.5 shows an example of instance segmentation and object segmentation.

EXHIBIT 9.5 Instance and object segmentation

Instance Segmentation   Object Segmentation

Identification is the computer vision task involving identifying specific instances of an object, usually over time. One example of a common identification task is facial recognition. Object identification could be used to track a product and keep a record of which workstations it passed through. This might allow better tracing of products that fail quality control as well as insight into bottlenecks in the production process.

Although there are many more computer vision tasks with differing state-of-the-art models, a few other notable mentions include optical character recognition, pose estimation, depth mapping, and 3D reconstruction. Optical character recognition (OCR) could digitize paper workflows and allow seamless transitioning between digital and physical tracking systems. Pose estimation may be used for predicting and preventing human–robot collision by determining 3D orientation and the position of the robot, human, or both. Depth mapping is more prevalent in the world of self-driving, but it can provide a uniform replacement or augmentation for radar or lidar data. 3D reconstruction is fairly new but could allow the creation of 3D models, automatically speeding up prototyping and design stages.

Building a Model

Once you know what you want to build, the next step is creating a data set and model. This section addresses the process of building a supervised learning model for object detection, but the same process will apply elsewhere. The basic process involves selecting data, cleaning and annotating data, training or fine-tuning a model, evaluating the trained model, and continual improvement.

While not glamorous, selecting and cleaning the data are the most time-consuming and important steps in machine learning. Biases or faulty data can easily poison a model, reducing accuracy by a noticeable margin. At this stage, you should consider how much you are going to spend on data collection and annotation on a weekly, monthly, or yearly basis. Unfortunately, models often require regular retraining to handle issues like concept drift, where the statistical properties of a target change as the environment in which the model works changes. For example, if you build a model to detect safety vests, but the company changes the color of all safety vests, your model likely will perform worse because it hasn't seen the new color before. When selecting data, you want to make sure you get a representative set from all the different camera angles and conditions you expect the model to see when deployed. Ideally, there should also be a balance of classes that you expect to see, without any one class being over- or underrepresented. You should also make sure to be aware of any biases or shortcomings in the data set, preferably documenting them alongside the model. The siloed nature of data at larger companies can interfere with the ability to get widespread and useful data selected. Where possible, long-term investment in a single data lake can make this process faster, cheaper, and more agile.

After data selection and some analysis, cleaning and annotation of the data pose a number of unique challenges. Before starting annotations, you should write up a guide for how to annotate the data. This minimizes communication errors and allows more uniform results, which in turn leads to better model performance. While you may need to annotate the data internally, external annotation services likely offer better guarantees for annotation accuracy. Internally, you will likely have to train team members and double-check annotations, since everyone will have a different idea about annotating initially.

Once you have your annotated data, you should split it into three sets: training, validation, and testing. At least 10% of your data should be set aside for testing and around 20% should be set aside for validation. This leaves the remaining 70% for training. These data sets are kept separate to ensure that you have adequate data to estimate accuracy of your model. If there is overlap between these data sets, you will have no way of performing unbiased evaluation of your model or ensuring that no overfitting occurred. Overfitting is when the model performs better on the training data than on the test/validation data. This is usually because it conforms too closely to the data, effectively memorizing it rather than generalizing. You want to stop training before you overfit the data. The purpose of a separate validation set is to allow you to experiment with parameters of different models without selecting a model that best fits the test set. This becomes more important if you are testing a large range of models or parameters but is good practice either way.

Now that you have data, you can start experimenting with models. While state-of-the-art models may offer higher accuracy, they often are more expensive to train and deploy because they are more complex. One good way to evaluate model complexity is to look at inference time. The longer it takes to perform inference on one or many images, the more complex it is and the longer it will take to train. If you can, start with simple initial models such as MobileNet or an SSD-based model and scale up if you need better performance. You can fine-tune (train only the last few layers of a model) or train the entire model, although the latter requires significantly more data to do without overfitting. In general, fine-tuning will be enough for most purposes. Like model complexity, start simple and see whether you need more.

Once you have a trained model, you should check its performance on the validation set. This will allow you to see how well it performs on data it hasn't seen before. From here, repeating the training step with different models or different parameters can produce better results. When you are happy with the results, evaluating on the test data set should give an indication of the performance you will see for the model when deployed.

From here, the model can be updated with whatever frequency is necessary to obtain good performance. One of the greatest benefits of machine learning is that more data and computation tend to produce better results so there is a straightforward improvement process. Over time, you can build up a data set allowing a better model to be made. This data set can also be a valuable asset to any company, providing a benefit over your competitors.

Machine Learning Pipelines

Given the continuous nature of machine learning, it may make sense to have a pipeline, much like you would for data science. Pipelines may also offer cost-saving techniques such as active learning, which are beyond the scope of this chapter. Machine learning pipelines can easily become a greater time and maintenance sink than the models they were used to build.[8] If your needs are met by one of the many existing free or paid pipelines, these will almost always offer more features than a pipeline you build yourself.

Having a pipeline can also be beneficial by allowing integration of automated tests to check whether the model is performing well. Organizing and tracking model performance can help diagnose issues and regressions before they cause major issues.

Issues with Artificial Intelligence

While artificial intelligence can be extremely powerful, it is important to understand the ethical implications of it. Ethical issues with artificial intelligence are not limited to some far-off reality with superhuman artificial intelligence or complete societal automation; plenty of existing machine learning–based systems cause real harm every day. It is your job to consider the implications of your use of artificial intelligence and avoid creating more harm through your application of artificial intelligence to manufacturing.

At their core, most machine learning systems look to optimize an objective; these algorithms will often replicate any systemic bias they are trained on. These biases can lead to critical failures of systems incorporating the compromised model. In a now-famous example, Amazon set out to build an automated hiring algorithm, hoping to mitigate sexism in the hiring process; despite not being fed data on the gender of candidates, it learned to discriminate against women.[9] One potential cause of such behavior is a proxy, where seemingly unrelated data can infer protected attributes such as race, age, or gender. This can have potential legal ramifications, especially as laws evolve to manage artificial intelligence.

Beyond biases, artificial intelligence systems often introduce a level of opacity into their design. While it may be amusing the first time your object detection model misses detecting a person because they wore an unexpected color of T-shirt, a Tesla crashing into a parked car or a robotic arm hitting an assembly line worker show us that the stakes can be deadly serious. What is worse is that in many of these cases, there is no way to be sure the problem is fixed. Sure, you can improve the training data, rerun training, and validate that the original scenario no longer occurs, but you can't verify that the model won't fail under another novel circumstance. This is one of the major downsides to machine learning; even if you achieve over 99% accuracy, there is no predicting what kind of failure you may experience in the remaining cases.

These hidden failures can also lead to real-world attacks on AI systems. Adversarial attacks rely on modifying input to models to intentionally disrupt the output. This may be as simple as placing tape on the right parts of a stop sign to prevent it from being recognized.[10] As artificial intelligence becomes more widespread, businesses will need similar cybersecurity practices for adversarial activities to what they now have for physical security.

It is also important to recognize that most artificial intelligence systems will work alongside humans. Artificial intelligence will drive robots and cobots. It may even optimize manufacturing floor layouts and worker scheduling. While automation can reduce the dangerous and repetitive tasks done by humans, it also runs the risk of harming humans, especially where AI and humans interact. Done improperly, artificial intelligence can optimize warehouses and manufacturing in ways that are harmful to workers.[11]

This isn't to discourage artificial intelligence, but it is important to be aware of the ethical issues when implementing these systems. Especially when thinking about integrating machine learning, applying different ethical decision-making frameworks can be useful. Potential ethical decisions can be assessed by recognizing, researching, evaluating, testing, and reflecting.[12] In this process, you want to recognize decisions that may cause harm to individuals or groups of people. Then you should research the relevant information, including your options for action. This should include consultation with possible stakeholders who may be impacted or harmed by the decision. While some concerns may be more or less important in the decision making, missing concerns may result in poor decision making. Next, you should evaluate the different options, preferably under different ethical frameworks. The easiest approach to conceptualize is usually a utilitarian approach of maximizing good and minimizing harm. After arriving at a decision, you should test that the decision addresses the issue and concerns raised in the research phase. Then, after implementing the decision, you should return to the issue to reflect on the implementation and ensure that it was implemented in an appropriate way for all stakeholders. Following this process should help ensure that you have a positive impact with your artificial intelligence application.

At the time of this writing, regulation of artificial intelligence is limited, but it is likely to increase in the next 10 years, with multiple agencies signaling interest in regulatory oversight. Currently, the National Artificial Intelligence Initiative Act of 2020 and General Data Protection Regulation (GDPR) are most influential when it comes to artificial intelligence uses, but neither has significant restrictions on artificial intelligence use. It is likely that the European Union will pass some form of the Artificial Intelligence Act in 2022, which may be much more restrictive. Recent industry initiatives have pushed for fairness, accountability, and transparency in machine learning; algorithms and machine learning should not be an excuse for companies causing harm.

Like much of machine learning, AI ethics is an extraordinarily fast-moving field. I would highly recommend checking out additional information on the topic, including researchers such as:

- Kate Crawford
- Timnit Gebru
- Guillaume Chaslot

and websites like:

- https://exploreaiethics.com/
- https://www.fatml.org/
- https://www.fast.ai/2018/09/24/ai-ethics-resources/

Conclusion

Artificial intelligence, machine learning, and computer vision are a gold mine to Smart Manufacturing, driven by the proliferation of Big Data and the low cost of computing. Innovation in these areas can drive process improvements by automating tasks and reducing human error as well as assisting workers with dangerous or repetitive tasks. While knowledge of code is useful, it isn't required to build machine learning systems. Advances in no-code systems enable anyone to build machine learning models, reducing the barrier to entry and enabling experts to apply their domain knowledge to building useful artificial intelligence systems. While there are potential issues with artificial intelligence, careful decision making prevents future harm and avoids costly mistakes. As seen in the case studies, computer vision is already revolutionizing Smart Manufacturing.

Sample Questions

1. What is the relationship between artificial intelligence, machine learning, and deep learning?
   a. There is no relationship.
   b. Machine learning and deep learning are subsets of artificial intelligence, but completely separate from each other.
   c. Deep learning and artificial intelligence are subsets of machine learning.
   d. Artificial intelligence contains machine learning. Deep learning is a subset of machine learning.
2. Artificial intelligence experienced how many winters?
   a. Zero
   b. One
   c. Two
   d. Three
3. Which of the following have enabled the rise of machine learning?
   a. Big Data
   b. Moore's law
   c. Specialized hardware
   d. All of the above
4. Which model architecture revolutionized computer vision by outperforming other techniques?
   a. Faster R-CNN
   b. Convolutional Neural Networks
   c. AlexNet
   d. Reinforcement learning
5. What is the difference between supervised and unsupervised learning?
   a. Supervised learning requires constant human interaction; unsupervised learning solves more problems.
   b. Unsupervised learning works on data without labels, while supervised learning requires labels.

    c. Unsupervised learning requires an interactive environment which an agent explores, while supervised learning requires labeled data.

    d. All of the above

6. Which of the following is *not* an example of image classification?

    a. Finding a person in an image

    b. Checking whether an image is a cat or a dog

    c. Determining whether a machine in an image is on or off

    d. Classifying whether the image is taken on the East or West Coast.

7. What should you ensure that you do with training, validation, and testing data?

    a. Mix the data to ensure that you have an even distribution of each when training and evaluating your model.

    b. Keep the data separate and don't use testing data to evaluate different models.

    c. Train until the model performs significantly better on training data than validation or test data.

    d. Set aside at least 50% of your data for validation and testing.

8. When building an artificial intelligence system, you should always:

    a. Consult with stakeholders to determine potential consequences of the system.

    b. Develop and deploy systems quickly to determine whether they can accelerate your process.

    c. Test systems by implementing them in your manufacturing facility.

    d. Avoid artificial intelligence and machine learning at all costs.

9. Which of the following is an example of an adversarial attack?

    a. A sticker is placed on a machine to prevent it from properly detecting the machine state.

    b. A model fails to recognize a person in an image because of a change to safety vest color.

    c. A model learns to detect machine state by image brightness because the training data features machines in their "off" state at night.

    d. A cobweb obscures the camera's view and causes poor model performance.

10. Who is responsible for a machine learning algorithm that causes harm?

    a. Nobody; the algorithm works in obscure ways.

    b. It is difficult to say because there is little legal precedent, but the company or anyone involved in the deployment of the system may be liable.

    c. The person harmed by the algorithm.

    d. The company.

Notes

1. Karpathy, A. (November 11, 2017). Software 2.0. Medium. https://karpathy.medium.com/software-2-0-a64152b37c35.

2. IBM Cloud Education. (May 1, 2020). What is deep learning? IBM. https://www.ibm.com/cloud/learn/deep-learning.

3. Shreyas, S. (March 7, 2019). Relationship between artificial intelligence, machine learning and data science. Medium. https://medium.com/@shreyasb494/relationship-between-artificial-intelligence-machine-learning-and-data-science-15a87e2cc758.

4. Anyoha, R. (August 28, 2017). The history of artificial intelligence. Science in the News, Harvard University. https://sitn.hms.harvard.edu/flash/2017/history-artificial-intelligence/.

5. Houlsby, N., and Weissenborn, D. (December 3, 2020). Transformers for image recognition at scale. Google AI Blog. https://ai.googleblog.com/2020/12/transformers-for-image-recognition-at.html.

6. Cat. Pixabay. https://pixabay.com/photos/cat-kitten-pet-striped-young-1192026/. Dog. Pixabay. https://pixabay.com/photos/bulldog-dog-puppy-pet-black-dog-1047518/

7. Chafik, A. Unsplash. https://unsplash.com/photos/2_3c4dIFYFU.

8. Sculley, D., Holt, G., Golovin, D., et al. (2014). Machine learning: The high-interest credit card of technical debt. *SE4ML: Software Engineering for Machine Learning* (NIPS 2014 Workshop).

9. Goodman, R. (October 12, 2018). Why Amazon's automated hiring tool discriminated against Women. ACLU. https://www.aclu.org/blog/womens-rights/womens-rights-workplace/why-amazons-automated-hiring-tool-discriminated-against.

10. Eykholt, K., Evtimov, I., Fernandes, E., et al. (June 2018). Robust physical-world attacks on deep learning visual classification. In *Proceedings of the IEEE Conference on Computer Vision and Pattern Recognition* (pp. 1625–1634).

11. Ghaffary, S. (October 22, 2019). Robots aren't taking warehouse employees' jobs, they're making their work harder. *Vox.* https://www.vox.com/recode/2019/10/22/20925894/robots-warehouse-jobs-automation-replace-workers-amazon-report-university-illinois.

12. Markkula Center for Applied Ethics. (2015). A framework for ethical decision making. https://www.scu.edu/ethics/ethics-resources/ethical-decision-making/a-framework-for-ethical-decision-making/.

CHAPTER 10

Networking for Mobile Edge Computing

Jeff Little

Introduction

Smart Manufacturing is powered by mobile computing. Without mobile technology, Smart Manufacturing would not be practical. Mobile devices are the platforms by which manufacturing workers and managers can connect easily to the Cloud. The Industrial Internet of Things (IIoT) generates massive amounts of data with connected devices. By combining mobile's ability to provide networks with the data generated by the IIoT, manufacturers and distributors have powerful new sources of information to improve operations and eliminate paper-based practices.

In this chapter, you will be introduced to modern networking. The chapter covers basic networking architecture concepts and introduces you to the extensive networking vocabulary. At the end of this chapter, you should be able to discuss, at a high level, the capabilities and required features of the Edge networks needed to support Mobile and Edge Computing with network engineers and IT support personnel. At the end of this chapter is a list of sources where additional information can be found about network design, implementation, deployment, and maintenance. Finally, a list of commonly used acronyms is provided.

Brief History of Networking

Mobile communications starts with analog radio telephony. Mobile technology is usually designated by the number of the generation, or XG, number; thus 4G or 5G. The first mobile communications is designated as 0G and generally thought to start with the car phone, introduced in 1946 by the Bell System. Hand-held portable two-way radios called walkie-talkies date back to the late 1930s, but these were only able to transmit on one radio channel and all walkie-talkies in the area used only that one channel. There was no notion of either circuit or packet switching. The car phone, however, was able to use multiple channels and could connect to the public switched telephone network so that you could call any other telephone from a car phone.

Wireless cellular telephony starts with 1G and today is using 4G and 5G technologies, with a 6G in early development and planned to launch sometime after 2030. The technical aspects of each of these technologies (3G, 4G, and 5G) are discussed later. Beginning in the 1980s, the first generation of wireless analog cellular phones was introduced. These are called the 1G generation. Starting in 1991 in Finland, the first commercially available digital cellular phones were introduced, creating the 2G generation. While 2G has now been superseded by newer generations, it is still used in most parts of Europe, Africa, Central America, and South America. In the United States, Japan, Australia, and other countries, various carriers have announced phase-out plans or have already shut down 2G services. In some other parts of the world, such as the UK, 2G is still widely used for so-called dumbphones and IOT devices to avoid high licensing costs.

Beginning in 2001, the first commercial launch of 3G was in Japan. In the United States, the first commercial launch was in 2002. However, since the increasing use of 4G, 3G has been in decline and operators around the world have already or are in the process of shutting down their 3G services. Beginning in 2009, 4G was commercially launched in Sweden and Norway. In the United States, the launch was in 2010. Then in 2019, 5G technology was launched almost simultaneously in South Korea and the United States and the rollout is now expanding around the world. Finally, as mentioned earlier, 6G is expected to launch commercially sometime after 2030.

The history of the evolution of electronic networking generally begins with the invention of the telephone in the 1870s. Different telephones need to be connected together to support a call and the connection needs to be stable for the duration of the call. The telephone companies made these connections manually with human operators working at plug boards in centralized offices. This type of networking is called circuit-switched networking because a dedicated circuit is established for the duration of a call between two "stations" (telephones). Circuit-switched networking requires a lot of resources, particularly as the size of the network grows and can be expensive to build and maintain. Throughout the twentieth century, telephone companies developed ever more elaborate switching and transmission equipment to both expand the capacity and lower the cost of maintaining the so-called public switched networks for telephone and, eventually, other kinds of telephony-based communication such as FAX and telephone modems.

Beginning in the 1960s, a different approach called packet switching was developed. In this approach, information in the form of video, sound, images, data, computer files, text messages, and so on is digitally coded and divided into small digital packets that can then be sent into a fixed network of permanently established connections. The packets are routed through the fixed network based on an addressing scheme that allows the addressed packets to arrive at their destination and be reassembled, in the proper order, and then be delivered to recreate the original video, sound, data, text message, and so on. The bandwidth of these networks is such that even a continuous stream of a large flow of packets such as what would be needed for video or audio can be delivered and reordered as necessary to recreate the video or audio. A key factor of a packet-switched network is that the network connections do not change but are fixed. And on these connections, a stream of packets of many different information flows can travel intermingled. The addressing scheme is what allows these intermingled flows to be received properly at their intended destination so that there is no need for a dedicated circuit. It is generally recognized that the world's first packet-switched network was the ARPANET (with the first connections made in 1969), a special research network funded by the Advanced Research Projects Agency (ARPA), an agency of the US Department of Defense. The purpose of the project was to develop a robust fault-tolerant networking technology that could survive World War III.

Throughout the 1970s, 1980s, and 1990s, a number of competing packet-switched network technologies were designed and implemented for various purposes. These included technologies like ARCNET, Token Ring, SONET, ATM, ISDN, Frame Relay, FDDI, and others. More on these early technologies later.

Basic Networking Concepts, Architecture, and Capabilities

A fundamental concept is that of topology, or the layout of a network with its nodes, stations, and links. Stations are the entities that want to communicate with each other. Examples are computers, laptops, programmed logic controllers (PLCs), sensors, and servers. We can also consider telephones, cell phones, and cameras to be stations. Nodes are entities that connect between links. Examples are routers, switches, and hubs. We examine these in greater detail a little later. Links are the connections between stations and nodes. They can be one-way or two-way connections.

Exhibit 10.1 shows the simplest (but not very useful) network. This is a simple link between two stations.

EXHIBIT 10.1 Simplest network

The next stage in complexity involves connecting through a node, which may provide some kind of protocol conversion and/or may be used to increase the distance between the two stations beyond what a single link could provide.

Exhibit 10.2 shows the addition of a node to a network.

EXHIBIT 10.2 Network with a node

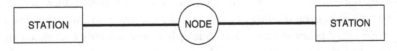

But such a network is subject to failure if a link or the node were to go down. Connectivity between the stations would be lost. When links were an expensive item, many early networks used a ring topology as a way to provide a network that can continue to operate even in the event of a link or node Failure.

Exhibit 10.3 shows the addition of a ring topology to a network. The two-way ring link connections are shown in gray.

EXHIBIT 10.3   Network with a ring topology

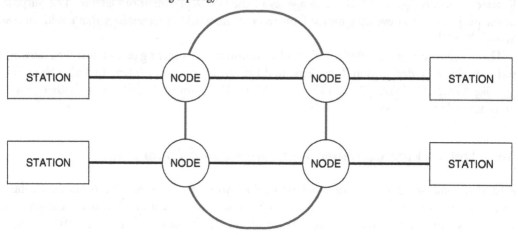

Now if one of the ring links goes down, or one of the nodes goes down, the stations still have a pathway to reach every other station, providing a protection against single faults in the network. With the availability of much cheaper links in recent years, ring topologies have largely disappeared and been replaced by mesh topologies.

Exhibit 10.4 shows a mesh topology consisting of a simple four-node mesh network.

EXHIBIT 10.4   Mesh topology

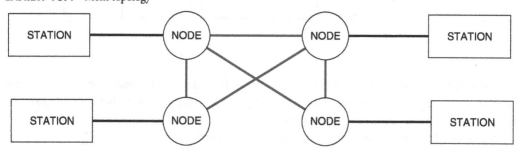

This topology is even more robust than the ring. With each node connected to every other node by a two-way link, even the loss of two red links will not prevent each station from communicating with every other station. If a node is lost, the stations connected to it will lose communication but the other stations in the network will still have communication to all the other stations on all the other nodes. Modern network technologies, notably Ethernet and the Internet, use this topology.

Another fundamental concept is that of the local area network (LAN) and wide area network (WAN). For stations that are located within close proximity to each other, such as in a home or building, these stations can be interconnected by a communication technology that emphasizes speed, lower cost, and simplicity. Such a network is called a LAN. In a LAN, the nodes and links are often passive, meaning that addressing is controlled by the stations themselves and the network merely acts as a medium to carry the packets. For stations that are geographically separated, such as across a city, a country, or halfway around the world, different communication technologies are employed that emphasize operation over a distance, increased reliability, error checking, and active routing and addressing of the packets being carried. Such a network is called a WAN. The most common forms of a LAN are the Ethernet or a Wi-Fi network. The Internet (TCP/IP), mobile phone cell networks, and cable TV are examples of a WAN network.

The Ethernet is, today, the most widely used packet-switched network technology used in LANs. Most people are now familiar with Cat 5 cables and hard-wired and Wi-Fi connections. Ethernet is a technology that depends on the transmission and reception of data packets in a shared media environment. A station wishing to send a packet or string of packets to another station attaches a unique address to those packets called a media access control (MAC) address. This is also sometimes called an equipment or physical address because it uniquely identifies the station to receive the packet(s). The sending station also attaches its own MAC address so the receiving station will know from where the packet came from. Special unique addresses have also been defined by the Ethernet standard to support a network broadcast of packets (with an address of all 1s) and selected multicasting of packets to some but not all stations. In its simplest configurations every station connected to the network can see all the packets being transmitted and selects out only those addressed to itself by the unique MAC address and the broadcast packets. MAC addresses are made up of six bytes or octets, and are generally assigned to the hardware of the station by the manufacturer of the hardware. They are therefore a unique identifier of that Station.

Exhibit 10.5 shows MAC addresses consisting of six bytes. They are generally displayed in the format where each digit is a hex digit (0–F). Lower-case letters instead of capital letters are also sometimes used.

EXHIBIT 10.5 Six-byte MAC addresses

00-00-CA-11-22-33

14-FE-B5-B9-26-97

00-E0-4C-68-17-FE

00-13-CE-32-F3-38

Data from the sending station is cut into packets of varying length (64 to 1,518 octets). These are then tagged with a preamble, the MAC address of the destination and the source, and a packet type designation. At the end of the packet, a cyclic redundancy check (CRC) is added that the receiving station can use to check for transmission errors in the packet. Messages, files, or data streams larger than what one packet can accommodate are cut into multiple packets.

Since all the stations on the Ethernet network share the media, it is possible for two stations to try to send their packets simultaneously, resulting in a garbled transmission and what is called a collision. The avoidance of collisions is accomplished through Ethernet CSMA/CD (carrier-sense multiple access with collision detection) rules. A simplified description of these rules is as follows:

- When a station is preparing to transmit a packet, it listens to its Ethernet input to see if anyone else is transmitting at that time (called CSMA). If another transmission is detected, the station waits until that transmission is over.
- Then the station begins to transmit its packet, while also monitoring the shared media (and its own transmission) to see whether a collision is occurring. The preamble that is at the head of the packet helps with this step by establishing the initial so-called "carrier wave," which is a string of 1s and 0s that can be seen by all the other stations. If no collision is detected during the entire transmission of the packet by the transmitting station, then the transmission is considered to be completed. As a double-check, the receiving station will check the CRC as well for any transmission errors and, if any are detected, will request a resend of the corrupted packet after a short wait time.
- If the transmitting station while monitoring its own transmission detects a collision (called CD), it immediately halts the transmission, waits for a randomly selected wait time, and then tries again.
- In the event of multiple collisions, a complicated scheme of randomly adjusting the wait times between transmission attempts helps prevent the Ethernet network from being tied up for long periods of time during periods of high traffic.

Ethernet thus is a system that does not support a 100% utilization of the total bandwidth available. But as high-bandwidth links are available, it can carry a significant amount of traffic without bogging down. This permits a relatively large number of stations to share the Ethernet media at reasonable cost.

Exhibit 10.6 shows the initial Ethernet installations used a single large cable with stations hanging off of multiple taps into that cable. This topology is often called a multi-drop topology.

EXHIBIT 10.6   Initial Ethernet installations

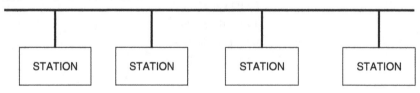

Today, for virtually all Ethernet hard-wired installations, this multidrop topology has been replaced by one that depends on two-way twisted-pair wiring (the Cat 5 cable) and a node device (called a hub) that replicates a packet coming into it on one link to all the other links going out of it.

Exhibit 10.7 shows a simple configuration.

EXHIBIT 10.7   Simple configuration with hub

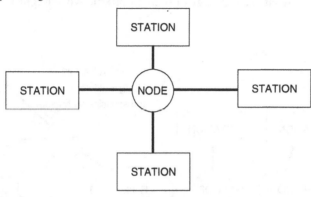

When the node is a hub, it is a very simple repeater device that replicates an incoming packet on all the other links attached to the node, creating a fully shared media environment. Nodes can also be more sophisticated and smarter. By monitoring the packets coming in on a particular link and noting the source MAC address(es), the node can determine over time what stations are attached to that link. It can then identify packets coming in on another link that are addressed to a station on that particular link and replicate that packet out only on that link. This is called packet-switching and such a node is called a switch. Packet-switching can greatly reduce the amount of meaningless traffic passing by a station and increase the apparent bandwidth of the overall network. Modern switches are able to receive and switch multiple packets simultaneously as long as they are coming in on and being transmitted out on different links.

These examples are simple ones. Real LANs, particularly large ones with lots of stations, can have a number of hubs and switches connected together to create complex LAN network topologies. Modern switches are also capable of a number of other functions. They can manipulate packets, monitor traffic statistics, and do other tasks. When dealing with a real network, it is important to map out its topology showing the hubs, switches, stations, links, and so on in order to be able to understand and maintain it properly.

As LANs grow in size and complexity, it may be desirable to divide the topology up into many different subnetworks, networks using different technologies or protocols, or other kinds of divisions. This in effect creates distinct separate LANs that can then be interconnected. This is the essence of internetworking. While Ethernet is the most common LAN technology, it is not the only one. In the early days of networking there were many competing network technologies and protocols with different transmission technologies and addressing schemes. To interconnect stations using different technologies and protocols, interconnecting these separate LANs is required. To interconnect separate LANs, a special node is required that can translate the contents of a packet to show different source and destination addresses, different data formats, different transmission technologies, and different network protocols. Such a node is referred to as a router.

Exhibit 10.8 shows what the topology of such an internetwork might look like.

EXHIBIT 10.8 Internetwork topology

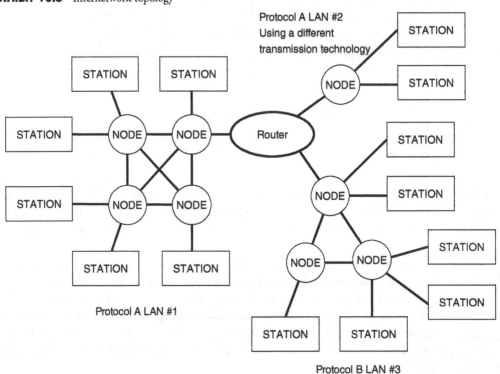

In Exhibit 10.8 we see three separate LANs connected by a router. LAN #1 and LAN #2 use the same networking protocol but use different kinds of transmission technology. In LAN #3, a different networking protocol is in use. The router now reconfigures the packets traveling from one LAN to another to match the protocol and transmission technology of the destination LAN. With the predominance of the TCP/IP protocol in recent years, this kind of protocol conversion is not as common as it once was. But examples still exist, such as cable and digital subscriber line (DSL) modems. Such routers are also sometimes referred to as gateways. When a larger network of the sort described by Exhibit 9.8 covers a wide geographic area such as a city, a nation, or worldwide, it is called a WAN or wide area network. The Internet, the widest global network, is usually shown as a cloud symbol. This is a shorthand way of showing the wider Internet without having to show any of the detail of how it is put together. Often, the actual topology of the Internet in a local area is not clearly understood by those connecting to it, and it is not necessary that they know this information.

Exhibit 10.9 shows a typical example of a LAN connected to the Cloud (aka the Internet).

EXHIBIT 10.9 LAN connected to the Cloud

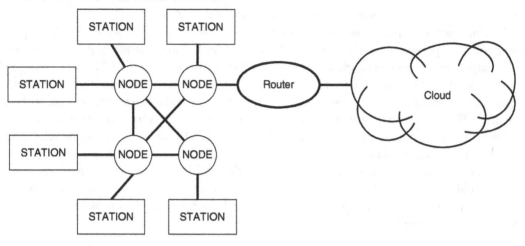

Protocol A LAN #1

Network Address Management

Now let's explore how modern network addressing works in the TCP/IP protocol. The most common internetworking protocol and the one used by the Internet. As discussed earlier, Ethernet uses fixed unique MAC addresses to identify stations. In the early days of networking, when there were many different LAN protocols and technologies, there were many different ways to identify/address stations. The TCP/IP protocol defines a generic Internet or IP address that can be used to address a station across different LANs and across different LAN protocols. In our examples we focus on Ethernet as the LAN protocol, sometimes referred to as the data link layer.

The MAC addresses, referred to as physical addresses, are assigned to the station hardware at the time of manufacture, although some of these interfaces do allow the manual setting of the MAC address by a user. To support internetworking where different LAN protocols may use different physical address formats, a logical address called an IP address is assigned to the station to permit its packets to move across network boundaries between different LAN protocols and different logical addressing schemes. In TCP/IP this assignment is handled by a protocol called address resolution protocol (ARP). Now a station is addressed in two ways. There is the physical address (in Ethernet this is the MAC address) used on the LAN and the logical IP address used both on the local LAN and the wider WAN network. As packets come into the router from the WAN, the router uses a lookup table to translate the IP address on the packet into the physical address that will be recognized by the receiving station. It does the reverse translation on packets going from the LAN station to the WAN network. In this way, a station need only know the logical IP address of the station it wants to send and receive from on the wider WAN. The physical addresses and protocol details are largely hidden, and management and translation is handled by the router. The station can also use the IP address to establish a connection to a station on the local LAN without having to know the physical address up front. The TCP/IP protocol permits a station to use an IP address to locate and discover the physical address of the station with that IP address so that the two stations can establish a connection to each other through the LAN network. Again the station need only know the IP address of the station it wants to communicate with on the LAN.

Having logical addresses also permits special unique addresses to be assigned that support functions like private networks, network classes, loopback, broadcast (or multicast), subnetting, supernetting, and network maintenance. One of the goals of TCP/IP is to create an environment that handles many of the tasks of setting up network addressing automatically through suitable functions and protocols. The rules governing the definition, use, and assignment of IP addresses are complex and we explore only a few very important ones here. Additional information about IP addresses, ARP, and TCP/IP can be found in the references listed at the end of the chapter.

TCP/IP supports two versions of IP addresses; the initial and most common are the IPv4 version addresses. They take the form of a four-byte binary number and are usually shown in what is known as "dotted-decimal notation." Several examples are shown in Exhibit 10.10. More on the comments later.

EXHIBIT 10.10 IPv4 addresses

IP Address	Binary	Comment
10.0.0.0	00001010.00000000.00000000.00000000	Example of a private address
127.0.0.24	11111111.00000000.00000000.00011000	Example of a loopback address
196.168.0.20	11000100.10101000.00000000.00010100	Example of a private address, often used in home networking
216.58.195.78	11011000.00111010.11000011.01001110	A globally recognized Internet IP address. This one happens to be Google.com.
255.255.255.255	11111111.11111111.11111111.11111111	Direct broadcast address, recognized by all stations on the network

Starting in 1995, the Internet Engineering Task Force (IETF) introduced a new IP address format to replace IPv4. This was because as a 32-bit binary address, IPv4 only permits approximately 4.3 billion addresses to exist. Due to the way the classification of the address space was defined for IPv4, many of the address values are wasted. It was assumed that Internet growth would eventually exhaust this address space, so IPv6 is defined with a 128-bit binary address that supports approximately 3.4×10^{38} addresses, which is a huge increase in the total address space. Since IPv6 was introduced, its adoption has been slow because of the cost and effort to update the infrastructure of the overall Internet. Also, IPv4 has introduced a number of features that reduced the need to assign a unique globally recognized IPv4 address to each station. These new features include classless addressing, the use of the Dynamic Host Configuration Protocol (DHCP) for address allocation in a dynamic way, and network address translation (NAT). This is why today IPv4 is still the most commonly used IP-addressing protocol. But IPv6 provides some other improvements that will become more important in the future as the Internet continues to evolve. These include an improved header design, new options and the allowance for extensions, support for resource allocation, and support for better security. More on IPv6 can be found in the references listed at the end of the chapter. Meanwhile we focus on IPv4.

The original definition for IPv4 included the notion of address classes. While this proved useful when the Internet was small, by the 1990s it was causing problems.

Exhibit 10.11 shows how the classes are defined, where "X" is any value between 0 and 255.[1]

EXHIBIT 10.11   Definitions of classes

Class	IP Address Range	Number of Blocks	Address Size of a Block	Comment
A	0.X.X.X to 127.X.X.X	128	16.8M	For large organizations
B	128.X.X.X to 191.X.X.X	16K	65.6K	For medium-size organizations
C	192.X.X.X to 223.X.X.X	2.1M	256	For small organizations and home use
D	224.X.X.X to 239.X.X.X	1	4096	For multicasting
E	240.X.X.X to 255.X.X.X	1	4096	Reserved for future use

However, starting in 1996, classless addressing was introduced that largely rendered this class system obsolete. Now blocks of addresses are of varying size and allocated through a central authority called the Internet Corporation for Assigned Names and Addresses (ICANN). ICANN does not normally assign names and addresses to individual organizations. Instead it assigns large blocks to Internet service providers (ISPs), and they divide up the addresses into smaller variable-size blocks, assign them to their customers, and manage them. This permits a much more efficient way of managing the total IPv4 address space. Still, there are special addresses that are defined for specific purposes.

Exhibit 10.12 shows a few of the most commonly used addresses and how they are applied to both class and classless address spaces.[2]

EXHIBIT 10.12   Common IP addresses

Host IP Address	Name or Purpose	Comment
0.0.0.0	The unspecified address	Used in several ways by several functions and protocols.
10.X.X.X	Reserved for use by a large private network	For a large private network, often behind a firewall or NAT router.
127.X.X.X	Loopback address	Used to perform a virtual loopback operation. The packet never leaves the station but loops back as if it had been received from the network.
192.168.X.X	Reserved for use by a small private network	For a small private network, often behind a firewall or NAT router.
224.X.X.X through 239.X.X.X	IP multicast addresses	Used for multicasting on the network.
255.255.255.255	Direct broadcast address	Used to broadcast a packet to all stations on the network.

Subnets

As mentioned earlier, for large Ethernet networks (as well as other protocols) it is advisable to be able to divide a network up into smaller subnetworks so that all of the stations do not have to see *all* of the network traffic (packets) flow by. Then the traffic can be divided (switched) between separate physical networks while still having the logical topology that permits any station to talk to any other station. As mentioned earlier, this can be accomplished through the use of switches and routers. But TCP/IP also provides another mechanism called subnetting that can do the same thing. Subnetting a large network topology can become quite complicated, but it is possible to understand the general theory from a simple example. This makes use of the concept of the mask.

Each IP address has associated with it a second value called the mask. The mask is also a four-byte, 32-bit binary number that uses the same dotted-decimal notation as the IP addresses seen in Exhibit 10.10. The purpose of the mask is to act as a filter on the IP address so that certain "blocks" of IP addresses may be carved out to create the subnets. A combination of an IP address with a mask allows parts of the IP address to "flow through" to a new address value that can be used by the router for managing the flow of packets. Thus by selectively blocking off parts of an IP address, the routers placed between subnets can selectively either block or allow groups of packets to pass through. With allocation of the right IP addresses on a subnet, combined with the right set of masks in the routers and the stations, the IP addresses of a large network can be divided into suitable blocks such that a subnet need only see the traffic coming from and to its stations. The routers can now route packets based on the subnet blocks to only those subnet networks where the addressed stations are located and avoid sending all the traffic over the entire network. While such a setup can at first seem confusing, TCP/IP protocols allow such subnetted networks to be created and managed almost automatically through the use of appropriate software and protocols.

Exhibit 10.13 shows an example subnetted network in diagram form.

In Exhibit 10.13, while all the traffic to and from the network is routed across Subnet #1, Subnets #2 and #3 see only their own traffic to and from their own stations. This greatly reduces the traffic on those networks. If the connections from Routers B and C to their respective subnets are long-distance WAN links, then the required bandwidth and cost of those links can be greatly reduced. Such a topology would be very advantageous for an organization that has outlying satellite locations, such as branch offices. Subnet #1 and Routers A, B, and C are located at corporate headquarters and Subnets #2 and #3 and Routers D and E are located at distant locations, such as in other cities or states.

Network Address Translation

As ISPs became more prominent in the 1990s, it also became apparent that a simple mechanism was needed to permit local networks to be separated from the much larger Internet. The local stations did not want to see all of that traffic intended for someone else and setting up a large number of subnets or using some other method was too cumbersome. There was also a desire for added simple security in these smaller networks found in small businesses, homes, and so on. Internet service providers were also becoming important in connecting to small businesses and homes, which made the problem even more acute. So in 1993, the concept of

EXHIBIT 10.13 Subnetted network diagram

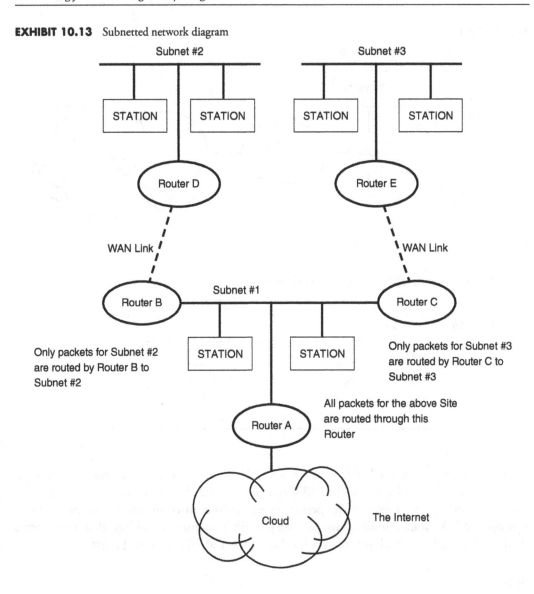

network address translation (NAT) was introduced. A NAT router connects to the larger Internet on one side (often through an ISP) and to the smaller "private network" on the other.

Exhibit 10.14 shows a NAT router connecting the Cloud to a private router.

The NAT router acts as an IP address translator. On the right side, facing the ISP Cloud, the ISP issues a single IP address to the router for it to communicate with the Cloud. On the left side, the NAT router translates packets flowing to and from the stations on the private network to this single ISP-facing IP address. On the private network, each stations and the left-facing connection of the NAT router have different IP addresses assigned by the NAT router or manually. These addresses are assigned from special blocks of addresses that have been defined as private network addresses. There are three groups.

EXHIBIT 10.14 NAT router

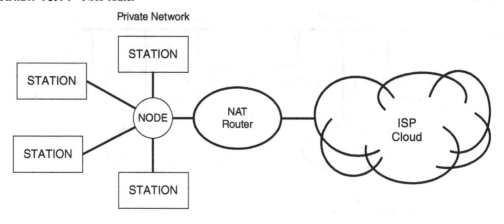

Exhibit 10.15 shows IP addresses in table form.[3]

EXHIBIT 10.15 IP addresses

Range	Total Addresses Available
10.0.0.0 to 10.255.255.255	16.8 million
172.16.0.0 to 172.31.255.255	1.0 million
192.168.0.0 to 192.168.255.255	65,500

The most commonly used address range is 192.168.X.X. This is most commonly found in use with home networks where the NAT router is included with a cable modem or DSL modem from the ISP provider. To permit many stations to communicate to the Internet through the NAT router while using only one globally recognized IP address that is supported by the ISP Cloud, the NAT router uses another concept of TCP/IP called ports.

Ports

TCP/IP permits multiple connections to exist in one station to different applications running on that station. For example, you might have an email application, a browser, and a videostreaming application all running simultaneously in the same station. To keep them separate, TCP/IP permits packets to be identified by subaddresses called port addresses. Packets being sent back and forth for an application that is running on a station can be identified by these port addresses so that they will be delivered to the correct application inside the station and in the right order. Port addresses are 16-bit quantities and usually shown in decimal notation.

Exhibit 10.16 shows some of the commonly seen port addresses that have been assigned to specific applications by the Internet Assigned Numbers Authority (IANA).[4]

EXHIBIT 10.16 IANA port addresses

Port	Assignment
20	File Transfer Protocol (FTP)
25	Simple Mail Transfer Protocol (SMTP) for email routing
53	Domain Name System (DNS) service
80	Hypertext Transfer Protocol (HTTP) used in the World Wide Web
110	Post Office Protocol (POP3)
161	Simple Network Management Protocol (SNMP)

A NAT router also makes use of port addresses to separate the different connection flows for different stations on the private network that flow back and forth to the ISP Cloud. It does this through a special address translation table that it establishes and maintains inside itself. A NAT router may also provide various security services. For example, it can prevent an external station from outside the private network from logging in, or even seeing the private network stations. It prevents monitoring the traffic on the private network that does not go to the ISP Cloud but flows only between the private network stations on the private network. Much more information about NAT can be found in the references listed at the end of this chapter.

Autoconfiguration of Networks

One of key concepts that was developed as the Internet evolved was that of autoconfiguration. As the Internet grew from its original ARPANET instantiation into the global network we know today, manually configuring and maintaining things like address and domain name tables was impossible. As network engineers would say, it did not scale. So a number of special protocols were developed over the years to handle the tasks of configuration automatically without manual intervention, or even knowledge of the configuration by the average user. Many of these protocols such as Open Shortest Path First (OSPF) and Border Gateway Protocol (BGP) perform this function in the global Internet but do not really affect LANs or private networks. Here we outline two of these autoconfiguration networks that you will encounter in a LAN or private network.

The first is Dynamic Host Configuration Protocol (DHCP). DHCP is a dynamic configuration protocol that replaced the earlier Bootstrap Protocol (BOOTP) that depended upon manually set up tables in the early days of the Internet. Put simply, a DHCP server is given an allocation of IP addresses for it to manage. When a new station (called in DHCP a client) is added to the network, it initiates a negotiation with the DHCP server on the network to get an IP address assigned to it. A number of messages are passed back and forth between the DHCP server and the client to accomplish this negotiation. The IP address may be assigned dynamically, meaning that it will expire after a time and the client will need to request a new one, or, statically, meaning that it will be fixed and not expire for as long as the client is on the network. Clients can also release and renew IP addresses upon request to the DHCP server. In this way, when a new station is added to a network such as a LAN or private network, it does not need to have its IP address set manually, nor does any other station on the network need to have its IP address manually added to its table. It is all handled automatically.

The other protocol is Domain Name System (DNS). DNS is the system that maps IP addresses to domain names, because people are better able to keep track of sites and stations on the Internet using names rather than long numbers like an IP address. When a station wishes to communicate with another station using a domain name, a DNS server will map that name to its database of names and IP addresses and inform the station what the IP address is for that domain name. Thus a station may want to communicate with the station called Google.com if it wants to do a search from a browser. DNS will map the name Google.com to an IP address. Domain name spaces have been defined hierarchically, meaning that to better keep track of what, who, and where they are, labels have been defined that can be added to the domain name, separated by dots.

A number of predefined domain name labels have defined for the Internet, called dot-labels. Some relate to a site's function or purpose, and some relate to location.

Exhibit 10.17 shows some common examples of predefined domain labels.[5]

EXHIBIT 10.17 Predefined domain labels

Dot Label	Description
.com	For commercial organizations
.edu	For educational organizations
.gov	For government institutions
.net	For network services centers
.org	For nonprofit organizations
.us	Indicates the site is located in the United States
.ru	Indicates the site is located in Russia
.uk	Indicates the site is located in the United Kingdom
.gr	Indicates the site is located in Greece
.ng	Indicates the site is located in Nigeria

Today, ICANN manages the top-level development and architecture of the Internet domain name space. It authorizes domain name registrars, such as HostGator or Bluehost, to assign and register domain names such as google.com, PBS.org, microsoft.com, harvard.edu, wikipedia.org, and so on.

Security and Reliability

Network security is a very complex and complicated subject that cannot be covered here in any real depth, although a few key concepts are discussed. More information can be found in the references listed at the end of this chapter.

Cryptography, the science of encrypting messages with codes and cyphers, has a long history. Today, encryption of messages on the Internet is accomplished by complex mathematical algorithms. A number of these algorithms are in use today but the two most commonly used are the following.

The Data Encryption Standard (DES) was designed by IBM and adopted by the US government as the standard encryption method for nonmilitary and nonclassified use. DES uses a 56-bit key number to encrypt text using a very complex multistep algorithm. The same key is used by the sender to encrypt the message and by the receiver to decrypt the message. This is known as shared-key or symmetric-key cryptography. To establish communications, both the sender and the receiver must agree on the key to be used before communication can start.

An alternate method called public-key cryptography has also been developed. In this system, there is a public key and a private key. The private key is kept by the receiver. The public key, which is publicly announced, is used by the sender. In this system, any sender can use the public key to send a private message to the receiver without having to first make an agreement as to what key to be used. There can be many senders using the public key and all of their messages will arrive securely and privately to the receiver because only the receiver can decrypt them successfully. Pretty Good Privacy (PGP) is an example of the use of public-key encryption, although it also supports symmetric-key encryption.

Now that messages and data can be encrypted, emails, files, and even data streams such as audio and video can be sent and received securely. Encryption, particularly symmetric-key encryption, can be used to guarantee that a message or data is being sent from a known secure source and received only by a known and secure source.

What about someone trying to send harmful messages or packets into a network or station? Or someone snooping on the traffic passing across a LAN or private network? To provide this kind of security, a device called a firewall is employed to filter out harmful or suspect packets from passing into a LAN or private network, or to prevent packets from flowing out from the LAN or network to destinations that are suspect. It accomplishes this function using the information it can see in the packets themselves, which includes source and destination addresses, codes that show the purpose of the packet, or even, in more advanced firewalls, the nature of the data being transmitted in the packet. Much more information about firewalls can be found in the references listed at the end of this chapter.

Introduction to the OSI Model

Now that we have discussed a number of the aspects of TCP/IP networks, this section introduces an architectural concept that you may encounter when talking or reading about networks in general. This concept is called the *Open Systems Interconnection model*, or, as it is widely known, the OSI model. Now that we have discussed some recognizable aspects of the TCP/IP network protocols and technologies, these can be fitted into this architectural model. Other concepts and technologies found in this chapter are also related to their layer of this model.

The OSI model is a conceptual model that can be used to describe the functions of a telecommunications system independent of the internal structure or technologies used. In other words, it can be applied to describe any number of telecommunications and computer network systems and architectures.

Exhibit 10.18 shows the OSI model divided into seven layers.[6]

EXHIBIT 10.18   Layers of the OSI model

Layer No.	Layer Name	Function
1	Application	To allow access to network resources
2	Presentation	To translate, encrypt, and compress data
3	Session	To establish, manage, and terminate sessions
4	Transport	To provide reliable process-to-process message delivery and error recovery
5	Network	To move packets from source to destination; to provide internetworking
6	Data link	To organize bits into frames: to provide hop-to-hop delivery
7	Physical	To transmit bits over a medium: to provide the mechanical and electrical specifications

The following discusses each layer in a little more detail and relates various TCP/IP functions and other network concepts and technologies found in this chapter to the layer.

Physical Layer (7). The physical layer describes the transmission media over which the data bit stream is sent. This involves the physical and electrical specifications for the transmission media. Examples are the radio signals used in Wi-Fi, the electrical signals used by the Ethernet, and the nature of the wiring, cables, and connectors, and so on as defined in Standard IEEE 802.3, which defines things like 10BASE-T (Ethernet over twisted pair) or 1000BASE-X (Gigabit Ethernet over fiber).

Data Link Layer (6). The data link layer makes the raw transmissions more reliable by managing various aspects of the transmission such as framing (e.g. dividing the data into packets), physical addressing, link-level flow and error control, and access control. The functions, such as synchronous transmission, parity checking error control, and Ethernet MAC addressing, provide these kinds of functions.

Network Layer (5). The network layer is responsible for source-to-destination delivery of the data packets. This can apply across one or multiple network links. Whereas the data link layer provides delivery of a packet between two nodes across one link in a network, the network layer makes sure the packet gets from its original source to its final destination. The TCP/IP Protocol functions of routing control, source and destination addressing, multicasting, and packet sequencing provide this layer of function.

Transport Layer (4). The transport layer is responsible for the process-to-process of an entire message. This means between applications running on the source and destination nodes (hosts). The TCP/IP Protocol functions of sockets, ports, and so on provide this layer of function. Other functions that are end-to-end in nature as opposed to across one link are also provided by the transport layer. ACK (acknowledgment) or negative acknowledgment (NAK) of successful transmission and retransmission of packets or messages not received or with nonrecoverable errors are examples of this layer of function.

Session Layer (3). The session layer is the dialog controller. It establishes, maintains, checkpoints, and terminates communication between the source and destination nodes (hosts). In modern TCP/IP systems, the session layer as part of the TCP protocol is considered by some to be nonexistent or simply included with TCP.

Presentation Layer (2). The presentation layer handles the syntax and semantics of the data exchanged between the source and the destination. This includes data format changes, encoding (ASCII vs. EBCDIC coding, or HyperText Transfer Protocol, HTTP), compression, and encryption/decryption that are typically done at this layer (HTTP Secure, or HTTPS).

Application Layer (1). The application layer enables the user, whether a human or software, to access the network. Examples of the applications at this layer are mail services (email), directory services, browsers, file transfer utilities, and remote logon or virtual terminal utilities. These applications generally have no means to see or access resources of the network.

Much more detail about the OSI model can be found in the ISO/EIC 7498 Standard.[7] The initial version of this Standard was first released by ISO (International Organization for Standardization) in 1980.

Basic Wi-Fi Concepts, Architecture, and Capabilities

The term *Wi-Fi* stands for wireless fidelity. It is also sometimes referred to as WLAN, or wireless local area network. Wi-Fi is basically a technology that uses radio to create a local shared media network that can be used by a variety of different kinds of stations such as laptops and desktop computers, printers and other peripheral devices, and mobile devices such as cell phones, tablets, and so on. It allows these stations to exchange data with each other without wires when they are located anywhere within the signal range of the Wi-Fi. When the Wi-Fi is connected to the appropriate router or modem connected to the Cloud, the stations can also connect to the Internet. Wi-Fi is a trademark of a nonprofit organization called the Wi-Fi Alliance. The Wi-Fi Alliance restricts the use of the term "Wi-Fi Certified" to products that have completed special interoperability tests successfully. This ensures that anything that is Wi-Fi Certified will work with any Wi-Fi network that it encounters.

Wi-Fi uses multiple parts of the IEEE 802 family of protocol standards. The IEEE 802 standards are a set of standards established by the Institute of Electrical and Electronic Engineers (IEEE) for LAN and other kinds of networks. The intention of Wi-Fi is for it to interwork seamlessly with wired Ethernet networks. Since the notion of wireless networking was introduced in the early 1970s, and since the founding of the Wi-Fi Alliance in 1999, there have been many generations of the technology. Today, there are three basic generations that are in use with modern equipment, called Wi-Fi 4, Wi-Fi 5, and Wi-Fi 6. However, older equipment may still use the older standards and there are a number still in use today. All of these standards have a high degree of backward compatibility so that they can all interconnect with each other. Advanced features of the newer generations are only available to that generation and to later ones. But earlier generations will still connect.

Exhibit 10.19 is a table outlining the most basic features of each generation.

EXHIBIT 10.19 Features of Wi-Fi generations

Generation	Max. Data Rate	Year Adopted	Radio Frequency	Comments
Wi-Fi 6E 802.11ax	600 to 9608 Mbits/second	2019	2.4/5/6 GHz	Extended Wi-Fi 6
Wi-Fi 6 802.11ax	600 to 9608 Mbits/second	2019	2.4/5 GHz	
Wi-Fi 802.11ac	433 to 6933 Mbits/second	2014	5 GHz	
Wi-Fi 4 802.11n	72 to 600 Mbits/second	2008	2.4/5 GHz	
802.11g	6 to 54 Mbits/second	2003	2.4 GHz	Sometimes informally called Wi-Fi 3
802.11a	6 to 54 Mbits/second	1999	5 GHz	Sometimes informally called Wi-Fi 3
802.11b	1 to 11 MBits/second	1999	2.4 GHz	Sometimes informally called Wi-Fi 2
802.11	1 to 2 Mbits/second	1997	2.4 GHz	Sometimes informally called Wi-Fi 1

Wi-Fi stations communicate by sending data packets over a shared radio connection. As with all radio, this is accomplished by modulating and demodulating radio carrier waves. Wi-Fi stations come programmed with globally unique 48-bit MAC addresses, just like with Ethernet. Packets use the MAC addresses to identify the source and destination of the packet and each station receives a packet and only passes it on to the computer/tablet/application when that packet has the station's MAC address as the destination. In this way it operates exactly like Ethernet. This also extends to using the same carrier-sense multiple access with collision avoidance (CSMA/CA) mechanism used by the Ethernet to permit a number of stations to share a shared radio connection.

Wi-Fi standards also permit the use of separate channels within a frequency band. For example, in the United States, channels 1–11 are defined for the 2.4 GHz band and a much larger number of channels for the 5 and 6 GHz bands. The channels are allowed to overlap each other. Channels can be shared between different networks, but only one station can transmit on a channel at a time. Some complicated rules about how to share groups of channels have been developed over time to prevent one Wi-Fi network from interfering with another in high-density areas where a number of different Wi-Fi networks may be closely placed, such as an apartment building, a residential neighborhood, or an office complex. Check with a network engineer or expert when deciding what channels to configure in your Wi-Fi network.

Exhibit 10.20 shows a typical Wi-Fi LAN topology. The Wi-Fi station that creates the Wi-Fi network is called an access point. This is often coupled with a NAT router into one piece of equipment to permit a wired connection to the Internet or WAN network, with the access point acting as the source and control for the Wi-Fi network.

EXHIBIT 10.20 Typical Wi-Fi LAN

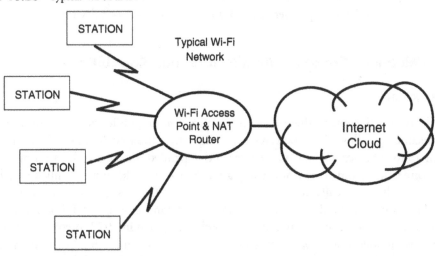

A Wi-Fi access point often also includes provisions for security of the wireless network. So-called public Wi-Fi networks such as those found in coffeeshops and public spaces permit any Wi-Fi enabled device to connect to the Wi-Fi network. But for homes and offices, and places like hotels, it is often advisable to restrict access to the Wi-Fi network to authorized stations only. This was originally accomplished through a protocol called Wired Equivalent Privacy (WEP). WEP was originally designed to prevent snooping on wireless networks by outsiders through the use of encryption. But WEP encryption keys proved to be too vulnerable to various hacking tools and so WEP is now considered obsolete. The Wi-Fi Alliance introduced Wi-Fi Protected Access (WPA) to correct the vulnerabilities in WEP. But eventually even WPA was shown to have vulnerabilities to improved hacking tools. In 2004 the more secure WPA2, using an advanced encryption standard, was introduced. In 2007, a new feature called Wi-Fi Protected Setup (WPS) was added to the Wi-Fi standards that allowed WPA and WPA2 security to be bypassed in many situations. The only remedy was to turn this new WPS feature off. Then in 2017 a flaw in WPA2 was discovered that allowed a new hacking tool to gain access. To be fully secure across a Wi-Fi network, adding a further layer of security in the form of a Virtual Private Network (VPN) protocol or secure Hypertext Transfer Protocol (HTP) over Transport Layer Security (HTTPS) is often recommended. Today, WPA2 is considered secure, provided a strong passphrase is used. WPA2-PSK uses a passphrase to authenticate and generate the initial data encryption keys. These are then dynamically and randomly varied. WPA2-PSK is considered the most secure version of WPA2. In 2018 WPA3 was announced as a replacement for WPA2 with further increased security. However many current Wi-Fi networks still use WPA2 or WPA2-PSK. It is also true that many Wi-Fi access point products default to an encryption-free open access mode of operation. Novice users who believe they are benefiting from a zero-configuration access point that can be used out of the box do not realize that their new Wi-Fi network is fully exposed, meaning it has zero security.

Wi-Fi is not without its other problems. The 2.4 GHz band in particular is very close to the resonant frequency of the free water molecule. This is also the frequency that most microwave ovens operate at. Wi-Fi equipment operating on this band may experience interference from microwave ovens, cordless telephones, USB 3.0 hubs, and Bluetooth devices.

Mobile Cell Phone Concepts, Architecture, and Capabilities

Mobile Generations

Early in the twentieth century, there were attempts to create a mobile telephone. Improvements in radio technology allowed the Bell System to introduce the first Mobile Telephone Service (MTS) in 1946. This service was not cellular and depended on the intervention of operators to make connections; and since it was using a radio link, it could support only a very limited number of simultaneous calls. It was also very expensive. The first mobile cellular phone was developed by Motorola and demonstrated in 1973. The handset weighed 4.4 pounds. The first commercial automated cellular network was launched in Japan in 1979. A commercially available handheld mobile phone was not available until 1984. Since these early days, cellular technology has advanced through several generations. Exhibit 10.21 outlines those generations and the dates of their introduction.

EXHIBIT 10.21   Mobile generations

Mobile Generation	Introduction Date	Type of Service	Comments
0G	1946	Noncellular, analog radio-based, and operator support required. Used VHF radio bands.	Starting with Bell System's MTS, this analog radio service went through several improvements over the years until cellular technology made it obsolete. In recent years it has been referred to as 0G.
1G	1979	Analog cellular technology with a maximum frequency of 1.0 GigHz.	First demonstrated in 1973, the first cellular network introduced in 1979 in Japan, mobile analog cellular service introduced in 1983.
2G	1991	Digital cellular technology supporting multiple services, with a maximum frequency of 2.75 GigHz.	First introduced in Finland, supports digital encryption of phone conversations, text messages, and picture messages. Additional services that came to be supported are Internet access and digital photography.
3G	2001	Digital cellular technology using the Wideband Code Division Multiple Access (WCDMA) protocol, supporting multiple services, with a maximum frequency of 3.95 GigHz.	First introduced in Japan, supports a number of services including mobile Internet access, fixed wireless Internet access, wireless voice telephony, video calls, and mobile TV.
4G	2009	Digital cellular technology supporting multiple services, with a maximum frequency of 4.9 GigHz.	First introduced in Norway and Sweden, supports a number of services including mobile Internet access, Internet Protocol (IP) telephony, gaming services, HDTV, and videoconferencing.
5G	2019	Digital cellular technology supporting multiple services, with a maximum frequency of 39 GigHz.	Began worldwide deployment in 2019 and deployment is ongoing as of 2022. Uses three different radio bands, the highest band giving download speeds in the gigabit/second range.

The way radio bands are allocated for digital cellular mobile phones is extremely complicated and controlled by a number of international organizations. The same can be said for the protocols used. Different countries and different mobile service suppliers use different protocols that can be incompatible with each other. For our purposes it is enough to understand that a modern 4G or 5G mobile phone has a degree of connectivity that is unmatched by any other technology at this time.

There are two basic types of mobile phones, smartphones and feature phones. Smartphones are, in effect, small personal computers of ever-increasing power. They are capable of a number of applications and services, a list that is constantly growing. So-called tablets that have 4G or 5G connectivity are also considered smartphones. Typical smartphones can support voice and video telephony, Internet connectivity, contain cameras for still photography and video, can receive and display mobile TV, contain a Global Positioning System (GPS) receiver to determine the location of the mobile phone, and can download and run applications, games, and so on. They also typically use a mobile operating system that shares common traits across devices from different suppliers or may be proprietary to one supplier.

Feature phone is a term used to describe a mobile phone device that is limited in its capabilities when compared to a smartphone. Feature phones typically provide voice telephony and text messaging capabilities and may or may not include a camera for pictures and video. They also typically use custom-designed software and a user interface that is unique to the feature phones provided by a particular supplier.

Since the introduction of 1G cellular mobile phones, mobile telephony has used the same basic concepts. It is the details that have changed with each generation. These basic concepts are:

- The creation of cells to divide up the service area
- The use of radio frequency bands
- Offloading mobile traffic when alternate services, such as Wi-Fi, are present

Cell Architectural Concepts

A mobile cellular service provider divides up their service area (a city or geographic region) into separate zones, or cells, that are served by cell towers that are erected within the cells. Usually, there are three cell towers in each cell, located in different locations within the cell. Each cell uses a set of radio frequency bands or channels for two-way communication to the mobile devices located in the cell. The set of bands used is different from those used by the neighboring cells so that there is no interference. Due to bandwidth limitations, the number of cell phones that can be handled in one cell is limited. Generally, an average cell tower can handle 30 simultaneous users for voice calls and 60 simultaneous users of 4G data. Cells are sized according to the expected density of mobile devices. In large cities where the density of population and the density of mobile devices are higher, cells are sized smaller than cells in small towns or in rural areas. Cell towers are then interconnected to each other and to the Internet and public telephone network by wired connections. A mobile phone thus contacts a cell tower to alert it of its presence over the radio bands and the cell tower can then connect that mobile phone to the Internet or public telephone network through the wired connections. As the mobile phone moves around and crosses the boundary from one cell to another, the mobile phone will "hand off" to the new cell it is moving into. It automatically disconnects from the previous cell and connects to the new cell to give the best reception possible. If the mobile phone was in the middle of a call or data upload/download, this handoff is usually so smooth that the mobile phone users do not notice it happening. When a mobile phone detects an alternate infrastructure, such as Wi-Fi, it will hand off to the new infrastructure to reduce the load on the cell network whenever practical. This will usually be for Internet connections, text messaging, and other nonvoice services.

Mobile Networking Security and Reliability

Starting in the early 1990s, mobile phones began to contain subscriber identity modules, or SIM cards. A SIM card is a small microchip that contains a secure international mobile subscriber identity (IMSI) number, which uniquely identifies every user of a cellular network. It is stored as a 64-bit field and is usually shown as a 15-digit number. To prevent eavesdropping and tracking of the mobile phone user, the IMSI is sent as rarely as possible. Upon entering a new geographic area, the mobile phone will send the IMSI to the cellular network. The cellular network then randomly assigns a Temporary Mobile Subscriber Identity (TMSI) number that the mobile phone can use as long as it remains in the geographic area. To further prevent eavesdropping and tracking of the mobile phone user, the cellular network can change the TMSI on a regular basis. However, while these measures can prevent the most casual eavesdropping, snooping, and tracking, a mobile device is really not secure unless encryption for the actual phone call is enabled.

Mobile phone security includes the following requirements to truly protect mobile phone users and their data.

- Security and integrity of the identity of the phone, and thus the privacy of its user
- Security of the voice or data transmission to and from the phone
- Integrity of the voice or data transmission to and from the phone
- Security of personal and business data stored on the phone
- Control of location information from the phone that can be used to track the user
- Prevention of limiting or denying access to the user of a phone, that is, protecting availability

To achieve these requirements it is important to understand the nature of the threats mobile phones face. There are a number of threats that attackers exploit; some of the most common are:

- BotNets: Malware that can attack multiple phones. Users generally acquire the malware through email attachments, infected applications, and malicious websites. The malware gives attackers control of the phones, in which they perform various kinds of harmful acts.
- Malicious applications: Applications that are infected with malware or are written as malicious applications that an attacker uploads as games or applications to various mobile phone application marketplaces or stores. Often the intention of the malware is to steal personal information and data from the phone and send it via backdoor communication channels to the attacker. Other kinds of malicious behavior are also possible, including having the phone silently record conversations and send those recordings to the attacker.
- Malicious links: Are found on social networks to spread malware as spyware, backdoors, and so-called Trojan horses.
- Spyware: A special form of malware that allows attackers to monitor phone calls and text messages and track the phone, and thus the user's identity and location.

The attackers themselves are made up of four different groups:

1. **Professionals** who may be affiliated with a government or a commercial business. They steal data from the general public for use in advertising and commerce and engage in government and/or industrial espionage.
2. **Simple thieves** who profit from the sale of data or identities that they have stolen.
3. **Black hat hackers** who usually attack availability with the intention of causing damage to the mobile phone or its contents, such as deleting or scrambling files, records, and address books. They may also attempt to scramble files, records, and so on to hold them hostage in order to extract a ransom from the mobile phone user. This kind of attack is now commonly called ransomware.
4. **Gray hat hackers** whose goal is not to cause harm or steal data but to expose the vulnerabilities of the mobile phone, its applications, or its software.

The type and number of threats to mobile phones is a list that is growing all the time. In many ways, a smartphone is vulnerable to many of the same threats faced by computers. Smartphones are in fact a special subset of computer, with the same kind of features and capabilities. Because of their very portable nature, and the fact that they are constantly using widely shared media for communication, mobile phones are also vulnerable to some threats that are unique. These threats include:

- **Juice jacking.** USB ports serve dual purposes, as a charging port and a data port. Mobile phones are vulnerable to having data stolen from them, or having malware installed on them, when using malicious charging ports set up in public places or hidden in normal charging adapters.
- **Password cracking.** Attackers watch over the shoulder of a mobile phone user to determine the password they use to open their phone or determine the password from a pattern of smudges on the phone's screen (called a smudge attack).
- **Eavesdropping.** Eavesdropping occurs through the use of a fake cell tower commonly called an IMSI catcher. The fake tower catches the secure IMSI number as the mobile phone sends it out to establish its initial connection with what it believes is a legitimate cell tower. The fake tower can now issue a TMSI number to the phone and then eavesdrop on all the communication with the phone and initiate other malicious actions.
- **Tracking.** A mobile phone can be tracked in two ways. One is by hacking the phone and having readouts from an internal GPS device sent to the attacker through a backdoor communications channel. The other can be performed passively by noting the cell in which the phone is currently located. It is even possible to utilize signals from the different cell towers in a cell to locate the phone within the cell using differences in the timing of the signals. These last two techniques require the support of the mobile phone service provider and are usually used by government entities and law enforcement. Using these two techniques, any mobile phone can be tracked and hacking or compromising the phone is not required.

To deal with these threats, countermeasures are also constantly evolving. They involve many of the same countermeasures used with computing, including the following:

- **Antivirus software.** Special software applications that monitor for and detect known malware.
- **Firewalls.** An application or process running on the mobile phone to detect and prevent unwanted outside intrusions into the mobile phone.
- **Notifications.** Warning messages posted to the mobile phone user when unusual activity is detected, such as a note-taking application attempting to use the phone's camera.
- **Biometric identification.** Using fingerprint readers or eye or face recognition to validate the identity of a user of the mobile phone.
- **Encryption.** Using techniques to encode voice or data traffic being exchanged across the cell phone network, or to protect files and other data stored on the phone itself. Encryption techniques are evolving all the time, and today's modern methods are much more secure than those of even a few years ago.
- **Resource monitoring.** Detecting and reporting unusual or heavier than normal usage of phone resources, indicating suspicious behavior.
- **User awareness.** Actions taken by the user to limit potential threats.

This last countermeasure may be one of the most important. Many threats are made possible by the carelessness of the mobile phone user. The following is a short list of actions the user should remember to take to better protect their phone and its contents.

- Lock the phone when it is not in use.
- Don't leave the phone unattended, particularly in public places.
- Enable encryption whenever it is available.
- Disconnect peripheral devices that are not currently in use.
- Don't ignore security warnings and notifications.
- Maintain control of permissions given to applications. Give apps rights to use only the resources they need to do their job. For example, an application for playing music does not need access to GPS services.

Future Evolution of Mobile Networking

As mentioned earlier, mobile networking technology is constantly evolving and will continue to advance into the foreseeable future. The next big step will be the introduction of 6G, the sixth-generation standard for wireless mobile communication. This successor to 5G is planned to be even faster than 5G and will support services and applications beyond anything seen so far. It is currently under development and the radio frequency bands it will use are under discussion. A launch date is not yet set but some experts think that 6G will be available around 2035. As mobile networking continues to evolve, we can expect to see further improvements in security, accessibility, coverage, and added services.

IT and Telecommunications Networking Convergence

Convergence of the Internet and Telephony

Since the development of Voice over Internet Protocol (VoIP) in the early 1990s, the eventual convergence of the Internet and voice telephony of all kinds has been predicted. So-called landline phones are still supported in many countries, including the United States. Landline phones depend on analog wired connections. These very old services are expected to be obsolete and shut down in the 2020s. Going forward, they will be replaced by packet-oriented voice communications sharing the same digital media of the Internet and mobile telephone networks. Already the long-distance backbone of the public switched telephone network (PSTN), the network owned and operated by the telephone companies, is being converted from older Synchronous Digital Hierarchy (SDH) and Synchronous Optical Networking (SONET) technologies to optical TCP/IP networks. These trends will continue and in recent years a number of older now-obsolete telephony and networking technologies have or are disappearing. In addition to the technologies named above, some others that you may encounter are:

- **ARCNet.** An earlier computer networking technology.
- **Token Ring.** An earlier computer networking technology created by IBM.
- **Integrated Services Digital Network (ISDN).** A digital telephony standard, now largely replaced by digital subscriber line (DSL) systems.
- **Asynchronous Transfer Mode (ATM).** An earlier digital telephony transmission technology.
- **Frame Relay.** An earlier packet-switched telecommunications technology.
- **Fiber Distributed Data Interface (FDDI).** An earlier optical fiber data transmission standard.

Capabilities and Benefits of Mobile Edge Networking

Setting up a Mobile Edge Network is necessary to support Mobile Edge Computing. Such a network is in effect a private network, such as the example shown in Figure 10.15. It may include both hard-wired TCP/IP connections to various nodes and stations as well as Wi-Fi local area networks connected to the private network. Then through a firewall and NAT router, secure connections can be made to various services in the Cloud. For Smart Manufacturing, this kind of topology can offer many benefits over depending on connection services provided by ISPs, public mobile phone networks, and so on. This section outlines only some of those benefits. Only where absolutely necessary should any station to be connected to the Mobile Edge Network utilize a connection through a mobile telephone service. For security purposes, such connections will need careful engineering including use of virtual private network (VPN) connections, encryption where needed, special identity and resource control, and so on.

One major benefit will be enhanced performance and control. Some of the key aspects of performance and control are listed below.

- **Traffic Offloading.** The Mobile Edge Network will offload your multi-access Edge computing (MEC) traffic to a separate private network where it will not interfere with or be interfered with by other public traffic and other business traffic on other networks within the business or factory.
- **Better Control.** Running a separate private network for MEC will permit better control of the stations and devices connected to that network and keep them free from interference from other public or business traffic on other networks.
- **Congestion Reduction.** Careful network topology design through the use of switching, Wi-Fi, and subnets will permit better control of high-volume traffic, allowing control and mitigation for traffic congestion.

- **Latency Reduction.** In a similar way, careful network topology design will limit latency and improve overall performance.
- **Bandwidth Improvement.** Where necessary, the use of switching and subnets will allow better control of bandwidth and its use by the various stations connected to the Mobile Edge Network.

Better control and performance will permit enhanced operations, such as the following.

- Maintenance of the Mobile Edge Network will be simplified and more focused.
- Upgrading the Mobile Edge Network will be easier and less likely to affect or be affected by the public or other networks in the business.
- Being able to monitor, control, and upgrade the topology of the Mobile Edge Network will provide increased flexibility.
- All of these attributes can be expected to result in reduced operations expenses of the Mobile Edge Network.

Other benefits to setting up and using a Mobile Edge Network are:

- Active device location and tracking will be simplified and can also be made more accurate.
- Video capture and analytics will be simplified, and the extensive video traffic can be handled more efficiently with the improved performance and bandwidth that good topology design can provide.
- Security, which is a critical factor in how the Mobile Edge Network is designed and controlled, will be enhanced through the better control and more focused security measures applied to the Mobile Edge Network. The network will also be less vulnerable to the many threats that face the public mobile network or the other networks in the business or factory.

Summary

This chapter introduced you to some basic concepts, architecture, and capabilities of modern networking, including the concepts of topology, circuit and packet switching, Wi-Fi, mobile and cellular telephony, and some of the vocabulary used in modern networking. It also reviewed some of the extensive history of the development of networking – where it has been, where it is today, and where it may be going in the future. Armed with this knowledge you should now be able to discuss at a high level the capabilities and required features of the Edge networks needed to support Mobile Edge Computing.

Sample Questions

1. What is the difference between circuit switching and packet switching?
   a. Circuit switching is now obsolete and was used only for analog communications (e.g. telephones).
   b. Circuit switching uses dedicated connections that are established only for the duration of the communication, whereas packet switching uses a fixed media topology and does not require dedicated connections.

    c. Circuit switching requires the intervention of a human operator to set up the connection.

    d. Packet switching is used only for radio connections such as Wi-Fi and mobile telephone.

2. What was the main reason early networks used ring topologies?

    a. A simple ring topology was easier to set up and maintain.

    b. Tracing faults in a ring topology was faster and easier.

    c. The failure of a single two-way link in a ring allowed all stations on the ring to still communicate with each other despite the failure.

    d. Links were expensive and there just were not that many available.

3. Why is mesh topology preferable to a ring topology?

    a. Mesh topology is more robust than ring and can tolerate multiple link failures.

    b. Using a mesh topology means that each node has its own link.

    c. Mesh topology can support wide area networks, whereas rings cannot.

    d. Mesh topology is inherently cheaper than ring topology.

4. The Ethernet uses what form of switching?

    a. Circuit switching

    b. Packet switching

    c. Both circuit and/or packet switching, depending on the topology

    d. None of the above

5. What is the difference between a hub and a switch in an Ethernet network?

    a. The switch sets up a dedicated temporary connection between stations while the Hub does not.

    b. A hub and a switch are basically interchangeable.

    c. The switch can alter packets passing through it to connect two different kinds of networks with different protocols.

    d. The switch repeats packets it receives only out of the link that it knows the destination station is connected to. A hub simply repeats packets it receives out of all the links.

6. What is the original basic purpose of a router in a TCP/IP network topology?

    a. To connect the local network to the Cloud

    b. To support multiple stations connected to a network

    c. To translate the format of packets from one protocol to another when local networks using different protocols/technology are to be connected

    d. To provide security for the network by checking packets as they flow by

7. IP addresses were originally assigned to classes to accomplish what purpose?

    a. To separate different sets of addresses so they do not conflict with each other

    b. To help ease the management of the large IP address space across multiple entities including government, commercial businesses, educational organizations, and so on

    c. To control the number of addresses that can be used at one time

    d. To create separate address blocks to help manage security

8. What is a NAT router most often used to connect a local network to?

    a. An Internet service provider (ISP)

    b. A special security monitor router

    c. Networks using protocols other than TCP/IP

    d. Long-distance links to the Cloud

9. Wi-Fi is used for what purpose?
   a. To connect to a wide area network so that wireless links can be made across large distances
   b. To create a local area network that selected stations can securely connect to through wireless radio links rather than wires
   c. To permit other users who happen to be in the local area to connect to the Internet without any restrictions
   d. To create mobile phone cells
10. In mobile telephony, the location of cell towers and the size of a cell are determined by what factor(s)?
   a. The geographic location of the cell
   b. The ruggedness of the terrain of the cell (i.e. how many mountains, large buildings, etc. are present to block radio signals)
   c. The anticipated density of cell phone users in the location
   d. The generation of the cell phone technology (i.e. 3G vs. 4G vs. 5G)
11. What may be the most important countermeasure to achieve security in the use of a mobile phone?
   a. Strong passwords
   b. Encryption of the data flowing across the cell network
   c. Resource monitoring and control of mobile phone apps
   d. User awareness. The user not ignoring warnings, setting strong passwords, and maintaining physical security for the phone are some examples.
12. Telephone networking and the Internet have converged over the recent years for what reasons?
   a. Lower cost of packet-switched networks
   b. Reduced cost of maintenance of packet-switched networks
   c. Improved redundancy of mesh networks over circuit switching and rings
   d. All of the above

References

Forouzan, Behrouz A. (2006). *TCP/IP Protocol Suite*, 3rd ed. New York, NY: McGraw-Hill Higher Education.

Comer, Douglas E. (1999a). *Internetworking with TCP/IP*, Vol. 2, 4th ed. Upper Saddle River, NJ: Prentice-Hall.

Comer, Douglas E. (1999b). *Internetworking with TCP/IP*, Vol. 3, 4th ed. Upper Saddle River, NJ: Prentice-Hall.

Comer, Douglas E. (2000). *Internetworking with TCP/IP*, Vol. 1, 4th ed. Upper Saddle River, NJ: Prentice-Hall.

Comer, Douglas E. (2001). *Internetworking with TCP/IP*, Vol. 3, Linux/Posix Sockets ed. Upper Saddle River, NJ: Prentice-Hall.

Perlman, Radia. (1999). *Interconnections: Bridges, Routers, Switches, and Internetworking Protocols*, 2nd ed. Reading, MA: Addison-Wesley.

Popular Acronyms Used in Networking and Mobile Computing

Term	Definition	Added Comments
AES	Advanced Encryption Standard	Used in WPA-2.
AP	access point	Another term for a station in a Wi-Fi network that provides a connection to the Wi-Fi.
API	Application Programming Interface	A system of tools and resources in an operating system, enabling developers to create software applications.
ARP	Address Resolution Protocol	An IP protocol mapping logical addresses to physical addresses. In IPv4 this is a mapping of IPv4 addresses to Ethernet MAC addresses. Also, a command line interface (CLI) command that can be used manage the logical-to-physical mapping table in an IP-enabled device.
Base Station		Term for an access point (device) that connects a local Wi-Fi to a larger network, often the Internet.
CDMA	code-division multiple access	Channel access method for shared-medium cell phone networks developed by Qualcomm and used in 2G and 3G cell phones. Not compatible with GSM cell phones. Goes obsolete with 4G. In the United States, Sprint, Verizon, and US Cellular used CDMA on 2G/3G.
DHCP	Dynamic Host Configuration Protocol	A protocol that allows a station with a domain name to request the assignment of an IP address, often on a temporary basis. A DHCP server must be present in the network. Without DHCP, the IP address must be manually assigned to enable TCP/IP communication.
DMZ	demilitarized zone	A DMZ, sometimes referred to as a perimeter network or screened subnet, is a physical or logical subnetwork that contains and exposes an organization's external-facing services to an untrusted, usually larger, network such as the Internet. The DMZ functions as a small, isolated network positioned between the Internet and the private network.
DNN	deep neural network	
DNS	Domain Name System	A service that translates domain names (like Eddy.com) to IP addresses, sometimes called the phonebook of the Internet.
ERP	enterprise resource planning (or planner)	
ETSI	European Telecommunications Standards Institute	An independent, not-for-profit standardization organization in the field of information and communications. ETSI supports the development and testing of global technical standards for ICT-enabled systems, applications, and services.
E-UTRA	Evolved Universal Mobile Telecommunications System (UMTS) Terrestrial Radio Access	The LTE RAN for 4G that provides higher data rates, lower latency, and is optimized for packet data. It uses OFDMA radio access for the downlink and SC-FDMA on the uplink.
GERAN	GSM EDGE Radio Access Network	The radio part of GSM/EDGE together with the network that joins the base stations and the base station controllers. The core of the GSM network.

(Continued)

Term	Definition	Added Comments
GSM	Global System for Mobile Communications	A standard developed by ETSI to describe the protocols for second-generation (2G) digital cellular networks used by mobile devices such as mobile phones and tablets. Also used in 3G but goes obsolete in 4G. Not compatible with 2G/3G CDMA cell phones. In the United States, AT&T and T-Mobile use GSM on 2G/3G. The rest of the world uses GSM too.
HSPA	high speed packet access	
IIoT	industrial IoT	Industrial Internet of Things
IP	Internetworking (or Internet) Protocol	IP defines how computers and IP-enabled devices send packets of data to each other.
IP stack		A collection of programs, applications, and code that permits a computer or IP-enabled device to communicate with other computers and IP-enabled devices. Also supports various network maintenance activities with the computer or device.
ISV	independent software vendor	Just what it says.
ITC	information and communications technology	An extensional term for information technology (IT) that stresses the role of unified communications and the integration of telecommunications (telephone lines and wireless signals) and computers, as well as necessary enterprise software, middleware, storage and audiovisual, that enable users to access, store, transmit, and manipulate information.
JSAN	JavaScript Archive Network	
JSON	JavaScript Object Notation	JSON is a standard text-based format for representing structured data based on JavaScript object syntax. It is commonly used for transmitting data in web applications (e.g., sending some data from the server to the client so it can be displayed on a web page or vice versa). JSON is used for serializing and transmitting structured data over network connection. It is primarily used to transmit data between a server and web applications. Web services and APIs use JSON format to provide public data. It can be used with modern programming languages.
KPI	key performance indicator	
L-DNN	Lifelong (or Continual) Deep Neural Network	
Link		As used in this publication, a communication connection that may be wired or wireless, one-way or two-way, and connects stations and nodes together.
LTE	Long Term Evolution	The project to develop a high-performance air interface for cellular mobile communication systems and the last step in the development of fourth-generation (4G) radio technologies.
MAC address	media access control address	A unique identifier address assigned to a NIC for use as an address in communications. Used in Ethernet, Wi-Fi, and Bluetooth networks.
MEC	mobile Edge computing	

Term	Definition	Added Comments
MNO	mobile network operator	MNO, also known as a wireless service provider, wireless carrier, cellular company, or mobile network carrier, is a provider of wireless communications services.
MRP	manufacturing resource planning (or planner)	
NAT	network address translation	A technology that permits the creation of private networks that use private IP addresses to communicate to the Internet through a set of one or more global addresses.
NIC	network interface controller	A component that provides networking capabilities for a computer or network-enabled device.
Node		As used in this book, a communication entity that connects network links together.
OEE	overall equipment effectiveness	
OFDMA	orthogonal frequency-division multiple access	A multi-user version of the popular orthogonal frequency-division multiplexing (OFDM) digital modulation scheme.
OPEX	operations expense	The cost of running your network, app, service, etc.
OT	operational technology	
OTT	over-the-top	Over-the-top (OTT) services are any type of video or streaming media that provides a viewer access to movies or TV shows by sending the media directly through the Internet. An OTT player would be a streaming service such as AppleTV, iTunes, Netflix, Hulu, and so on.
P2P	peer-to-peer	A P2P network is a network in which interconnected stations ("peers") share resources among each other without the use of a centralized administrative system. P2P is used to share all kinds of computing resources such as processing power, network bandwidth, or disk storage space. However, the most common use case for P2P networks is the sharing of files on the Internet. Examples are Skype, BitTorrent, and Kazaa.
PING	acket Internet groper	Named for the "ping" of submarine sonar. Use this CLI command to ask an IP address to respond to see if it is there and to provide a latency time.
PLC	programmable logic controller	A PLC is an industrial computer control system that continuously monitors the state of input devices and makes decisions based on a custom program to control the state of output devices.
PSTN	public switched tele-phone network	The land-based analog telephone switching network such as those maintained by AT&T, Verizon, and so on. Your analog home phone connects to the PSTN.
QoE	quality of experience	A purely subjective measure from the user's perspective of the overall quality of the service provided.
QoS	quality of service	Measures key network performance metrics.
RAN	radio access network	Part of a mobile telecommunications network, it is the radio access technology. Examples are GSM, GERAN, UTRAN, and E-UTRA.

(Continued)

Term	Definition	Added Comments
SC-FDMA	Single-carrier frequency-division multiple access	A frequency-division multiple access scheme, also called linearly precoded OFDMA (LP-OFDMA). Like other multiple access schemes (OFDMA), it deals with the assignment of multiple users to a shared communication resource.
Station		As used in this book, an entity that communicates across a network to other stations. Examples are computers, laptops, peripheral devices, mobile phones, and IoT-enabled devices.
Station		This definition, while not used in this book, is sometimes used as a term for an access point in a Wi-Fi network that provides a connection to the Wi-Fi.
TCO	total cost of ownership	Original cost-plus ongoing costs like maintenance, fuel, insurance, and so on.
TCP	Transmission Control Protocol	A standard that defines how to establish and maintain a network connection through which applications can exchange data. TCP works with the Internet Protocol (IP).
TDMA	time-division multiple access	Channel access method for shared-medium cell phone networks used in GSM.
UTRAN	UMTS Terrestrial Radio Access Network	Collective term for the network and equipment that connect mobile handsets to the public telephone network or the Internet.
WPA-2	Wi-Fi Protected Access – Version 2	

Notes

1. Comer, Douglas E. (2000). *Internetworking with TCP/IP*, vol. 1, 4th ed., p. 65. Upper Saddle River, NJ: Prentice-Hall.
2. Forouzan, Behrouz A. (2006). *TCP/IP Protocol Suite*, 3rd ed., pp. 96, 99. New York, NY: McGraw-Hill Higher Education.
3. Ibid., p. 99.
4. Ibid., p. 259.
5. Comer, *Internetworking with TCP/IP*, p. 388.
6. Forouzan, *TCP/IP Protocol Suite*, p. 29.
7. International Organization for Standardization. (November 15, 1989). ISO/IEC 7498-4:1989 – Information technology – Open Systems Interconnection – Basic Reference Model: Naming and addressing. ISO Standards Maintenance Portal. ISO Central Secretariat.

CHAPTER 11

Edge Computing

Vatsal Shah and Allison Yrungaray

Introduction: What Is Edge Computing?

In Smart Manufacturing we see technology buzzwords come and go, but Edge computing is only increasing in importance as companies work to digitally transform. Let's start with a few simple definitions.

- **The Edge** includes any device or asset at the Edge of the network in the field or on the factory floor.
- **The intelligent Edge** includes all those devices that are smart and connected, thus producing valuable data.
- **Edge computing** processes the vast amounts of data from these devices more efficiently next to the asset, instead of sending all of it across a network to data centers or the Cloud.

The Edge brings computing as close to the data source as possible, down to the factory floor. Running fewer processes in the Cloud and enterprise systems and moving them closer to the devices that generate data transforms the way data is handled, processed, and delivered for immediate business benefits.

What kind of data is captured at the intelligent industrial Edge? Data points are almost endless but might include temperature, speed, humidity, vibration, sound, or video. Edge data has a great deal of power when harnessed, providing the power to help companies make better business decisions to improve asset availability, performance, and quality.

Today, 50% of assets at the industrial Edge are connected, according to Gartner's *2019 Industrial Connectivity Market Report 2019–2024*, with the percentage expected to increase steadily year over year.[1]

Exhibit 11.1 shows the rise of industrial connectivity from 2018 to 2024.[2]

EXHIBIT 1.1   The rise of industrial connectivity (2018–2024)

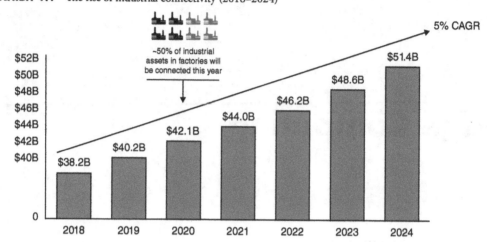

Source: From IoT Analytics.

Edge computing is exploding as data volumes and use cases expand. Gartner noted in 2018 that they expected 75% of data to be processed at the Edge, outside the traditional data center or Cloud, by 2025.[3]

Benefits of Edge Computing

Many companies have taken a Cloud-first approach to Smart Manufacturing, without properly considering the importance of the Edge. The growth in the number of IoT devices and data being sent to the Cloud means increased latency and costs. Processing data locally with Edge computing not only complements the Cloud by offsetting some of those challenges, but also activates key use cases and applications that are better hosted on-premises. Following are the key benefits of Edge computing:

- **Faster insights.** The Cloud has its place for long-term analysis, but the value of industrial Edge computing lies in making use of the data at the asset, where it has the greatest impact and zero latency.
- **Immediate action.** Edge computing allows companies to collect data, analyze it at the Edge, and then take immediate action to solve maintenance problems, increase efficiency, improve production, and more.
- **Preserve bandwidth.** Networks and infrastructure haven't quite caught up to the data explosion, so utilizing Edge computing to send only the data that is needed to the Cloud saves time and money.
- **On-Premises Use Cases.** The Cloud makes sense for machine learning and other long-term or Big Data analysis, but the Edge activates on-premises use cases like condition-based monitoring and operational equipment effectiveness (OEE) for a quick return on investment (ROI).

The vast importance of the Edge is beginning to come to light as more industrial use cases are enabled. Industrial Edge computing powers preventive maintenance, asset condition monitoring, OEE, vision systems, quality improvements, and more. Edge data also powers more advanced use cases like artificial intelligence and machine learning in the Cloud.

Top Use Cases for the Edge in Smart Manufacturing

The desired outcomes for most Smart Manufacturing initiatives are surprisingly similar. Most companies aim for better insight into asset health and performance, improved OEE, predictive maintenance, and improved processes and efficiency from machine learning. Following are four of the most common use cases for Edge computing in Smart Manufacturing.

1. **Asset Condition Monitoring.** Condition-based monitoring is often the first use case for Smart Manufacturing. Many companies are reactive, solving problems as they see them with their own eyes. Condition-based monitoring allows them to move to the next step of intelligence by collecting and analyzing machine data to understand how assets are performing. Creating KPIs and alerts for basic machine data such as temperature, vibration, and velocity provides intelligence that can be acted on in real time to ensure that all machines run as planned.
2. **OEE.** Overall equipment effectiveness is a discrete manufacturing best practice that measures productivity based on three factors: availability, performance, and quality. OEE evaluates how effectively a manufacturing operation is utilized by comparing fully productive time to planned production time. Modern Edge technology offers readily available, real-time data to fuel OEE measurements and then helps companies determine how they can improve operations over time.
3. **Predictive Maintenance.** Anomalies in the production line often lead to unplanned downtime and costly maintenance. Edge computing can enable predictive maintenance to reduce the cost of failure. The ideal platforms come with key performance indicators (KPIs) for asset utilization, uptime/downtime, and more. Use cases often start with reactive maintenance, move to predictive, and eventually prescriptive maintenance to reduce repair costs, minimize downtime, maximize output, and improve remediation efficiency.
4. **Machine Learning and AI.** Machine learning and AI start with connectivity at the Edge. Connecting devices, collecting and normalizing production data, and making that data available to Big Data and machine learning systems allow companies to develop the machine learning models needed to improve operations. Being able to rapidly deploy and run these new models at the Edge completes the feedback loop and provides the Edge-to-Cloud intelligence needed for continuous optimization.

The Data Challenge

Machine data from the Edge powers these and other advanced use cases both on-premises and in the Cloud, but the applications are only as good as the Edge data that feeds them. Achieving complete access to factory data is the biggest challenge most companies face. Although many individual shop floor assets may have connectivity, there is still a massive data problem in manufacturing for the following reasons:

- **Closed Operational Technology (OT) Infrastructure.** Legacy industrial systems were built to be closed and secure, relying on proprietary protocols and typically having no connection beyond the four walls of the factory. Securely accessing a closed OT infrastructure and bringing it online can be time-consuming and costly without the right tools.
- **Mass Data Fragmentation.** Data is the fuel for digital transformation, but data fragmentation is a roadblock. Unstructured data combined with silos across locations leads to an incomplete view of machines and processes.
- **Mix of Legacy and Modern Systems.** The typical factory includes disparate systems including sensors, machines, connectivity, storage, networking, applications, and more.

   Add in information technology (IT) systems also chosen without collaboration and the result is a complicated spider web of technologies.
- **Difficulty Ingesting All Data.** Data must be ingested before it can be analyzed, and transporting data from all sources is not easy. An incomplete data picture can lead to misleading conclusions and bad decision making, so companies need to collect all data from all machines.

Very few factories have true and complete access to all of their assets and systems at the Edge. They don't have a middleware platform in place to connect to every machine, collect data, normalize it, and then distribute it to any person or application who needs access to that data for Smart Manufacturing. If they do have some of these things in place, they often are not able to scale the solution to multiple lines and factories.

Smart Manufacturing, simply put, is where the data tells us what to do. Getting the right Edge data to the right people at the right time can mean the difference between never getting Smart Manufacturing initiatives off the ground and truly transforming a factory into an efficient, data-driven operation.

Deployment Challenges

In addition to the data challenge, Smart Manufacturing initiatives with an Edge computing component face several other deployment challenges. In fact, a 2018 McKinsey survey found that 84% of companies deploying a Smart Manufacturing solution are stuck in pilot mode, 28% of them for more than two years.[4] Capgemini surveyed 1,000 manufacturers and in its 2019 report found that "deployment and integration of digital platforms and technologies" was the number-one challenge impeding the progress of smart factory initiatives.[5]

The phenomenon is called "pilot purgatory," when a small-scale test or proof of concept drags on for too long and can slow or even stop the Smart Manufacturing progress. Pilot purgatory happens primarily for one of the following reasons: deployment is slow and expensive, factory systems and data are complex, or the project lacks clear ROI.

An astonishing number of Smart Manufacturing pilots last years, which not only slows progress but also increases costs. By the time a project is ready to deploy, business priorities may have evolved, budgets have changed, and new technologies may even be on the horizon.

Leaders are often slow to deploy new solutions, holding tight to legacy solutions and failing to see the importance of operating quickly and nimbly to adjust to changing market conditions and increased competition. Many Smart Manufacturing initiatives also try to do too much at once, adding needless functionalities that escalate costs and complexity.

The second primary reason many deployments fail is complex systems and data. The typical factory has multiple disparate systems at the Edge, hundreds of heterogeneous devices, and is often managed by an operational technology team with legacy technologies that do not work with IT effectively. Instead of designing an enterprise technology convergence strategy where both teams have a vested interest in the project, local teams often drive siloed approaches without the right expertise to balance Edge and Cloud computing, and the end result is a deployment that can't scale across the enterprise, or work with Big Data and Cloud applications.

Finally, many companies fail to clearly define success at the beginning of the project. The first misstep is that implementations are often prompted by "shiny" new technology rather than a value-led implementation prompted by a business use case. Success is initially defined as a successful pilot instead of a measurable business goal such as improving OEE, optimizing a specific production line, or reducing scrap. Without clear objectives, Edge deployments fail because they never reach an unmeasurable goal. A successful deployment is dependent on clear metrics from the beginning and designing ways to track the progress of initiatives with both IT and operational technology (OT) involved. Every deployment needs greater accuracy, clarity, and precision when it comes to measuring progress.

Solving Deployment Challenges with an Edge Computing Platform

Manufacturers need two things to solve the data problem, prevent pilot purgatory, speed up deployment time, reduce complexity, and increase ROI: first, an Edge platform purpose-built for Smart Manufacturing with a specific set of capabilities to accelerate time-to-value and reduce cost, complexity, and risk; and second, a deployment rollout plan that focuses on defining clear ROI through business use cases, which is described in detail later.

Earlier we defined Edge computing, but what exactly is an Edge computing platform? An Edge platform bridges the gap between industrial devices at the Edge and advanced analytics in the Cloud. An Edge computing platform handles the collection of large volumes of data from all assets, analyzes it at the Edge, and makes it available in a ready-to-use format to any Cloud or enterprise system. Then, the platform takes the data models created in the Cloud and puts them to work next to the equipment on the shop floor to improve operations and realize the full value of Smart Manufacturing.

While an Edge platform handles a lot of information from several sources, and hence its value, there are four core capabilities it should handle seamlessly: device connectivity, Edge analytics, data integration, and application enablement.

1. **Device Connectivity.** Device connectivity has four essential parts or functions when it comes to Edge computing platforms: device drivers, data collection, data normalization, and data storage. Basic data collection from a select few industrial machines does not meet the requirements for enterprise Smart Manufacturing. The ideal platform offers rapid data

connectivity to all modern and legacy industrial systems with just a few clicks, enabling data collection and structuring data to be used by any Edge or enterprise application. The way platforms offer rapid device connectivity is with a large number of prebuilt drivers to connect to both legacy and modern devices and systems quickly.

2. **Edge Analytics.** The platform should be able to perform analytics at the Edge and enable advanced analytics and machine learning in the Cloud. Prebuilt data visualizations and analytics at the Edge for common KPIs such as condition-based monitoring and OEE allow customers to realize rapid time-to-value. The ability to run data models created in the Cloud back at the Edge provides closed-loop Edge-to-Cloud operations.

3. **Data Integration.** The next building block for an Edge computing platform is prebuilt integration with leading third-party Cloud or Big Data systems for machine learning and advanced analytics. Once connectivity and analytics are enabled, the platform should be able to stream ready-to-use data to any Cloud or enterprise system. The ability to send data bidirectionally between OT and IT systems provides the anywhere-to-anywhere integration needed to enable enterprise-scale Smart Manufacturing.

4. **Application Enablement.** Last, the Edge computing platform should be able to host and access public or private applications in a centralized repository, with the ability to rapidly and securely deploy and then run applications at the Edge. The platform should also stream normalized and structured data to any prebuilt or custom application at the Edge and run those same applications as close to the data source as possible.

Exhibit 11.2 shows an intelligent Edge computing platform.

EXHIBIT 11.2   Intelligent Edge computing

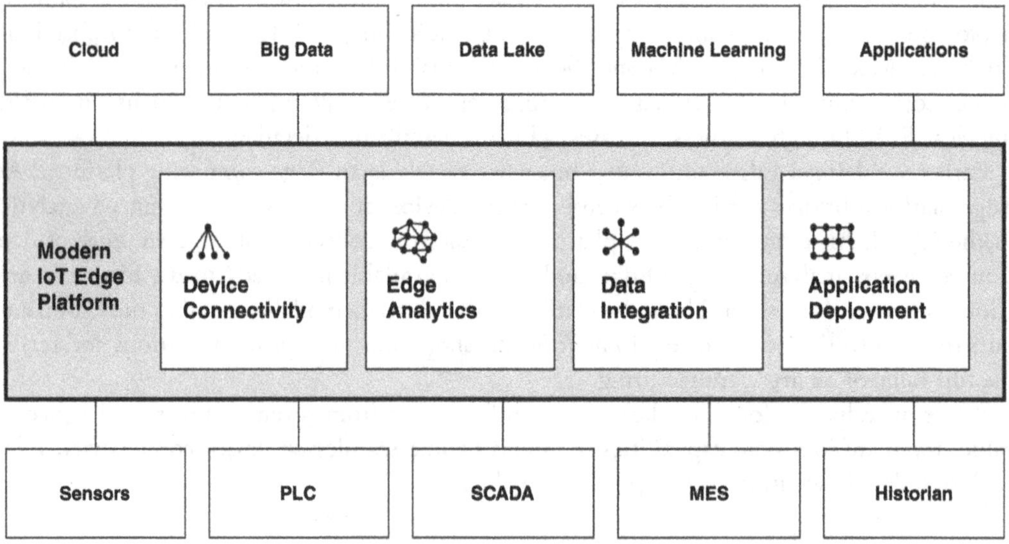

This type of intelligent Edge platform provides the common infrastructure to move data up to IT and back to the Edge with seamless cooperation. Now everyone in the enterprise can access the same tools, dashboards, data models, and KPIs so they are all on the same page for Smart Manufacturing success.

The Edge Computing Platform Landscape

With dozens of Edge computing and Smart Manufacturing platforms on the market, choosing a solution that checks all of the right boxes can be challenging. See the following list of vendors that have appeared in platform reports published by leading analyst firms, rated based on the completeness of their solution. The top-rated vendors have the most comprehensive solutions with all of the must-have Edge computing platform building blocks as outlined – device connectivity, Edge analytics, data integration, and application deployment. Each has its own set of core capabilities, but here is a list of some of the leading platform vendors:

- ABB
- Altizon
- AWS
- Bosch
- Braincube
- Clearblade
- Davra
- Eurotech
- Exosite
- Flutura Cerebra
- Foghorn
- GE Digital Predix
- Google
- Hitachi
- HPE
- IBM
- Inductive Automation
- Litmus
- MachineMetrics
- Microsoft Azure IoT
- Oracle
- OSIsoft
- PTC
- QiO
- RootCloud
- Samsung SDS
- SAP
- Siemens
- Software AG
- Telit

Edge-to-Cloud Computing

We've mentioned both Edge and Cloud computing in the same sentence a few times now, and that's because for large-scale manufacturing companies to transform and embrace Smart Manufacturing, they really need both. Edge computing is needed to enable use cases on the factory floor, but also to feed Cloud applications with good, complete data. Let's take a look at that balance.

At the bottom of the stack shown in Exhibit 11.3, we have the Edge. Edge devices include programmable logic control devices (PLCs), computer numerical control (CNC) machines, and factory floor systems like Supervisory Control and Data Acquisition (SCADA) or Historian (a software program that records a computer's process data). The Edge infrastructure layer is where the Edge platform resides, connects to all of the devices, and collects data. The data is then sent to the Cloud infrastructure where advanced applications reside and where data models can be built.

Exhibit 11.3 is an example of a technology stack with Edge devices and Edge and Cloud infrastructure.

EXHIBIT 11.3 Technology stack with Edge and Cloud infrastructure

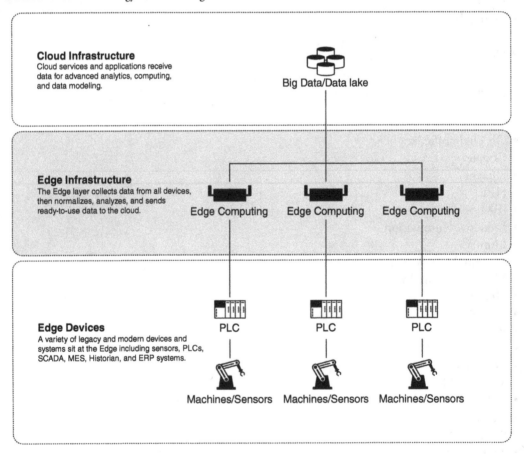

Many companies are either sending all of their data to the Cloud or keeping it at the Edge; they don't know how to effectively do both. Edge and Cloud technologies need to work together. Figure 11.4 clarifies how these two systems work together. With an Edge platform in place, assets are connected, data is collected securely at the Edge, and the people who need it are empowered to take action in real time against that data. At the same time, data is analyzed and sent to the Cloud to build, run, and train data models that can be run back at the Edge in production to drive change in the entire organization at scale.

Exhibit 11.4 is an example of a technology stack showing Edge infrastructure detail.

EXHIBIT 11.4   Technology stack with Edge infrastructure detail

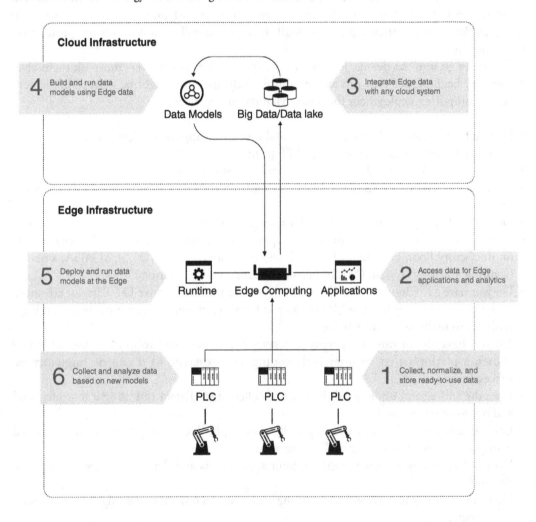

Edge-to-Cloud projects often start with simple data visualization and local analytics, then over time move to more advanced machine learning and continuous optimization use cases. No matter which Cloud platform a company chooses to standardize on – Amazon Web Services (AWS), Azure, or Google Cloud Platform – an Edge computing platform can collect, normalize, store, and integrate valuable data into the Cloud. As a result, companies can seamlessly move data from the Edge to the Cloud, and back, to run models and improve operations at scale.

How a Successful Edge Computing Rollout Works

The right Edge platform provides the ideal foundation for deployment success. With true flexibility and scalability, manufacturers can develop a proof of concept to make sure the solution checks all the important boxes, then enter the pilot phase to ensure that it satisfies every requirement for deployment. The manufacturer can quickly derive value and show ROI with instant KPIs, dashboards, and visualizations while they test the technology. No more two-year pilots stuck in pilot purgatory – this type of pilot will show its value in less than 30 days and then scale to a small deployment followed by multiple production lines and sites.

Before we go into the details, let's discuss the questions manufacturers must ask themselves before they begin. Once they can answer affirmatively, the time might be right to implement an Edge computing deployment for Smart Manufacturing success.

- Do you have a dedicated team, mandate, and executive support for the initiative?
- Do you have buy-in from both OT and IT teams?
- Have you agreed on a single business goal and data stream for the pilot?
- Have you agreed on measurable business goals, objectives, and outcomes for the entire initiative?
- Do you have an agile plan that allows for change over time?
- Do you have a way to easily connect a complex mix of modern and legacy OT data sources on the factory floor, including PLCs, a distributed Cloud system (DCS), SCADA, a manufacturing execution system (MES), Historian, databases, and sensors?
- Do you have IT, Cloud, or Big Data systems set up that can receive OT data for advanced analytics and machine learning? If not, do you have a platform that can provide instant Edge analytics to make use of data at the source?
- Do you have the in-house expertise necessary for a successful rollout? If not, have you chosen a platform that requires little coding and expertise so it can be implemented successfully?
- Does the entire team understand the value of collecting OT data and using it to understand and optimize operations?
- Do you have a plan for how to easily scale to multiple Edge deployments with centralized management and control over all Edge devices?
- Do you have a way to deploy purpose-built applications and deployment templates across all sites?
- Do you have a way to run machine learning and analytics models at the Edge for continuous improvement?

A successful deployment is agile, encouraging continuous evaluation and improvements that allow OT and IT teams to add data points and use cases as needs evolve over time. It is important to include both IT and OT teams in the entire process. Too often OT teams choose a solution, and then enterprise-wide rollout fails because IT was never involved in the process.

Exhibit 11.5 shows the steps to ensure a successful Smart Manufacturing deployment.

EXHIBIT 11.5 Deployment steps

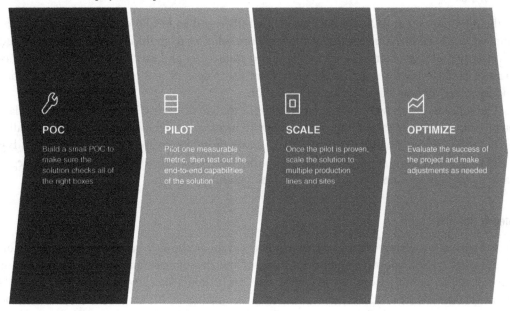

POC

Build a small POC to make sure the solution checks all of the right boxes

PILOT

Pilot one measurable metric, then test out the end-to-end capabilities of the solution

SCALE

Once the pilot is proven, scale the solution to multiple production lines and sites

OPTIMIZE

Evaluate the success of the project and make adjustments as needed

Step 1: Proof of Concept

Start with a clear vision of what the Edge computing and Smart Manufacturing initiative needs to accomplish. Determine the business goal – is it to gain visibility into the shop floor, to power a Big Data platform with a complete data picture, or perhaps to create dashboards for the maintenance team? Develop a checklist of things the product must be able to do to reach the goal, and look for an all-in-one platform that can collect, analyze, manage, and integrate all data from every asset on the factory floor.

Develop a proof of concept that starts with the platform on the shop floor to see how it behaves and choose one or two data points to collect that fit into the business goal. At the same time, consider future scaling needs by adopting a platform that will grow with the project rather than requiring a myriad of solutions. The point of the proof of concept is to make sure the solution checks all the important boxes, and to gain enough confidence in the product to implement it and move to the pilot stage.

Step 2: Pilot

Once a solution is chosen, define a specific use case with clear metrics and goals to measure the success of the pilot. Start by solving one business problem, such as reducing downtime or scrap on a specific line. Determine the data points needed to detect anomalies and trends, who needs access to the data, and where the data needs to go. Implement all aspects of the platform, from data collection to local analytics to integration with Big Data and enterprise applications.

A complete OT-IT deployment at the pilot phase is critical to understanding the technical requirements needed for success. The pilot is designed not only to help test the capabilities of the platform, but also to provide the opportunity to understand how to integrate new technology into day-to-day operations. The right technology should uncover new insights and new behaviors.

Step 3: Scale

If the pilot shows that ROI and the features are fully tested and explored, it can confidently be applied across all machines in the plant. Scale the solution to multiple production lines, and ultimately multiple plants. Create a template to scale safely and easily with proven best practices. The Edge platform can handle any number of devices or sites with centralized Edge management tools and application orchestration.

The solution grows with the customer, so once the initial deployment is proven, more data streams and more use cases can be added safely to increase productivity, efficiency, and quality across the business. Analyze larger data sets, create more dashboards, and integrate with Big Data or machine learning platforms for even more ROI. Continue to evaluate and adjust regularly to get more business value out of the deployment.

Step 4: Optimize

In the evaluation phase, OT and IT stakeholders evaluate the success of the project based on the success criteria adopted at the beginning of the deployment and make adjustments as needed. Think about lessons learned, what else can be done, and how to build on success. Is more data needed? Should the project expand to new use cases? If changes are needed, redefine the scope quickly and deploy again in a matter of days. An agile smart factory that can adjust based on how things are going will pay dividends in the long run.

A modern Edge platform is flexible enough to make changes on the fly, create new dashboards, add data points, send data to a new application, and more. Taking the time to stop and measure success, then improve upon the process, takes discipline but is well worth the effort, especially for a Smart Manufacturing deployment that thrives on continuous improvements.

Summary

Reaching true Edge-to-Cloud continuous optimization and machine learning is the end goal for many manufacturers. Others simply want to visualize and analyze operations in real time. No matter where the journey begins, the path is the same. It starts with connecting and collecting data from every machine at every location. An Edge platform provides the foundation needed for a complete data picture to be shared across the enterprise and achieve operational excellence at the Edge and beyond.

Sample Questions

1. Edge computing can be defined as:
    a. Processing data next to the asset
    b. Any device or asset at the Edge of the network in the field or on the factory floor
    c. Connected devices producing data
    d. All of the above

2. What kind of data is captured at the intelligent industrial Edge?
   a. Humidity
   b. Temperature
   c. Sound
   d. All of the above
3. What percentage of industrial assets are connected today?
   a. 50
   b. 25
   c. 75
   d. 100
4. What are some of the benefits of Edge computing?
   a. Faster insights
   b. Short-term analytics
   c. Preserve bandwidth
   d. All of the above
5. True or false: The typical factory includes a mix of modern and legacy systems, which is the foundation for the data challenge.
6. What is the number-one reason Smart Manufacturing deployments fail?
   a. Budget
   b. Complex systems
   c. Pilot purgatory
   d. Failure to define success
7. What percentage of companies deploying a Smart Manufacturing solution are stuck in pilot mode?
   a. 28
   b. 84
   c. 42
   d. 10
8. What are the four core capabilities an Edge platform should handle?
   a. Device connectivity, Edge analytics, data integration, and application enablement
   b. Cloud analytics, building applications, machine learning, and lifecycle management
   c. Machine maintenance, resource planning, and data storage
   d. None of the above
9. How should companies handle Edge versus Cloud computing?
   a. Companies need to choose one or the other
   b. Companies should keep all data at the Edge
   c. Companies should all adopt the Cloud
   d. Companies must balance the Edge and the Cloud
10. What is the order for a Smart Manufacturing deployment?
    a. Pilot, Proof of Concept, Scale
    b. Scale, Pilot, Proof of Concept
    c. Proof of Concept, Pilot, Scale
    d. Pilot, Scale, Proof of Concept

Notes

1. IoT Analytics. (2019). *Industrial Connectivity Market Report 2019–2024.* https://iot-analytics.com/product/industrial-connectivity-market-report-2019-2024/.
2. Wopath, M. (August 13, 2019). 5 Industrial connectivity trends driving the IOT convergence. *IoT Analytics.* https://iot-analytics.com/5-industrial-connectivity-trends-driving-the-it-ot-convergence/.
3. Gartner. (2018). What edge computing means for infrastructure and operations leaders. https://www.gartner.com/smarterwithgartner/what-edge-computing-means-for-infrastructure-and-operations-leaders/.
4. de Boer, E., and Narayanan, S. (April 16, 2018). Avoid pilot purgatory in 7 steps. McKinsey & Company. https://www.mckinsey.com/business-functions/organization/our-insights/the-organization-blog/avoid-pilot-purgatory-in-7-steps.
5. Petit, J.-P., Brosset, P., Puttur, R. K., et al. (2019). *Smart factories @ scale.* Capgemini Research Institute. https://www.capgemini.com/wp-content/uploads/2019/11/Report---Smart-Factories.pdf.

CHAPTER 12

3D Printing and Additive Manufacturing

Bahareh Tavousi Tabatabaei, Rui Huang, and Jae-Won Choi

Introduction

Additive manufacturing (AM) is a technology used to create an object from a computer-aided design (CAD) model by adding materials in a layer-by-layer manner (Ngo et al. 2018). This chapter aims to provide a brief introduction to the basic concepts of additive manufacturing, the history and current state of this technology, and a description of the methods, materials, and application of this technology in different industries.

The first commercialized system for additive manufacturing, known as the stereolithography apparatus (SLA), was developed by Charles Hull in 1986 (Ngo et al. 2018). This system incorporated an ultraviolet (UV) laser to solidify a photo-curable liquid polymer (Bose and Bandyopadhyay 2019). Remarkable developments such as fused deposition modeling (FDM), inkjet printing, and other techniques have been made since the advent of SLA (S. H. Huang et al. 2013).

Rapid prototyping (RP) and other terms had been used to represent the various AM processes until the American Society for Testing and Materials (ASTM) Technical Committee agreed on the new term *additive manufacturing* (ASTM International 2012). RP was initially regarded as a process to fabricate a model quickly for the purpose of visualizing a design or creating a representative example of a final product, and this process has helped engineers to create and analyze the designs they have in mind (Wong and Hernandez 2012). However, since the advent of significant advancements in AM processes, materials, and applications, AM technology is being used not only for the fabrication of prototypes but also for creating final products (Ngo et al. 2018; Vaezi et al. 2013).

AM processes are advantageous in their minimal waste of material, capability for printing parts with complex geometry, and customization features that allow them to surpass traditional manufacturing methods (Ngo et al. 2018), where traditional manufacturing methods require deep analysis for process planning to determine how to create complex features and what tools are needed. AM technology simplifies the process for fabricating objects with complex geometry, as AM does not require such complex, costly, and time-consuming steps to create a part. AM technology merely requires three-dimensional information about the part to be built and a basic understanding of how the AM machine works (Gibson et al. 2021).

In an AM process, each part is created by adding material layer by layer, and each layer is a horizontal cross-section of the part. As illustrated by Exhibit 12.1, building the part using thinner layers will make the output match more closely to the original. Various AM processes use different methods to form the layers. Some processes use heat energy to bind the layers together. Other processes spray a binder or a solvent for this purpose (Gibson et al. 2021; S. H. Huang et al. 2013; Gibson et al, 2021).

EXHIBIT 12.1 CAD model and slicing: (a) CAD model; a finished part using (b) a thicker layer thickness and (c) a thinner layer thickness.

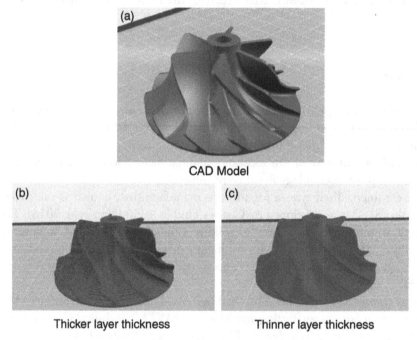

CAD Model

Thicker layer thickness   Thinner layer thickness

In general, AM processes include four major steps (Exhibit 12.2):

1. A solid model is created in CAD software and converted to a stereolithography (STL) file format. (Note: 3D scanning can also be used to create a solid model.)
2. The STL file is transferred to an AM machine, where it is possible to manipulate the size, position, and orientation of the part.
3. Once the machine is set up, it starts to build the part one layer at a time.

EXHIBIT 12.2 Example of a general AM process cycle

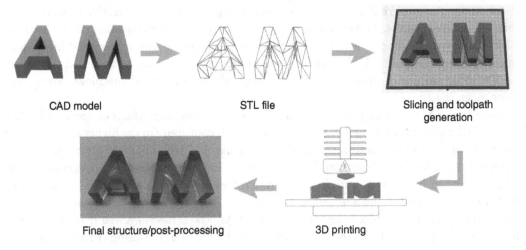

CAD model STL file Slicing and toolpath generation

Final structure/post-processing 3D printing

4. In some cases, the part may need to be cleaned once the building process is complete, or support structures that were built along with the part will need to be removed. In general, post-processing of the part is required to achieve an acceptable finish.

Many approaches have been introduced for the classification of AM processes. One approach is based on the type of material used for printing (Chua and Chou 2003; Chua et al. 1998). Another popular classification is based on the technology used, such as classifications for processes that use extrusion technology or laser technology (Burns 1993; Kruth et al. 1998). In this chapter, we apply a classification method according to the building style and material transformation principles (a version of Pham's classification, as described in Pham and Gault, 1998). Based on this approach, the following seven categories can be used to accommodate all AM processes:

1. Vat photopolymerization (VPP)
2. Material extrusion (MEX)
3. Material jetting (MJT)
4. Binder jetting (BJT)
5. Powder bed fusion (PBF)
6. Sheet lamination (SHL)
7. Directed energy deposition (DED)

These seven processes are described in greater detail in subsequent sections. It should be noted that development of other technologies, such as computer-aided design (CAD) and laser technology, as well as advances in material science, played a significant role in the development of rapid manufacturing processes (Cooper 2001; Kruth 1991; Noorani 2006; Wong and Hernandez 2012). Novel materials and methods have made it easy to adopt AM for new applications. Some recent advances have drastically decreased the cost of three-dimensional (3D) printing so that these machines can be used in schools, homes, and other settings.

Despite the progress made in developing AM technologies, there is still a need in some cases to use computer numerical control (CNC) machining, which is a subtractive process rather than an additive process. There are still some limitations in the materials that can be used for AM. At the current time, it is not possible to print all materials commonly used in product manufacturing (Kruth 1991). In addition, size limitations are also an issue, as AM cannot be used to print parts with dimensions that are larger than existing AM machines (Wong and Hernandez 2012).

Nowadays, AM technology is being used for many purposes, as products customized by AM eliminate the need for mold making and tooling, which contribute to the higher costs for parts produced in traditional manufacturing. Architects use AM technology to create prototypes rapidly and at a low cost (Ngo et al. 2018). Construction is another industry benefiting from AM: inexpensive houses were built in China by WinSun in less than 24 hours (Wu et al. 2016). AM is also being used in the biomedical field due to its ability to produce patient-customized products (Stansbury and Idacavage 2016). Wohlers Associates has suggested that the demand for customized products is driving the expansion in the use of 3D printing (Berman 2012).

History

Early Stages of Additive Manufacturing

When looking at the historical development of additive manufacturing, determining its exact origins is difficult. Although Charles Hull was named as the first person to introduce a commercialized AM machine in the 1980s (Bose and Bandyopadhyay 2019), numerous research efforts in the 1950s through the 1970s were conducted that relate to the 3D printing of objects (Gibson et al. 2021).

In 1951, Otto John Munz suggested a system using technology that is similar to the current process for stereolithography. Munz's system used a cylinder with a piston. An adequate amount of photo-curable emulsion is added when the piston goes down and layers are created by fixing the agent. An image of the object is formed after exposure to light. The object can be carved out photochemically or manually to produce the final solid object. A schematic diagram of this system is shown in Exhibit 12.3a (Bourell et al. 2009).

In 1968, Wyn Kelly Swainson developed a system in which a plastic model is built using a process in which two laser beams selectively crosslink a photosensitive polymer (Swainson 1977). This system is shown in Exhibit 12.3b.

In 1971, Pierre Alfred Leon Ciraud invented a powder-based system that is very similar to the direct deposition AM method (Ciraud 1972). In this process, particles of the material are applied to a matrix in a variety of ways, such as by using electrostatics and a nozzle. The particles become attached to each other and form a layer when they are heated selectively by a laser beam, an electron beam, or any other heat source. The scheme is shown in Exhibit 12.3c.

In the late 1970s and early 1980s, Ross F. Housholder (Housholder 1981), Hideo Kodama (Kodama 1981), and Alan Herbert (Herbert 1982) also proposed processes for making three-dimensional objects by adding layers of materials (Wohlers and Gornet 2014).

Exhibit 12.4 shows early 3D patterns that were printed using the systems by Housholder, Kodama, and Herbert.

EXHIBIT 12.3   Schematic diagrams showing (a) the Munz system, (b) the Swainson system, and (c) the Ciraud system

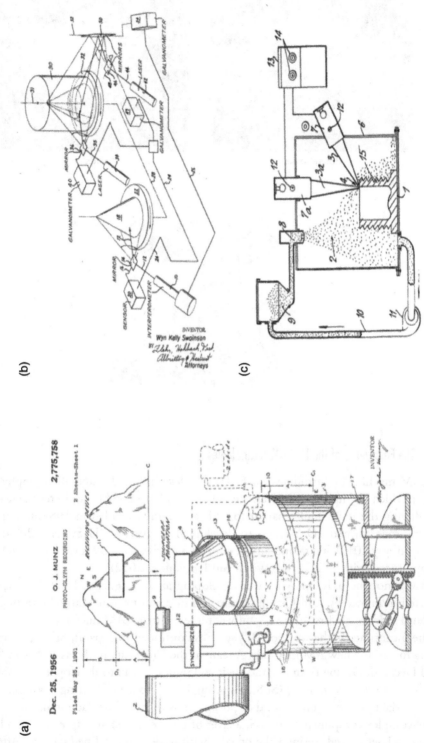

(a)

(b)

(c)

Sources: (a) Bourell, Beaman, Leu, and Rosen (2009); (b) Swainson (1977); (c) Ciraud (1972).

EXHIBIT 12.4   3D patterns using the Housholder, Kodama, and Herbert systems

Householder   Kodama

Herbert

1980s – The Emergence of the First AM Technologies

In the AM machine he introduced in 1983, Charles Hull used a stereolithography apparatus (SLA), in which UV light is selectively shined onto a photo-curable liquid resin, creating layers by solidifying the resin. Hull's patent for this apparatus was filed in 1984 and approved in 1986. He then formed 3D Systems to develop SLA systems (3D Systems 2014; Bose and Bandyopadhyay 2019). This company developed STL files to make it easy for an AM machine to communicate with a CAD file (Wong and Hernandez 2012).

Hull was not the only one who was developing an AM process. Parallel work was being conducted in France by Jean-Claude Andre and in Japan by Yoji Marutani around the same time, but Hull was the first to successfully commercialize an AM machine.

In the meantime, as the work on 3D systems continued to expand, other researchers were working to develop new AM methods. A team at the University of Texas at Austin (consisting of Carl Deckard, Dr. Joe Beaman, and their colleagues) began to develop a new AM technology called *selective laser sintering* (SLS). Sintering is the process of heating a powdered metal or other material to coalesce into a solid or porous mass without liquefaction in the manufacture of more complex components. In SLS, a layer of powder material is spread onto a build platform, and a laser is used to sinter the powder with respect to a defined pattern. After successfully building the first SLS machine, Deckard and Beaman established a company called Nova

Automation, which was later became DTM Corp. before being acquired by 3D Systems in 2001 (Lou and Grosvenor 2012).

In 1986, Helisys Inc. commercialized a technology called laminated object manufacturing (LOM), in which layers of paper material are laminated in a layer-by-layer manner and then cut using a CO_2 laser to create a part. After bankruptcy, this company continued as Cubic Technologies Inc. (Gibson et al., 2021).

At the same time, Scott Crump and his wife, Lisa, developed a new AM process known as fused deposition modeling (FDM) while they were making a toy for their child (On3DPrinting .com 2017). In this process, material contained in a reservoir is forced out through a nozzle when pressure is applied. Scott Crump applied for a patent, and the couple formed Stratasys in 1989; the patent for the device was granted in 1992 (Crump 1992; Derdak and Pederson 2005).

1990s – Process and Innovation

In 1993, binder jetting (called *3D printing* at that time) was developed at the Massachusetts Institute of Technology (MIT). In this process, a binder is printed onto a powder bed to form part cross-sections. The MIT group licensed this technology to several companies; of these, Zcorp (which later was bought by 3D Systems) and ExOne were the most successful. Zcorp primarily concentrated on low-cost machines, while ExOne focused on powder-form metal and sandcasting applications (Sachs, Cima, and Cornie 1990).

Around the same time, Sanders Prototype Inc. developed a machine that deposits droplets of a wax material onto a plate (Wohlers 2004) using a process known as *inkjet technology*. This technique was originally demonstrated during the development of a process called *ballistic particle manufacturing* in the 1980s (Gibson et al. 2021).

At this time, most of the machines were developed to make objects using polymers. As a result, many companies began to develop a process in which an object can be made from metal or ceramic. Electro Optical Systems (EOS) GmbH in Germany was among the first companies to start working to make metal parts by using SLS. EOS built the first metal prototype in 1994 (EOS GmbH 2021).

In 1997, an AM technology known as *laser engineered net shaping* (LENS) was developed by Sandia National Laboratories to create metal parts. In the LENS machine, which was commercialized by Optomec Inc., metal powder is melted and deposited to form a layer (Sandia National Laboratories 1997).

2000s – Development of New Applications

Additive manufacturing continued to grow in the 2000s with the addition of new materials, processes, and applications. AM was being used to print metal parts through the application of methods such as micro casting and sprayed materials. Moreover, new research began in the biomedical field to fabricate functional artificial kidneys, prosthetic legs, and artificial blood vessels. The manufacturing industry started to use AM technology in different ways to make products in less time and with the minimum amounts of material and energy waste. At this time, the use of open-source hardware became more widespread, and people were able to make new products on their own. The Connex series of 3D printers was introduced by Objet Ltd., providing users with the ability to build a part using two different materials in a single print (Nano Dimension USA 2021).

AM Standards

Standards were developed for AM to help organizations and companies learn how to adopt AM technologies, reach confidence in this area, characterize materials and their properties, and adopt common terminology for communicating with suppliers and clients. The first attempt to develop standards for AM began in 2009, when Ian Gibson and his colleagues worked with ASTM International (formerly known as the American Society for Testing and Materials) to form the F42 committee. Other standards organizations, such as the International Organization for Standardization (ISO) and the American Society of Mechanical Engineers (ASME) began working on AM standards in subsequent years. In 2013, ISO and ASTM worked together to promote AM standards as an agreement among various organizations (ASTM 2016). Six sub-committees in ASTM F42 are responsible for developing standards on test methods, design, materials and processes, environmental health and safety, applications, and data in AM (Gibson et al. 2021). Various organizations are currently working on AM standards, which indicates that more comprehensive standards will be made available in the future. A summary of ASTM and ISO standards that are available as of 2020 are presented in Exhibit 12.5.

EXHIBIT 12.5 Summary of the ASTM and ISO standards in 2020

	ASTM Standard Designations	Title
Terminology	ASTM52900	Standard Terminology for Additive Manufacturing – General Principles – Terminology
Design	F3413	Guide for Additive Manufacturing – Design – Directed Energy Deposition
	ASTM52910	Additive manufacturing – Design – Requirements, guidelines and recommendations
	ASTM52911-1	Additive manufacturing – Design – Part 1: Laser-based powder bed fusion of metals
	ASTM52911 -2	Additive manufacturing – Design – Part 2: Laser-based powder bed fusion of polymers
	ASTM52915	Specification for additive manufacturing file format (AMF) –Version 1.2
	ASTM52950	Additive manufacturing – General principles – Overview of data processing
Materials and Processes	F2924	Standard Specification for Additive Manufacturing Titanium-6 Aluminum-4 Vanadium with Powder Bed Fusion
	F3001	Standard Specification for Additive Manufacturing Titanium-6 Aluminum-4 Vanadium ELI (Extra Low Interstitial) with Powder Bed Fusion
	F3049	Standard Guide for Characterizing Properties of Metal Powders Used for Additive Manufacturing Processes
	F3055-14a	Standard Specification for Additive Manufacturing Nickel Alloy (UNS N07718) with Powder Bed Fusion
	F3056-14e1	Standard Specification for Additive Manufacturing Nickel Alloy (UNS N06625) with Powder Bed Fusion
	F3091	Standard Specification for Powder Bed Fusion of Plastic Materials
	F3184	Standard Specification for Additive Manufacturing Stainless Steel Alloy (UNS S31603) with Powder Bed Fusion
	F3187	Standard Guide for Directed Energy Deposition of Metals

EXHIBIT 12.5 Summary of the ASTM and ISO standards in 2020 (*Continued*)

	ASTM Standard Designations	Title
	F3213	Standard for Additive Manufacturing – Finished Part Properties – Standard Specification for Cobalt-28 Chromium-6 Molybdenum via Powder Bed Fusion
	F3301-18a	Standard for Additive Manufacturing – Post-Processing Methods – Standard Specification for Thermal Post-Processing Metal Parts Made Via Powder Bed Fusion
	F3302	Standard for Additive Manufacturing – Finished Part Properties – Standard Specification for Titanium Alloys via Powder Bed Fusion
	F3318	Standard for Additive Manufacturing – Finished Part Properties – Specification for AlSi10Mg with Powder Bed Fusion – Laser Beam
	F3434	Guide for Additive manufacturing – Installation/Operation and Performance Qualification (IQ/OQ/PQ) of Laser-Beam Powder Bed Fusion Equipment for Production Manufacturing
	ASTM52901	Standard Guide for Additive Manufacturing – General Principles – Requirements for Purchased AM Parts
	ASTM52904	Additive Manufacturing – Process Characteristics and Performance: Practice for Metal Powder Bed Fusion Process to Meet Critical Applications
	ASTM52903	Material extrusion-based additive manufacturing of plastic materials Part 1 & 2
Test Methods	F2971	Standard Practice for Reporting Data for Test Specimens Prepared by Additive Manufacturing
	F3122	Standard Guide for Evaluating Mechanical Properties of Metal Materials Made via Additive Manufacturing Processes
	ASTM52902	Additive manufacturing – Test artifacts – Geometric capability assessment of additive manufacturing systems
	ASTM52921	Standard Terminology for Additive Manufacturing – Coordinate Systems and Test Methodologies
	ASTM52907	Additive manufacturing – Feedstock materials – Methods to characterize metallic powders

Additive Manufacturing Process

Seven processes for additive manufacturing are defined by the ASTM standard. This section describes the principles behind each process, the materials used for manufacturing parts, and the benefits and drawbacks for each process.

VAT Photopolymerization

Process

Vat photopolymerization (VPP) is a liquid-based method in which a liquid photo-curable resin is solidified when a light source is shined on its surface. Stereolithography (SLA), which was patented by Charles Hull in 1986 (Hull 1988) and commercialized by 3D Systems (Wong and

Hernandez 2012), was the first available VPP process. In SLA, as in other AM processes, a model designed using CAD software is translated to a file in STL format, and the model is then sliced to many layers. The UV laser scans each layer to solidify the resin and form cross-sections (Guo and Leu 2013).

Some VPP systems feature a configuration that uses a bottom-up approach in which the light source is located below the vat. After the current layer is fully formed, the build platform moves up by a thickness of one layer. In some VPP systems, a blade is employed to sweep the resin across the vat, while other systems do not have such a blade. Each new layer is scanned and attached to the previously built layer (Exhibit 12.6).

EXHIBIT 12.6 A schematic diagram of VPP

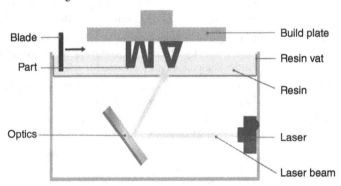

Other VPP systems use a configuration with a top-down approach in which the laser and optical system are located above the resin and a blade is generally used to flatten the resin surface (Gibson et al. 2021). Support structures incorporated into the design of the part are used to fix the printed part to the platform and to support any overhanging parts (S. H. Huang et al. 2013); consequently, a part printed by VPP may need be heated or exposed to UV light as a post-process in order to obtain the desired mechanical properties (Ngo et al. 2018).

VPP processes can be classified into different types based on the radiation source and layering method used in building the part.

Radiation sources such as gamma rays, X-rays, ultraviolet (UV) light, electron beams, and visible light can be used for photopolymerization in VPP, but most commercial machines use UV or visible light. Methods for layer fabrication include vector scanning, mask projection, and two-photon polymerization. In vector scanning, a laser scans the material and solidifies the resin point by point. For the mask projection approach, a pattern generator such as a digital micromirror device (DMD) or a liquid crystal display (LCD) is used to form a beam pattern; this pattern solidifies the entire layer at once, resulting in an increase in the printing speed. In the two-photon polymerization technique, two photons are required to enable the photosensitizer to emit light that solidifies a very tiny volume of a resin (usually less than 1 μm³), whereas the other two approaches require only a single photon. The parts fabricated using different VPP approaches may have different layer thicknesses or be printed at different resolutions (Wong and Hernandez 2012).

A variety of VPP processes has been developed to print with materials other than polymers or to print parts using more than one material. Some VPP units allow the creation of structures by using liquid monomers containing metallic or ceramic particles to form metallic-polymer or ceramic-polymer composites, respectively (Andrzejewska 2001; Crivello 1993; Hageman 1989; Travitzky et al. 2014). Other VPP units provide users with the ability to build parts using more than one material. In this process, the resin vat must be drained and loaded with a new material to be printed with the same layer or in a different layer, or multiple resin vats are required (Choi, Kim, and Wicker 2011; Crivello 1984).

Materials

Acrylic or epoxy-based monomers are commonly used as photopolymers in the VPP process (Ngo et al. 2018). The first photopolymers developed in the 1960s were made from acrylates and were highly reactive; however, the printed parts exhibited shrinkage (Wohlers 1991). Later, epoxy-based photopolymers were developed that were able to create a more durable and accurate part with less shrinkage (Crivello 1998). While the epoxy resins have lower reactivity (resulting in a slower cure speed), the produced parts are more brittle following curing (Lu et al. 1995). Nowadays, a combination that uses both epoxides and acrylate is favored for VPP printing, and ceramics such as zirconia, silica, and alumina materials can be used as a suspension in the liquid resin (Travitzky et al. 2014).

Advantages and Disadvantages

VPP machines are capable of creating parts with high accuracy and an excellent surface finish, making it possible to create fine features (Ngo et al. 2018), as shown in Exhibit 12.7. In addition, these machines can use a variety of light sources and scanning methods to produce parts

EXHIBIT 12.7   Examples of printed microstructures by VPP

Source: J. Lee et al. (2018).

in a wide range of sizes. As mentioned previously, it is now possible to create features in nanoscale by using two-photon photopolymerization (Gibson et al. 2021; Tofail et al. 2018).

The main drawback of using VPP technology is related to materials that are used for printing the parts. The materials used for VPP are limited to photopolymers with relatively poor mechanical properties that will become degraded over time (Gibson et al. 2021; Tofail et al. 2018); the slow printing speeds and higher cost are other disadvantages of using this process (Ngo et al. 2018).

Material Extrusion

Process

Material extrusion (MEX) is a widely used additive manufacturing process in which a material is pushed out through a nozzle and is solidified after being deposited onto the substrate (Gebhardt 2011; Tofail et al. 2018), as shown in Exhibit 12.8. Fused deposition modeling (FDM) is a trade name that is often improperly used to refer to material extrusion because FDM was commercialized as the first material extrusion machine by Stratasys Ltd. in the late 1980s (Crump 1991a, 1991b). A material extrusion–based machine includes a reservoir (which is heated to maintain the printing material in a molten state), an extrusion head with nozzles, and a build platform.

EXHIBIT 12.8   Schematic diagram of material extrusion

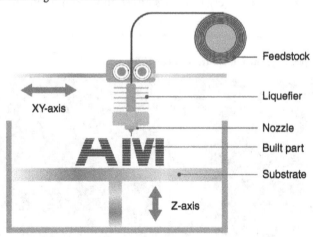

In the MEX process, the printing head having nozzles moves in an x–y plane, and the unit controls the starting and stopping of the flow of materials from the nozzles. When a layer is deposited, the vertical axis moves by one-layer thickness to make space for depositing the next layer (Gibson et al. 2021). This process requires support structures to hold disconnected or overhanging features (Gebhardt 2011). Two considerations should be taken into account when using the MEX process. First, the material must be solidified completely after deposition while still in its deposited form. Second, the deposited material must bond to the previous layer to create a solid part.

Two main approaches to solidifying materials are used in MEX machines. One approach is to control the material state by controlling its temperature, by using heater coils around the material reservoir to keep the temperature in the reservoir constant. This allows the material to flow easily through the nozzle and to bind with the next layer before solidification. It is important keep the temperature just high enough to ensure good flow, as high temperatures can lead to degradation for some materials. The second approach is a chemical-based solidification approach in which a curing agent or solvent is used to solidify and bond the material in a process that dries wet components. This approach is more commonly used for biomaterials (Gibson et al. 2021).

Depending on the material state, different loading mechanisms can be used for the material extrusion to ensure a continuous supply of material. For liquid printing materials, pumping is the proper method; for solid-state materials, the most effective method is to use filaments, pellets, or the powder form of the material. To aid in pushing the materials through the nozzle of the machine, the machine can use a screw, compressed gas, or pinch rollers.

A number of parameters can affect the printing and the part printed using a MEX unit. The shape and size of the extruded filament will depend on the shape and diameter of the nozzle. Material viscosity, nozzle geometry, and pressure drop will affect the flow of material through the nozzle (Bellini et al. 2004; Turner et al. 2014), while gravity and surface tension will affect the material shape after printing. The material can also undergo some changes when it cools or dries: shrinkage may occur during the cooling process, and this can be controlled by keeping the temperature difference between the reservoir and surrounding area to a minimum (Gibson et al. 2021). Thermoplasticity is another vital property for materials printed using MEX technology, as thermoplastic materials can be fused together during printing and solidified at room temperature to create a part. In the individual printing process, the layer thickness, the air gap within a layer or between layers, and the width and orientation of filaments are important parameters that can significantly affect the mechanical properties of the printed structure (Mohamed et al. 2015).

Materials

Polymers are the traditional materials used in material extrusion machines (Guo and Leu 2013), while polymer composites with ceramic or metal are new materials (R. Huang et al. 2021). ABSplus™, the most common polymer used for this technology, is a new version of acrylonitrile butadiene styrene (ABS). ABSi is another polymer for MEX, and it can be used to create translucent parts. Polycarbonate (PC)-based materials with higher tensile strength, or a mixture of ABS and PC, can also be utilized by material extrusion machines: products made from PC-ISO are used in food and medical applications, ULTEM° (polyetherimide, which has favorable characteristics against flame, smoke, and toxicity) is used for aircraft applications, and other polymers are used to manufacture high-end apparatuses (Gibson et al. 2021). There have been intensive research activities for developing ceramic or metal composites that incorporate alumina, lead zirconate titanate (PZT), silicon nitride (Si3N4), zirconia, silica, bioceramics (Agarwala et al. 1996; Allahverdi et al., 2001; R. Huang et al., 2021; Rangarajan et al. 2000), and metal (Singh et al. 2021).

Fiber-reinforced filaments are composite materials that can be used in material extrusion technology. Examples of composite filaments include Onyx™ (a microcarbon fiber–filled nylon), high-strength high-temperature (HSHT) fiberglass, and carbon fiber (CF) filaments.

Each of these materials has specific mechanical properties that are suitable for various applications (Markforged 2020).

Advantages and Disadvantages

MEX machines are widely used, as they are relatively inexpensive and have low maintenance costs. These machines can create 3D parts using a simple method (Chohan et al. 2017), and the parts do not require chemical-based post-processing (Cooper 2001; Noorani 2006). MEX can also be used to fabricate fully functional parts (Tofail et al. 2018). As discussed previously, different materials with different mechanical properties can be used in this technology to produce parts that can be used for various applications (S. H. Huang et al. 2013).

The main drawback of material extrusion technology is, in general, its low resolution. The distinct boundary between layers (S. H. Huang et al. 2013) and poor surface finish are other problems with this method that can result in poor mechanical properties for the printed product (Chohan et al. 2017). To create fine features, it is necessary to use a nozzle with a small diameter, which increases the build time significantly; consequently, this method is not a good approach to use for producing fine details in the part. All commercially available nozzles have round openings, which have limitations for creating sharp corners (Gibson et al. 2021).

Material Jetting

Process

Material jetting is another AM technology in which a 3D part is created through the deposition of material droplets on a substrate (Tofail et al. 2018), as shown in Exhibit 12.9. As the materials need to be jettable in the material jetting machines, any solid materials (e.g. waxy materials) must be heated to convert them to a liquid state (3D Systems 2020). If materials are flowable, viscosity is another factor to be considered. Materials with high viscosity are not

EXHIBIT 12.9 Schematic drawing of material jetting

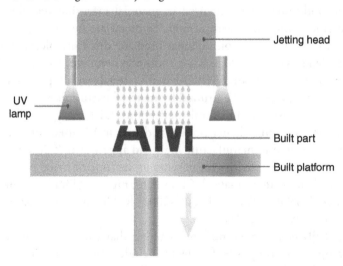

suitable for printing, so these materials require heating or the addition of solvents to lower their viscosity. The following paragraphs discuss research that is related to the development of material jetting technology materials or methods.

There are several reports for the printing characterizations. Gao and Sonin (1994) conducted an earlier study on the deposition and solidification of polymer droplets. In this study, the authors investigated three deposition modes (columnar, sweep, and repeated sweep) and used candelilla wax and microcrystalline petroleum wax for this purpose. Different frequencies and cooling rates were applied to study their effects on the columnar deposition of droplets. The authors determined that the contact angle where the drop solidifies, the sweep speed, and the droplet size are factors that affect the formation of solid lines, and smooth lines will be produced only within a small range of deposition frequencies. Reis et al. (1998) studied the effects of print head scanning speed and droplet velocity on the deposition of droplets. They reported that at a low droplet velocity, the droplets will be deposited discontinuously if the print head scans slowly, while droplets will be deposited continuously and form an unbroken line if the print head scans rapidly. At a high droplet velocity, the droplets splash upon contact with the substrate when the scanning speed is high, and bulges in the deposited material are formed when the scanning speed is low.

Despite all achievements and developments to date, there are still challenges that limit the growth of material jetting technology. One challenge is related to the requirement for liquid materials, and materials in other forms need to be dissolved in a solvent, suspended in a liquid carrier, or melted to be printable. In most cases, other components (such as surfactants) are added to obtain specific mechanical properties. Many factors such as droplet trajectory, impact, and substrate wetting are involved in the deposition of droplets. Another challenge is to control these factors to produce an acceptable print. Nozzle design and operation play an important role in the deposition characteristics by controlling droplet size and velocity (Bechtel et al. 1981; Pasandideh-Fard et al. 1996; Zhou et al. 2013).

Solidification is another phase of the material jetting process that needs to be controlled. It occurs when a melted material is hardened, the liquid portion of a solution evaporates, or a photopolymer is cured. Depending on the time and place the phase change happens, deposition and interaction between the droplet and substrate are determined (Orme and Huang 1997; Schiaffino and Sonin 1997). Another challenge in the material jetting process is related to printing droplets on the previously built layer and binding the printed layers (Gao and Sonin 1994; Tay and Edirisinghe 2001). Delamination is a potential problem in this area that can be solved by using a tool to plane the surface in an aperiodic manner (Gothait 2005; Sanders et al. 1996; Thayer et al. 2001, 2005). Nozzles often become clogged during printing because of the small size of their opening. Some approaches that can be applied in this case are discussed in Sanders et al. (1996). Some machines overcome this problem by scheduling a periodic cleaning cycle (Thayer et al. 2001) or by using sensors to detect and compensate for nozzle blockage and inconsistency (Bedal and Bui 2002; Gothait 2001).

The transformation of a liquid into discrete droplets is very important in material jetting technology, as the process is dependent on material characteristics, such that a small change related to the material can result in a significant change in droplet formation (Furbank and Morris 2004). There are two main types of droplets formation and expulsion methods: continuous stream and drop-on-demand, as shown in Exhibit 12.10.

EXHIBIT 12.10 Droplet formation and expulsion. (a) Schematic of continuous stream printing system (b) Schematic of drop-on-demand printing system

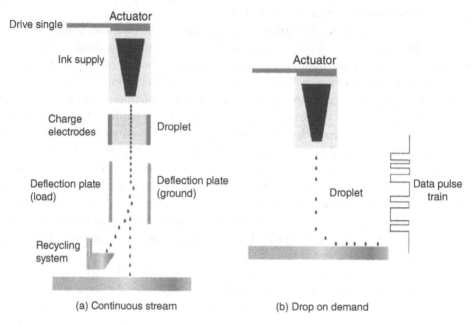

(a) Continuous stream   (b) Drop on demand

In continuous stream mode, the fluid in the reservoir is under a continuous and consistent pressure, causing a continuous stream of fluid to flow through the nozzle. As a result of Rayleigh instability, this stream is converted to small droplets, which are induced by an electrical field to gain charge. The charged droplets are directed by a deflection field to a desired location on the substrate or are caught so that they can be recycled and used for another print (Tay et al. 2003). Although this approach is widely used for food and pharmaceutical labeling (de Gans, Duineveld, et al. 2004), there are two problems associated with the use of this method: it is limited to materials with the ability to carry a charge, and the recycling of expensive materials can be problematic. The second method for droplet formation and expulsion is the drop-on-demand system, where different kinds of actuators, such as thermal, piezoelectric, and electro-static, are used to produce pressure pulses, making the fluid eject from the nozzle as droplets (de Gans, Duineveld, et al. 2004; Le 1998).

In general, the parameters that affect the quality of parts printed by material jetting technology are the pressure behind the material at which droplets are expelled; the distance between the substrate and the nozzle; the nozzle diameter; and the viscosity, density, and surface tension of the material.

Materials

ModelMaker, developed by Sanders Co. in 1994, was the first commercialized machine to use material jetting technology that used a wax material to print a part. In 1996, a wax-based apparatus called Actua 2100 was developed by 3D Systems. Later in 2001, 3D Systems introduced the ThermoJet, a new version of the Actua device. All of these machines use waxy thermoplastic materials to print structures (Wohlers 2004).

Polymers and acrylic photopolymers are used by some other commercial material jetting-based machines. Recently, researchers have focused on using acrylate photopolymers in material jetting technology, in which the print head moves in the horizontal plane and deposits photopolymer droplets while UV lamps simultaneously cure each deposited layer (Wohlers 2010). In 2000, Quadra (produced by Objet Co.) was the first machine introduced for printing with these materials. The Quadra featured a print head with more than 1,500 nozzles for jetting the photopolymers (Wohlers 2004).

Ceramic inks are used in material jetting technology to print advanced structures such as scaffolds for biomedical applications. The ink, in which ceramic particles (e.g., zirconium oxide) are suspended in a solvent, is injected from the nozzle and deposited onto the substrate. The ink is solidified by evaporation of the liquid part or, if a wax-based ceramic ink is used, the ink is melted and solidified by deposition on a cold substrate (Ngo et al. 2018). Research work on ceramic inks has focused on effects of the space between droplets, the substrate material, the space between lines, and other factors that may affect the deposition of ceramics (Tay and Edirisinghe 2001). Printing a suspension of ceramic particles in a wax carrier is another method developed to create 3D parts by material jetting; parts created by this method need to be sintered to achieve a higher ceramic density, and shrinkage may occur (Ainsley et al. 2002).

In the 2010s, Vader Systems introduced a process to print metal droplets using aluminum alloys, and XJet developed a method to print parts using metal or ceramic nanoparticles embodied in colloids. Metals are used in material jetting machines to print electronics, and a solder with a low melting point has been applied to this process (Q. Liu and Orme 2001a). Yamaguchi and coworkers used a metal alloy of Bi, Pb, Sn, and Cd to print parts using a material jetting machine (Yamaguchi 2003; Yamaguchi et al. 2000). Other research has focused on aluminum deposition (Q. Liu and Orme 2001b). However, for certain metals, the high temperatures needed to melt the metal can damage the printing system.

Advantages and Disadvantages

Material jetting technology has several advantages over other processes. It enables structures with high speed and high resolution to be fabricated, and it can be used to print materials in multiple colors (Petrovic et al. 2011; Singh 2011). As this printing method uses droplets, it can reduce material waste (Tofail et al. 2018). In addition, parts with multiple materials can be printed in a fast and cost-effective manner (Guo and Leu 2013; Ngo et al. 2018).

The main drawbacks of this process include nozzle blockage and costly ink cartridges (S. H. Huang et al. 2013). In the case of solution and dispersion deposition, inks with a volatile solvent can clog the nozzle as particles frequently precipitate in the nozzle (de Gans, Kazancioglu, et al. 2004), while low concentrations of materials suspended in solutions can lead to weak parts. Only a limited number of wax and photopolymer materials can be used, and only a small range of metals and ceramics can be printed in material jetting machines.

Binder Jetting

Process

Binder jetting is a powder-based additive manufacturing technology in which partial cross sections are built by jetting a liquid binder selectively on a thin layer of powdered material (Tofail et al. 2018). In this process, powder particles on the bed are attached together to build the layer as the binder (glue) is printed. This process is also known as *three-dimensional (3D) printing* in a number of sources (Butscher et al. 2012; Dimitrov et al. 2006; M. Lee et al. 2005; Melican et al. 2001; Sachs et al. 1990; Sachs et al. 1993; Seitz et al. 2005) due to its similarity to the inkjet process applied in two-dimensional printers (Cooper 2001; Halloran et al. 2011). Note that nowadays 3D printing is a switchable term for additive manufacturing, although it was used for binder jetting.

As mentioned in the history section, binder jetting technology was developed by a group at MIT in 1993 and was licensed to a number of companies, including ExOne and Z Corp. A new technique also was introduced by voxeljet AG that is capable of building parts of an infinite length in a continuous printing mode. In this technology, a sloping plane is used as a build bed (voxeljet AG 2020).

In the binder jet system shown in the schematic diagram in Exhibit 12.11, a powder feed piston controls the amount of the material needed for building each layer on the build platform. A leveling roller is used to spread the powder evenly over the bed. Binder is then applied by the print head to the powder. After printing one layer, a build piston lowers the building bed by a distance equivalent to a one-layer thickness, and a new layer of powder is added. After printing, the part must be separated from the building bed. Any remaining powder needs to be removed (Wong and Hernandez 2012) using a brush or air pressure; this unused powder material can be recycled and reused for another print (Gibson et al. 2021).

EXHIBIT 12.11   Schematic diagram of a binder jetting system

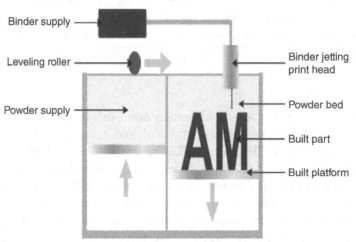

In the case of ceramic (Suwanprateeb et al., 2009; Tarafder et al. 2013) or metallic materials (especially when producing functional parts), the finished part will need to undergo extra post processes, including infiltration and/or sintering to reach the desired strength and mechanical properties (Bak 2003; Cooper 2001; Kruth 1991).

The quality of the parts built with a binder jetting system depends on the mechanical properties of the binder, the deposition velocity, the size and shape of the powder grains, the interaction between the powder and binder, and the post-processing methods (Utela et al. 2008; Wang et al. 2017). The wettability of the powder is another important factor for a successful printing: a powder with low wettability cannot be integrated strongly with the binder, while a high wettable powder will spread into other layers when binder is added (Amirkhani et al. 2012; Uhland et al. 2001).

Currently, a wide variety of binder jetting machines are available. Plaster-based powder and a water-based binder are used in low-cost machines. Similar to the systems used for material jetting, multiple nozzles can be applied to the print head, resulting in rapid deposition (Sachs et al. 1990). Binder jetting machines are inherently capable of printing color parts. In this case, some print head nozzles are allocated for printing colors while others are used to print the binder. Different binder materials can be used in binder jetting technology, making the process useful for various applications (Gibson et al. 2021).

Materials

Various groups of materials, including polymers (Cooper 2001; Halloran et al. 2011), metals, ceramics, and composites, are utilized in binder jetting technology to create parts with distinct characteristics (Guo and Leu 2013). Z Corp systems use starch-based powder and a water-based binder. Currently, commercial machines manufactured by 3D Systems use plaster-based powder and a water-based binder (3D Systems Co. 2020), and ColorBond, StrengthMax, and Salt Water Cure are materials provided by this company for infiltration (Gibson et al. 2021). voxeljet AG provides poly-methyl methacrylate powder and liquid binders for building parts (voxeljet AG 2020). Metal or sand powders are also materials used in machines manufactured by ExOne (2020).

Acrylic plastics and wax are examples of polymers that have been used in the binder jetting process (Ngo et al. 2018). When a polymer is applied for filtration, a dipping process is required. Polymers can also be used as binders for ceramic and metal parts. In this case, the fabricated part needs to be kept in a furnace for two or three cycles. The first cycle is a low-temperature cycle to burn off the polymer. In the second cycle, the metal grains are sintered together at high temperatures. The third cycle includes infiltration by an alloy that has a melting point lower than main material. The finished part will reach a density of greater than 90 percent after completing these three cycles (Gibson et al. 2021).

An example of using metallic materials in the binder jetting process is a process called *ProMetal*, in which stainless steel is used to fabricate dies and injection tools. As the parts created by binder jetting are highly porous, the printed part needs to be heated at a high temperature to fuse the steel with the binder (i.e. sintering), then infiltrated and heated with another metallic material having a low melting point (such as bronze powder) in a process called *infiltration* to increase the density and strength of the printed part (ExOne Co. 2010; Lipke et al. 2010; ProMetal RCT 2010).

Zirconia, silica, alumina, titanium silicocarbide (Ti3SiC2, with excellent mechanical and electrical characteristics), and bioceramics (such as hydroxyapatite) also are among the ceramic materials that have been investigated for use in binder jetting (Dimitrov et al. 2006; Sun et al. 2002). Metal–ceramic, polymer–metal, and ceramic–ceramic fiber-reinforced composites also have demonstrated their suitability for use in the binder jetting process (Rambo et al. 2005; Suwanprateeb et al. 2009).

Advantages and Disadvantages

Binder jetting has a number of advantages. Binder jetting machines can use a wide variety of materials to create parts (Guo and Leu 2013) and are capable of building large parts. The process does not generally require a support structure, since the powder bed supports the part during the building process – although in the case of metal parts (which require sintering), a support is required to stabilize the part while it is treated in the furnace (Gibson et al. 2021). Unlike other AM processes, there is no attachment between the part and build platform (Maleksaeedi et al. 2014). The binder jetting process is considered a high-speed printing process, as layers are formed by jetting binder droplets, which uses a small fraction of material and results in a shorter build time, although sometimes it requires post-processing. Moreover, print heads with multiple nozzles are used that cover a much larger powder bed area, which can also accelerate the building process (Bose and Bandyopadhyay 2019). Binder jetting is not considered to be an expensive AM process since low-cost materials can be used for building products (S. H. Huang et al. 2013) and no costly equipment such as lasers are required in this process (Sachs et al. 1990). Binder jetting technology, similar to material jetting, is able to print parts with different colors. More than one part can be fabricated at one time, and each part can be separated from the others by unbound powder. Another remarkable advantage of binder jetting is that the build material does not need a high temperature or any energy source to bond and form layers. In other words, the layers can be built at room temperature, which can eliminate the problems related to thermal effects such as crack and shrinkage (Gibson et al. 2021).

On the other hand, parts fabricated using this technology have high porosity, since friction between powder particles is high and there is no force to pack the particles (Maleksaeedi et al. 2014; Suwanprateeb et al. 2010). These porous structures are fragile and require post-processing that may be costly (Utela et al. 2008). This process creates a rough surface finish on the printed part.

Powder Bed Fusion

Process

As inferred from its name, powder bed fusion (PBF) is an AM process that uses a heat source to fuse powdered material spread on a fabrication platform to create parts (Tofail et al. 2018). As shown in Exhibit 12.12, the process occurs in a confined and heated

EXHIBIT 12.12　Schematic diagram of powder bed fusion

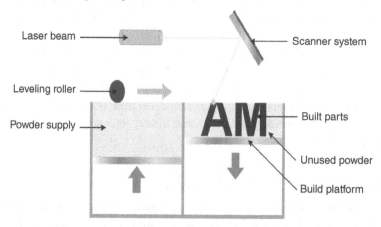

chamber that is filled with inert gas to protect the material from oxidation. Similar to the binder jetting approach, a powder feed chamber holds a build material, a leveling roller is used to distribute the powder evenly over the bed, and a build piston lowers the building bed a distance of a one-layer thickness after each layer is printed before adding a new layer of powder. A heater also is used to preheat and maintain the powder bed temperature a bit below the material melting point to prevent the part from becoming distorted and to help each layer bind better to the previously created layer (Pham and Gault 1998). At the end of the printing process, the newly fabricated part needs to remain in the building chamber to cool. This cooling period is necessary because the part can become deformed and degraded if it comes into contact with oxygen at room temperature immediately after the building process is complete (Bose and Bandyopadhyay 2019). After the part has cooled, it is removed from the building bed and any unused powder is removed (Wong and Hernandez 2012).

In some cases (e.g., when using ceramics), post-processing can include further sintering or/ and infiltration. The density of the part is an important factor when determining the method efficiency, as its density depends highly on the powder particle size and packing (Williams and Deckard 1998).

In PBF machines, any unbound powder can be recycled and reused for future PBF processes. However, since the powder properties might have changed as a result of being in a high-temperature environment and being in contact with atmospheric gases, the recycling process requires some consideration. Several recycling methods have been developed to address this issue; in one technique, a mixture of unused powder, overflow powder, and fabrication bed powder with a specific proportion is used as a build material powder (Gibson et al. 2021).

Selective laser sintering (SLS) is a powder bed fusion process in which a CO_2 laser is used to sinter the powder to build cross sections of a part. In SLS, the powder particles are not melted; they become fused when their surface temperature increases due to laser heating (Herzog et al. 2016; Matthews et al. 2017; Nie et al. 2015; Sheydaeian and Toyserkani 2018). 3D Systems in the United States and EOS GmbH in Germany are the most successful companies producing SLS machines (Guo and Leu 2013).

Selective laser melting (SLM) and electron beam melting (EBM) are two technologies derived from the SLS process. In SLM, powder grains are completely melted to bind them together for layer fabrication. Parts created using this method have relatively high density. On the other hand, heat control is the main challenge of the SLM process, as the focused high energy that is needed to melt the material can cause an increase in residual stress and layer deformation in the printed part (Abe et al. 2001; Kruth et al. 2004; Lu et al. 2000; Osakada and Shiomi 2006). In the EBM process, an electron beam is used as the energy source to melt material in a vacuum. As in SLM, the structures fabricated using this method are highly dense and strong, but the build speed in EBM is relatively faster than SLM, since electron beam energy is much denser than that of a laser beam, and the electron beam can scan faster than a laser (Cormier, Harrysson, and West 2004; Cormier et al. 2004; Harrysson et al. 2008; Heinl et al. 2007; Rännar et al. 2007).

The PBF process is controlled by many parameters related to the laser (or electron beam), the scanning process, the temperature, and the powdered material. In addition, there is a correlation between these parameters, so one parameter can be affected by another. For example, for a powder with a high melting point, a high-power laser must be used (Gibson et al. 2021).

Materials

As with the binder jetting process, a wide range of materials, including polymers, metals, ceramics, and composites, can be applied for powder bed fusion (Halloran et al. 2011; Salmoria et al. 2011; Tang et al. 2011). Polyamides (Dilip et al. 2017; Gong et al. 2017) such as nylon and styrene are among the polymers used in PBF, as this method can be used to create parts with characteristics that are similar to parts made by injection (Krznar and Dolinsek 2010; Tang et al. 2011). Polymers can also be used as binders for metal or ceramic powders, and they need to be removed by post-processing (Shahzad et al. 2014). Elastomeric thermoplastics are good candidates for fabrication of parts with high flexibility, as the mechanical properties of these materials do not change at high temperatures or when they come into contact with chemicals, and this makes them suitable for industrial applications. Polycaprolactone, polylactide, and poly-L-lactide are among the polymers that can be used for products with biomedical applications (Gibson et al. 2021).

Various metals and alloys can be used for fabricating parts using PBF processes, including stainless steels, titanium, titanium alloys, cobalt chromium (Guo and Leu 2013), and nickel alloys (Herzog et al. 2016). Sintering and infiltration parameters such as temperature and infiltrant volume need to be controlled to protect the metal parts from distortion (Gibson et al. 2021). It is important to mention that the materials used in SLM are limited to specific metals including steel and aluminum, while different polymers, metals, and alloys can be used in SLS (Ngo et al. 2018).

Powder particles are bound together to form layers by four main methods: solid-state sintering, chemically induced binding, liquid-phase sintering, and full melting (Kruth et al. 2005). In solid-state sintering, material particles are fused together instead of being melted to form layers; the sintering process depends highly on particle size, as larger particles require more time and heat to sinter. In chemically induced sintering, powder particles are bound together using a by-product from a chemical reaction between two different powders or between the powders and atmospheric gases. In liquid-phase sintering, binding occurs when a portion of the powder components melt while other portions remain solid; in this case, the melted particles bind the unmelted grains together. In full melting, which is used widely for metals and semi-crystalline polymers, all portions of the powder are melted when they are exposed to an energy beam; the printed parts are highly dense (Gibson et al. 2021).

Advantages and Disadvantages

In general, powder bed–based processes do not require support structures (except in case of metals), since unbound powder supports the part during the building process (Tofail et al. 2018). This will reduce build time, material, and energy waste (Gibson et al. 2021). A wide range of materials can be used, and any unbound powder can be recycled (Wong and Hernandez 2012). PBF can be used to fabricate parts with complex geometry and with good mechanical characteristics (Ngo et al. 2018).

The disadvantages of PBF include size limitations as well as poor surface finishes and low resolution, as this process is dependent on powder particle size (Wong and Hernandez 2012). Moreover, PBF requires a high-power energy source, which makes the process expensive, and

EXHIBIT 12.13 Schematic diagram of sheet lamination

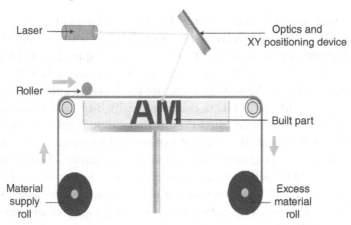

the building speed is also relatively slow (Ngo et al. 2018). Furthermore, the build time is long, as this process includes two time-consuming cycles: one cycle for preheating the build platform, and another cycle for cooling the printed part after finishing (Gibson et al. 2021).

Sheet Lamination

Process

In sheet lamination (SHL), which was among the first commercialized AM processes, a sheet or foil material passes over the build plate, and a section is cut from the sheet and bound to the previous layers to form a part. In one SHL process known as *laminated object manufacturing* (LOM), sheets of material are bound together by applying heat and pressure and are cut using a CO_2 laser to create the part (Vaupotič et al. 2006); the speed and the cutting depth of the laser are adjustable (S. H. Huang et al. 2013). As shown in Exhibit 12.12 and Exhibit 12.13, in the sheet lamination process, the sheet surface is coated with an adhesive that is activated when a high-temperature roller passes over the sheet (Liao et al. 2006; Mueller and Kochan 1999; Park et al. 2000; Weisensel et al. 2004).

Sheet lamination is accomplished by two different methods: form-then-bond and bond-then-form. As can be inferred from the names of the methods, the layers first are cut by a laser or mechanical cutter and then are attached together in the form-then-bond mechanism; in the bond-then-form method, the layer is first bonded to the substrate or previous layer and is then cut to create the part. Form-then-bond is more common for fabricating metallic or ceramic parts and, in some cases, post-processing (including heat treatment) is required to achieve certain mechanical properties (Gibson et al. 2021; Ngo et al. 2018).

Materials

Polymers, metals, ceramics, and papers are used for building parts in the sheet lamination process (Ngo et al. 2018). The first commercial LOM machine used butcher paper as a process material.

Advantages and Disadvantages

SHL has some advantages. The materials used for the sheet lamination process are not expensive (Tofail et al. 2018). The process can be used to build parts with different colors. The material can be handled without any difficulty, and no tooling is required to build the parts. Unlike most AM processes, SHL does not have issues in terms of part deformation or changes in mechanical properties during the build process, and it does not require any support structures or post-processing. In addition, SHL is considered a high-speed process since it does not require scanning of the entire cross-section – just the outline is cut to form the parts (Pham and Gault 1998). Furthermore, this method can be used to create large structures (Cooper 2001) as well as to produce functional parts and smart structures (Gibson et al. 2021).

Drawbacks of using SHL include material waste, as unused sheet material is not recyclable. It is difficult to create details and internal features in a part when using sheet lamination (Noorani 2006). In addition, the surface finish is not good, and the part requires post-processing (Ngo et al. 2018).

Directed Energy Deposition

Process

Directed energy deposition (DED) technology uses a heat source such as a laser, an electron beam, or a plasma arc to create focused energy to melt the material while it is being deposited through a nozzle to build the part. Laser deposition (Conduit et al. 2019), laser-engineered net shaping (Lewis and Schlienger 2000), electron beam melting, plasma arc melting, directed light fabrication (Lewis 1995), direct metal deposition (Atwood et al. 1998), 3D laser cladding (J. Liu and Li 2004), laser powder deposition (Costa and Vilar 2009), laser-based metal deposition (Dwivedi and Kovacevic, 2005, 2006), and laser freeform fabrication are other technologies that are based on DED. Laser characteristics such as laser type and power, material type, and the material delivery system are among the parameters that distinguish these technologies (Balla et al. 2009; Gasser et al. 2010; Hofmeister and Griffith 2001; Lewis and Schlienger 2000; K. Zhang et al. 2007).

In the diagram of the DED process shown in Exhibit 12.14, a material (in powdered form or a wire) is fed through a nozzle located near the energy source. The material changes to a molten state upon entering the molten pool created by the energy source. The molten material

EXHIBIT 12.14   Schematic diagram of directed energy deposition

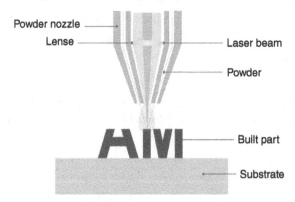

begins to solidify as soon as the machine head moves away. In this process, the cooling rate depends on the size of the molten pool and the scanning speed: the material solidifies rapidly when the molten pool is small and the print head speed is fast (Balla et al. 2008; Liao et al. 2006; Xiong 2009). A shielding gas (such as argon) is used for powder delivery to help to concentrate the material and protect it from oxidation (Bose and Bandyopadhyay 2019).

DED is also capable to print more than one material at the same time and by using multiple axes (Ngo et al. 2018).

Materials

Materials with acceptable stability in high temperatures can be used in the DED process. Metals are commonly used as build materials in DED machines; however, metals that conduct and reflect heat easily (such as gold) are not good candidate materials for DED, whereas metals capable of being welded properly are suitable. Different metallic materials, including stainless steel, tooling steel, titanium, Inconel (Ngo et al. 2018), and nickel-based alloys can be used as build materials (Cooper 2001). Ceramics are another group of materials that can be used for DED, but the process is not as easy as the one for metals. Only a limited amount of a ceramic can be melted in the molten pool, and cracks resulting from thermal change is a potential problem for the formed parts. However, ceramics in the form of a matrix composite or a metal matrix composite are suitable (Gibson et al. 2021).

Advantages and Disadvantages

Owing to the high cooling rate, DED technology can be used to create parts with unique grain structures, and DED is the best method for repairing defective industrial components. This technology can improve mechanical properties of parts by adding features and/or materials to an existing part (Tofail et al. 2018). Dense parts can be fabricated by this technology without the need for additional heat treatment (Griffith et al. 1999). Moreover, this technology provides designers with the flexibility to build a part with the desired properties by changing the process parameters such as the material composition, the feed rate, and the cooling rate (Gibson et al. 2021).

Despite these advantages, there are some drawbacks in using DED technology for the fabrication of parts. Parts created using this method are not accurate, and their surface roughness is relatively high. Low build velocity is also another disadvantage of DED (Tofail et al. 2018). Finally, this process is not a good approach for creating complex structures with fine features (Ngo et al. 2018).

Applications

Many different applications use AM technology. In this section, several interesting applications such as 3D-printed devices, 3D printing in construction, 3D bioprinting, and 4D printing are introduced.

3D-Printed Electronic Devices

Additive manufacturing of electronics such as sensors has attracted the interest of researchers in this field of study, and a considerable amount of research has been conducted to develop new methods and materials for this purpose (Espera et al. 2019). In the following paragraphs, we discuss the application of additive manufacturing for the fabrication of sensors.

Sensors play an important role in academic research studies, industrial projects, and devices we use in our daily lives. Different types of sensors have been fabricated for the various applications. In this section, we first present the basic concept of a sensor, then we categorize the 3D-printed sensors based on their applications. Finally, we review some studies on the fabrication of sensors by additive manufacturing.

Traditional manufacturing processes used for sensor fabrication such as lithography are costly, time-consuming, and less flexible (Niu et al. 2007; O'Neill et al. 2014). Additive manufacturing technology has demonstrated the capability to address these problems, as AM is an efficient tool for creating complex structures with high accuracy (Aremu et al. 2017; Tofail et al. 2018; Vilardell et al. 2019). 3D printing was used in for an electronics application in the year 2000, when a circuit was printed by inkjet technology (Sirringhaus et al. 2000). Sensors can be built by AM in two different ways: by printing a sensor with all its components or by embedding a sensor into a printed part (Khosravani and Reinicke 2020).

Sensors printed by AM technologies can be used for different engineering and medical applications. The multi-material 3D-printed pressure sensor developed by Emon et al. (2019) is an example of a highly sensitive sensor that was successfully fabricated by AM (Exhibit 12.15d). This sensor was built using a material extrusion process that employed three extruders. As illustrated in Exhibit 12.15c, the sensor consists of five stretchable layers: two insulator layers, two electrode layers, and one pressure-sensitive layer. Each intersection of two electrodes creates a sensing unit called a *taxel* (Exhibit 12.15a, b). This fabricated sensor demonstrated excellent performance and high reliability in different mechanical tests.

EXHIBIT 12.15 Schematics showing (a) top view of the 3D model, (b) side view of the 3D model, and (c) exploded diagram of the sensor. (d) Finished pressure sensor fabricated by multi-material extrusion

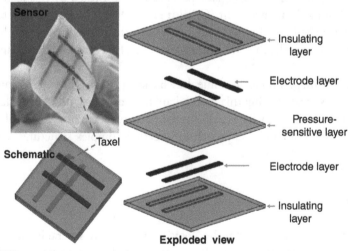

Source: Emon et al. 2019.

Particle sensors are another type of mechanical sensor that can be fabricated by AM. These sensors are used to detect specific particles in the atmosphere such as pollutants (Tang et al. 2017; Zusman et al. 2020). In Zhao et al. (2016), an impactor was designed and created for this purpose by using a material jetting process. In this case, the designers took advantage of material jetting technology to fabricate microchannels that enhance the performance of the sensor.

AM has also proved its potential for biomedical sensor applications. One example is a 3D-printed microfluidic device developed by Comina et al. (2015) to quantify glucose levels in the blood stream. This device uses the lens of a cell-phone camera for this purpose. In another study, Salvo et al. (2012) fabricated a medical electrode through the material jetting process to detect signals of electrical activity from the heart and brain.

The main challenge in the additive manufacturing of medical sensors is the limitation in the materials that can be used for printing, as the materials need to be biocompatible for most medical applications. In addition to the previously mentioned case studies, numerous other studies of the 3D printing of sensors have demonstrated the successful fabrication of other electronic components such as actuators (Zhang et al. 2019).

3D Printing in Construction

A recent report by Wohlers Associates indicates that architectural construction accounts for only a 3% share of the additive manufacturing industry (Wohlers 2017). Issues with traditional construction include workplace accidents, low efficiency in the workforce, and a lack of expertise in the workforce, and these factors are driving the demand for automated construction (Warszawski and Navon 1998). AM technology has proved its potential and capability for use in construction (Exhibit 12.16), especially in cases where creating complex structures using traditional methods is not possible. The first building that was constructed using fused deposition modeling technology was built in Amsterdam in 2014, and the use of AM technology in construction is still in its early stages. Nonetheless, AM technology is able to provide design flexibility and precise construction. Outstanding work in this area was

EXHIBIT 12.16   Examples of structural and nonstructural applications of 3D printing in construction

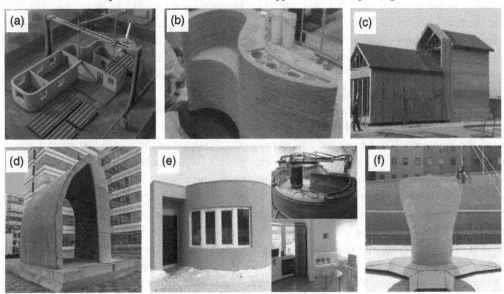

Source: Panda et al. (2018).

performed by Behrokh Khoshnevis, who developed a material extrusion–based method called Contour Crafting® and used it to construct buildings (Khoshnevis 2004). In this method, structures are made from cement and sand by robots on a crane at a construction site. In addition to construction for buildings and infrastructure on Earth, another potential application of Contour Crafting® technology is the construction of structures and habitats on the moon or other planets.

Lim et al. (2012) described some 3D printing approaches for building construction including Contour Crafting® (Khoshnevis et al. 2006), a D-shape industrial-scale 3D printer that uses a binder jetting process (D-Shape Enterprises LLC), and concrete printing that uses a material extrusion approach (Lim et al. 2009). In contrast to Contour Crafting®, parts for the D-shape and concrete printing techniques are printed off-site. Some challenges are associated with off-site projects: they require a large workforce or a large amount of work, and they generate a relatively high amount of waste material (Nadal et al. 2017). Cesaretti et al. conducted a feasibility study of the possible application of D-shape technology for construction in space (Cesaretti et al. 2014), and research in this area is ongoing.

In general, additive manufacturing technology has several advantages over traditional construction methods, as it requires less time, material, cost, and labor work to create a building, a bridge, or other structures. This technology also supports the use of recycled materials (Stoof and Pickering 2018). However, there are some limitations in applying AM to construction applications: AM can be used to build walls, roofs, and floors but not plumbing or electrical lines. Also, at this time, only a limited number of materials (including cement and hard plastics) can be used in this process (Ngo et al. 2018).

3D Bioprinting

3D bioprinting is an additive manufacturing process that uses bio-inks to fabricate 3D biological/biomimetic structures, functional tissues/organs, or biomedical devices in a layer-by-layer manner (Gu et al. 2015). The main difference between 3D bioprinting and conventional 3D printing is the material used for printing. The materials used in 3D bioprinting, which are known as *bio-inks* or *biomaterials*, include biological materials, biocompatible materials, living cells/tissues, growth factors, and biodegradable materials (Gungor-Ozkerim et al. 2018; Murphy and Atala 2014; Ozbolat 2016). 3D bioprinting has specific requirements to maintain cell viability and accurate cell deposition, including precise positioning, high resolution, and aseptic printing conditions (Murphy and Atala 2014).

As compared to conventional 3D printing, 3D bioprinting materials need to fulfill certain properties to enable successful printing to be achieved. First, the printability (i.e., the capability of the biomaterials to be printed and to maintain the shape of the printed structure), viscosity, gelation method, and rheological properties need to be studied to meet the conditions required for easy printing using a bioprinter (Naghieh and Chen 2021). Second is the biocompatibility in tissue engineering applications, where *biocompatibility* refers to the need for the implant biomaterials to coexist with endogenous tissue without causing any adverse local or systemic effects on the host (Alifui-Segbaya et al. 2017; Gungor-Ozkerim et al. 2018). The third is the mechanical/physical properties of the material, as the selection of a biomaterial should consider the desired properties of the 3D object, such as certain mechanical properties for resisting or producing force to maintain its functionality (Murphy and Atala 2014). Proper materials may include soft hydrogels for biocompatibility and thermoplastic polymers for intensity/strength (Holzl et al. 2016). More complex biopolymers and nanomaterials and composites are used in the fabrication of tissues and organs. Polysaccharides, polyesters, and composites based on proteins and nucleic acid have been printed and used as implants, artificial bone, artificial skin, and tissue scaffolds due to its nontoxicity, biodegradability, and biocompatibility (Gungor-Ozkerim et al. 2018). These materials enable the fabrication of scaffolds having high mechanical strength and porosity.

Exhibit 12.17 shows the overall process of 3D bioprinting of vascular model.

EXHIBIT 12.17   3D bioprinting vascular model

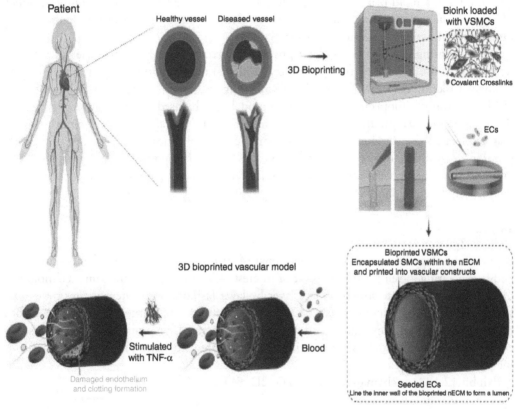

Source: Gold et al. 2021.

Several additive manufacturing technologies have been applied in 3D bioprinting, including material extrusion, material jetting, vat photopolymerization, and other processes (Hüseyin et al. 2017). Material jetting was the first technology applied for 3D bioprinting, where bioink droplets such as hydrogels with or without cells stored in a printing head were selectively dispensed onto a printing bed (Gibson et al. 2021). During the dispensing process, the dispensed droplets of biomaterial are controlled by pressure/force (generated by heat, piezoelectricity, acoustic waves, etc.). Material jetting has great advantages in 3D bioprinting due to its precise control over the dispensing of the bio-ink and the high resolution of the printed structures (Y. Huang et al. 2017; Vanaei et al. 2021).

Material extrusion can be used to print bio-ink materials with higher viscosity than those used for material jetting. In the material extrusion process, biomaterials are selectively dispensed via a nozzle with different extrusion methods, including extrusion with a screw, pneumatic extrusion, and extrusion with a piston (Mandrycky et al. 2016). Most hydrogel materials and high-density cell materials can be printed using the material extrusion method. However, the greater extrusion pressure will reduce the survival of cells and will limit the resolution of the print (Pati et al. 2015).

Vat photopolymerization is a type of 3D bioprinting in which the cells and hydrogel liquid photopolymers in a tank are used for printing and where the photopolymers are selectively cured by light (Ng et al. 2020). Vat photopolymerization allows the printing of parts at a high resolution and a rapid printing speed, and it also enables the fabrication of complex, hollow 3D objects such as structures that mimic native complex tissues/organs or mimic the branching systems of native vasculature (Holzl et al. 2016).

3D bioprinting can also be used for printing structures for regenerative therapies such as scaffolds, tissues/organs for transplantation, and stimulus-responsive regenerative scaffolds (Gungor-Ozkerim et al. 2018), and 3D cell culturing is another an important application area of 3D bioprinting. Bioprinted 3D organoid models can be used for drug screening, as functional tissue for therapy, and for other purposes (Knowlton et al. 2015). 3D bioprinting has also accelerated the development of organ-on-a-chip applications used in testing drugs by printing microfluidic devices that mimic the function and responses of native organs/tissue, precisely introducing cells into the microfluidic devices, and recreating the complex tissue microenvironment (Mandrycky et al. 2016).

4D Printing

A new technology known as four-dimensional (4D) printing has evolved from 3D printing and is generating a tremendous amount of interest. 4D printing uses the same technology as 3D printing to fabricate structures in a layer-by-layer fashion, but it incorporates time as the fourth dimension (Momeni et al. 2017). In 4D printing, the printed structures transform themselves such that they exhibit different forms, functions, or properties when exposed to environmental stimuli such as temperature, humidity, pressure, light, pH, water, or other stimuli (Choi et al. 2015).

Exhibit 12.18 is a schematic showing 1D, 2D, 3D, and 4D.

EXHIBIT 12.18   Schematic of four dimensions

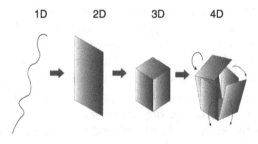

4D printing was first conceptualized in 2013 by Skylar Tabbis at MIT's Self-Assembly Lab, where a static 3D-printed structure using smart materials would transform its shape over time when immersed in water (Tibbits 2014). With the rapid development of 3D printing technology and the development of smart materials, it has become possible to print a broad range of materials that can be produced for self-assembly with 4D printing (Bajpai et al. 2020). Later, a more comprehensive definition of 4D printing was formulated to indicate a 3D-printed structure whose shape, properties, and function can change over time when exposed to predetermined stimuli, where the fourth dimension is time (Momeni et al. 2017).

Exhibit 12.19 shows 4D printing applications of hydrogels and composites stimulated by water (Kuang et al. 2019).

The most commonly used materials for 3D printing are polymers, metals, ceramics, and composites. However, the majority of these materials cannot be used for 4D printing because they lack a response to external stimuli. One of the very general methods to achieve 4D printing is to use 3D-printed smart materials or intelligent materials. These smart materials must be able to react to stimuli and be programmable so that they can form the desired shape (Chu et al. 2020). Shape memory polymers (SMPs), liquid crystal elastomers (LCEs), hydrogels, and multi-materials are the most widely used smart materials for 4D printing (Champeau et al. 2020).

SMPs are polymeric smart materials that are capable of shape transformation when stimulated by temperature, light, electricity, the presence of chemicals, or other stimuli (Kuang et al. 2019). For example, thermally induced SMPs can shift their shape in a glass-transition process. With reprogramming, the SMPs can be heated to a temperature above their transition temperature or melting temperature and recover to their initial shape. For LCEs, which are another class of smart materials, liquid crystal properties are combined with elastic properties, and the desired shape is programmed by controlling the orientation of the liquid crystal (Chu et al. 2020). When exposed to a certain stimulus (temperature, light, or another stimulus), LCEs undergo transformation and have the ability to transform and recover their shape rapidly, and the transition between the liquid crystal state and the isotropic state occurs in response to the stimuli. Another category of smart materials includes hydrogels, which are polymers capable of absorbing large amounts of water without being dissolved due to the presence of a crosslinking chain in its structure (Champeau et al. 2020).

SMPs materials that have a thermal response (i.e., they change shape in response to temperature) can be used as filaments in the material extrusion process (Nguyen and Kim 2020). SMP materials that respond to light by changing their polymer chains can be used to 4D-print materials in the vat photopolymerization method, as a response in terms of color, shape, or surface pattern can be triggered by light (Momeni et al. 2017). SMPs that have a response to water can be used as droplets in the material jetting method, and the printed 3D structures will fold and expand when they are immersed in water.

In 4D printing, stimuli or triggers are needed to start the transition of the shape, property, and function of the printed object. These stimuli (one or more) can be categorized into three categories: physical signals, chemical signals, and biological signals (Lui et al. 2019). Physical signals include temperature, light, humidity, water, an electrical field, a magnetic field, or other stimuli (Ma et al. 2019), and these stimulus signals determine which material(s) can be used for printing. For example, water stimulation is based on the principle that the response

EXHIBIT 12.19 4D printing applications

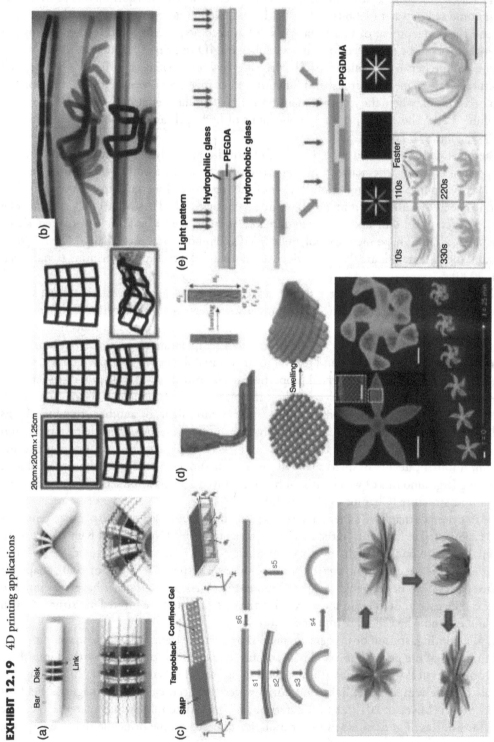

of a smart material to water can trigger a change in shape, driven by the different swelling ratios between smart (expandable) materials and rigid materials such that the smart materials will expand while the rigid materials will not (Tibbits et al. 2014). Examples of chemical signals include pH (acidity or basicity of a solution) and ionic concentration. For the pH stimulus, the smart materials will expand in response to a certain pH levels and shrink in response to other pH levels (Momeni et al. 2017). Biological signals such as the presence of certain enzymes or the presence or level of glucose can stimulus a change in shape or function of the 4D-printed material.

Summary

The advantages of additive manufacturing include minimal material waste, the capability for printing parts having complex geometries, and options for customization that make it surpass traditional manufacturing methods. This technology first appeared in 1986 with the development of stereolithography. Since then, other AM technologies using new materials and for new applications have been developed. Vat photopolymerization mainly uses photopolymers to create parts; while this method can be used to fabricate structures with high resolution, it has material limitations to photopolymers. Of the available AM technologies, material extrusion is most commonly used to build parts, as numerous low-cost (or desktop) machines are available in the commercial market. Since various materials with different mechanical properties can be used for this process, this technology has paved the way for creating parts for a wide range of applications. Low-cost extrusion machines are being widely used in schools, homes, and other sites in the community (such as libraries or maker spaces). Material jetting is among the fastest AM processes, since the use of a print head with multiple nozzles enables multi-material structures to be printed. Binder jetting is similar to material jetting, but it jets droplets of a binder onto the surface of a powdered material to create a layer (rather than jetting the build material). Powder bed fusion is another AM technology, in which a high-power infrared laser sinters or melts powdered materials; as this method does not require support structures for polymers, it reduces the build time, material use, and energy waste. In addition, this process is widely used with various types of metal powders. Sheet lamination technology, in which material sheets are cut and laminated in a layer-by-layer manner, benefits the manufacturing industry in a different way, as it uses inexpensive materials and does not require support structures. Finally, the directed energy deposition process, which is based on the melting and deposition of materials, is not only used to fabricate parts with high density but can also be used to repair existing parts that may be defective. Depending on the selected AM process, different forms of materials (including liquids, pastes, powders, filaments, sheets, and wire) can be used to create parts.

Nowadays, AM are being used for many purposes. Architects use this technology to create prototypes rapidly and at low cost. AM is being used in the biomedical field due to its capability for producing patient-customized products. Products customized by AM do not increase the cost of the product like traditional manufacturing methods do, since AM eliminates the need for molding and tooling for each individual design. Based on projections by Wohlers Associates, customized products are responsible for the rising trend in the use of 3D printing. AM technology has also proved its potential and capability to be used for aerospace, automotive, construction, and many existing industries.

Sample Questions

1. Who first commercialized AM machines?
   a. Charles Hull
   b. Scott Crump
   c. Yoji Marutan
   d. Gordan Moore
2. What is the official terminology for the 3D printing defined by ASTM/ISO?
   a. 3D printing itself
   b. Additive manufacturing
   c. Rapid prototyping
   d. Solid freeform fabrication
3. What is the process involving ultraviolet (or visible) light and liquid resin contained in a vat?
   a. Material jetting
   b. Vat photopolymerization
   c. Powder bed fusion
   d. Sheet lamination
4. Explain the process difference between material jetting and binder jetting.
5. What is the most widely used process for low-cost machines?
   a. Vat photopolymerization
   b. Directed energy deposition
   c. Material extrusion
   d. Sheet lamination
6. Which process is suitable for repairing defective parts?
   a. Directed energy deposition
   b. Powder bed fusion
   c. Material extrusion
   d. Vat photopolymerization
7. Which processes do not require support structures in general?
   a. Vat photopolymerization
   b. Material jetting
   c. Material extrusion
   d. Binder jetting
8. Explain 4D printing.
9. Which process is the most suitable for producing multi-material structures?
   a. Vat photopolymerization
   b. Sheet lamination
   c. Powder bed fusion
   d. Material jetting
10. What is the most suitable process for metal materials?
    a. Vat photopolymerization
    b. Powder bed fusion
    c. Material jetting
    d. Metal deposition

References

3D Systems. (2020). 3D printers. Available at http://www.3dsystems.com/3d-printers.

3D Systems. (2014). The journey of a lifetime: Celebrating 30 years since the first 3D printed part (3D Systems). 3D Systems, Rock Hill, SC. Available at http://www.3dsystems.com/30-years-innovation.

Abe, F., Osakada, K., Shiomi, M., et al. (2001). The manufacturing of hard tools from metallic powders by selective laser melting. *Journal of Materials Processing Technology* 111: 210–213.

Agarwala, M., Weeren, R. V., Bandyopadhyay, A., et al. (1996). Fused deposition of ceramics and metals: an overview. 1996 International Solid Freeform Fabrication Symposium.

Ainsley, C., Reis, N., and Derby, B. (2002). Freeform fabrication by controlled droplet deposition of powder filled melts. *Journal of Materials Science* 37: 3155–3161.

Alifui-Segbaya, F., Varma, S., Lieschke, G. J., et al. (2017). Biocompatibility of photopolymers in 3D printing. *3D Printing and Additive Manufacturing* 4: 185–191.

Allahverdi, M., Danforth, S., Jafari, M., et al. (2001). Processing of advanced electroceramic components by fused deposition technique. *Journal of the European Ceramic Society* 21: 1485–1490.

Amirkhani, S., Bagheri, R., and Yazdi, A. Z. (2012). Effect of pore geometry and loading direction on deformation mechanism of rapid prototyped scaffolds. *Acta Materialia* 60: 2778–2789.

Andrzejewska, E. (2001). Photopolymerization kinetics of multifunctional monomers. *Progress in Polymer Science* 26: 605–665.

Aremu, A. O., Brennan-Craddock, J., Panesar, A., et al. (2017). A voxel-based method of constructing and skinning conformal and functionally graded lattice structures suitable for additive manufacturing. *Additive Manufacturing* 13: 1–13.

ASTM International. (2012). Standard terminology for additive manufacturing technologies. ASTM F2792-12a. ASTM International, West Conshohocken, PA.

ASTM International. (2016). ASTM52901-16 standard guide for additive manufacturing – general principles – requirements for purchased AM parts. ASTM International, West Conshohocken, PA.

Atwood, C., Griffith, M., Harwell, L., et al. (1998). Laser engineered net shaping (LENS™): A tool for direct fabrication of metal parts. International Congress on Applications of Lasers and Electro-Optics. Laser Institute of America, E1–E7.

Bajpai, A., Baigent, A., Raghav, S., et al. (2020). 4D printing: Materials, technologies, and future applications in the biomedical field. *Sustainability,* 12.

Bak, D. (2003). Rapid prototyping or rapid production? 3D printing processes move industry towards the latter. *Assembly Automation.*

Balla, V. K., Bose, S., and Bandyopadhyay, A. (2008). Processing of bulk alumina ceramics using laser engineered net shaping. *International Journal of Applied Ceramic Technology* 5: 234–242.

Balla, V. K., Devasconcellos, P. D., Xue, W., et al. (2009). Fabrication of compositionally and structurally graded Ti–TiO2 structures using laser engineered net shaping (LENS). *Acta Biomaterialia* 5: 1831–1837.

Bechtel, S., Bogy, D., and Talke, F. (1981). Impact of a liquid drop against a flat surface. *IBM Journal of Research and Development* 25: 963–971.

Bedal, B. J., and Bui, L. V. (2002). Method and apparatus for controlling the drop volume in a selective deposition modeling environment. Google Patents.

Bellini, A., Gü, C., Eri, S. U., et al. (2004). Liquefier dynamics in fused deposition. *Journal of Manufacturing Science and Engineering* 126: 237–246.

Berman, B. (2012). 3-D printing: The new industrial revolution. *Business Horizons* 55: 155–162.

Bose, S., and Bandyopadhyay, A. (2019). Additive manufacturing: The future of manufacturing in a flat world. In: *Additive Manufacturing* (ed. A. Bandyopadhyay and S. Bose), 451–462. Boca Raton, FL: CRC Press.

Bourell, D. L., Beaman, J. J., Leu, M. C., et al. (2009). A brief history of additive manufacturing and the 2009 roadmap for additive manufacturing: Looking back and looking ahead. *Proceedings of RapidTech*, 24–25.

Burns, M. (1993). *Automated Fabrication: Improving Productivity in Manufacturing*. Prentice-Hall.

Butscher, A., Bohner, M., Roth, C., et al. (2012). Printability of calcium phosphate powders for three-dimensional printing of tissue engineering scaffolds. *Acta Biomaterialia* 8: 373–385.

Cesaretti, G., Dini, E., De Kestelier, X., et al. (2014). Building components for an outpost on the Lunar soil by means of a novel 3D printing technology. *Acta Astronautica* 93: 430–450.

Champeau, M., Heinze, D. A., Viana, T. N., et al. (2020). 4D printing of hydrogels: A review. *Advanced Functional Materials* 30.

Chohan, J. S., Singh, R., Boparai, K. S., et al. (2017). Dimensional accuracy analysis of coupled fused deposition modeling and vapour smoothing operations for biomedical applications. *Composites Part B: Engineering* 117: 138–149.

Choi, J., Kwon, O.-C., Jo, W., et al. (2015). 4D printing technology: A review. *3D Printing and Additive Manufacturing* 2: 159–167.

Choi, J. W., Kim, H. C., and Wicker, R. (2011). Multi-material stereolithography. *Journal of Materials Processing Technology* 211: 318–328.

Chu, H., Yang, W., Sun, L., et al. (2020). 4D printing: A review on recent progresses. *Micromachines (Basel)* 11.

Chua, C. K., and Chou, S. M. (2003). Rapid prototyping technologies and limitations. *Computer Aided and Integrated Manufacturing Systems. Volume 3: Optimization Methods*. World Scientific.

Chua, C. K., Leong, K. F., and Lim, C. S. (1998). Liquid-based rapid prototyping systems. *Rapid Prototyping: Principles & Applications in Manufacturing*. New York: Wiley.

Ciraud, P. A. (1972). Process and device for the manufacture of any objects desired from any meltable material. FRG Disclosure Publication 2263777.

Comina, G., Suska, A., and Filippini, D. (2015). Autonomous chemical sensing interface for universal cell phone readout. *Angewandte Chemie* 127: 8832–8836.

Conduit, B., Illston, T., Baker, S., et al. (2019). Probabilistic neural network identification of an alloy for direct laser deposition. *Materials & Design* 168: 107644.

Cooper, K. (2001). *Rapid Prototyping Technology: Selection and Application*. Boca Raton, FL: CRC Press.

Cormier, D., Harrysson, O., and West, H. (2004). Characterization of H13 steel produced via electron beam melting. *Rapid Prototyping Journal*.

Cormier, D., West, H., Harrysson, O., et al. (2004). Characterization of thin walled Ti-6Al-4V components reduced via electron beam melting. 2004 International Solid Freeform Fabrication Symposium, Austin, TX, August 2–4.

Costa, L., and Vilar, R. (2009). Laser powder deposition. *Rapid Prototyping Journal* 15 (4): 264–279.

Crivello, J. (1984). Cationic polymerization—iodonium and sulfonium salt photoinitiators. *Initiators – poly-reactions – optical activity. Advances in Polymer Science* 62: 1–48.

Crivello, J. (1993). *Latest Developments in the Chemistry of Onium Salts*. Barking, UK: Elsevier.

Crivello, J. (1998). Photoinitiators for free radical cationic and anionic photopolymerization. *Surface and Coatings Technology* 168.

Crump, S. S. (1991a). Fast, precise, safe prototypes with FDM. *ASME, PED* 50: 53–60.

Crump, S. S. (1991b). Fused deposition modeling (FDM): putting rapid back into prototyping. The 2nd International Conference on Rapid Prototyping. Dayton, OH, 354–357.

Crump, S. S. (1992). Apparatus and method for creating three-dimensional objects. Google Patents.

de Gans, B. J., Duineveld, P. C., and Schubert, U. S. (2004). Inkjet printing of polymers: State of the art and future developments. *Advanced Materials* 16: 203–213.

de Gans, B. J., Kazancioglu, E., Meyer, W., et al. (2004). Ink-jet printing polymers and polymer libraries using micropipettes. *Macromolecular Rapid Communications* 25: 292–296.

Derdak, T., and Pederson, J. P. (2005). *International Directory of Company Histories*, Vol. 67. St. James, MO: St. James Press.

Dilip, J., Miyanaji, H., Lassell, A., et al. 2017. A novel method to fabricate TiAl intermetallic alloy 3D parts using additive manufacturing. *Defence Technology* 13: 72–76.

Dimitrov, D., Schreve, K., and de Beer, N. (2006). Advances in three dimensional printing – state of the art and future perspectives. *Rapid Prototyping Journal* 12 (3).

Dwivedi, R., and Kovacevic, R. (2005). Process planning for multi-directional laser-based direct metal deposition. *Proceedings of the Institution of Mechanical Engineers, Part C: Journal of Mechanical Engineering Science* 219: 695–707.

Dwivedi, R., and Kovacevic, R. (2006). An expert system for generation of machine inputs for laser-based multi-directional metal deposition. *International Journal of Machine Tools and Manufacture* 46: 1811–1822.

Emon, M. O. F., Alkadi, F., Philip, D. G., et al. (2019). Multi-material 3D printing of a soft pressure sensor. *Additive Manufacturing* 28: 629–638.

EOS GmbH. (2021). 30 years of success: Our story. Available at https://www.eos.info/en/about-us/history.

Espera, A. H., Dizon, J. R. C., Chen, Q., et al. (2019). 3D-printing and advanced manufacturing for electronics. *Progress in Additive Manufacturing*, 1–23.

ExOne Co. (2010). 3D metal printing. Available at http://www.youtube.com/watch?v=i6Px6RSL9Ac&feature=related.

ExOne Co. (2020). https://www.exone.com.

Furbank, R. J., and Morris, J. F. (2004). An experimental study of particle effects on drop formation. *Physics of Fluids* 16: 1777–1790.

Gao, F., and Sonin, A. A. (1994). Precise deposition of molten microdrops: The physics of digital micro-fabrication. *Proceedings of the Royal Society of London. Series A: Mathematical and Physical Sciences* 444: 533–554.

Gasser, A., Backes, G., Kelbassa, I., et al. (2010). *Laser Additive Manufacturing: Laser Metal Deposition (LMD) and Selective Laser Melting (SLM) in Turbo-Engine Applications*. Wiley Online Library.

Gebhardt, A. (2011). *Understanding Additive Manufacturing*. Hanser Publications.

Gibson, I., Rosen, D. W., Stucker, B., et al. (2021). *Additive Manufacturing Technologies*. New York: Springer.

Gold, K. A., Saha, B., Rajeeva Pandian, N. K., et al. (2021). 3D bioprinted multicellular vascular models. *Advanced Healthcare Materials*, e2101141.

Gong, H., Dilip, J., Yang, L., et al. (2017). Influence of small particles inclusion on selective laser melting of Ti-6Al-4V powder. IOP Conference Series: Materials Science and Engineering, 012024. IOP Publishing.

Gothait, H. (2001). Apparatus and method for three dimensional model printing. Google Patents.

Gothait, H. (2005). System and method for three dimensional model printing. Google Patents.

Griffith, M., Schlienger, M., Harwell, et al. (1999). Understanding thermal behavior in the LENS process. *Materials & Design* 20: 107–113.

Gu, Q., Hao, J., Lu, et al. (2015). Three-dimensional bio-printing. *Science China Life Sciences* 58: 411–419.

Gungor-Ozkerim, P. S., Inci, I., Zhang, Y. S., et al. (2018). Bioinks for 3D bioprinting: An overview. *Biomaterials Science* 6: 915–946.

Guo, N., and Leu, M. C. (2013). Additive manufacturing: Technology, applications and research needs. *Frontiers of Mechanical Engineering* 8: 215–243.

Hageman, H. (1989). Photoinitiators and photoinitiation mechanisms of free-radical polymerisation processes. *Photopolymerisation and Photoimaging Science and Technology*. Springer.

Halloran, J. W., Tomeckova, V., Gentry, S., et al. (2011). Photopolymerization of powder suspensions for shaping ceramics. *Journal of the European Ceramic Society* 31: 2613–2619.

Harrysson, O. L., Cansizoglu, O., Marcellin-Little, D. J., et al. (2008). Direct metal fabrication of titanium implants with tailored materials and mechanical properties using electron beam melting technology. *Materials Science and Engineering: C* 28: 366–373.

Heinl, P., Müller, L., Körner, C., et al. (2008). Cellular Ti-6Al-4V structures with interconnected macro porosity for bone implants fabricated by selective electron beam melting. *Acta Biomaterialia* 4: 1536–1544.

Heinl, P., Rottmair, A., Körner, C., et al. (2007). Cellular titanium by selective electron beam melting. *Advanced Engineering Materials* 9: 360–364.

Herbert, A. J. (1982). Solid object generation. *Journal of Applied Photographic Engineering* 8: 185–188.

Herzog, D., Seyda, V., Wycisk, E. et al. (2016). Additive manufacturing of metals. *Acta Materialia* 117: 371–392.

Hofmeister, W., and Griffith, M. (2001). Solidification in direct metal deposition by LENS processing. *JOM* 53: 30–34.

Holzl, K., Lin, S., Tytgat, L., et al. (2016). Bioink properties before, during and after 3D bioprinting. *Biofabrication* 8: 032002.

Housholder, R. (1981). Molding process. *US Patent* 4,247,508.

Huang, R., El Rassi, J., Kim, M., et al. (2021). Material extrusion and sintering of binder-coated zirconia: Comprehensive characterizations. *Additive Manufacturing*, 102073.

Huang, S. H., Liu, P., Mokasdar, A., et al. (2013). Additive manufacturing and its societal impact: a literature review. *The International Journal of Advanced Manufacturing Technology* 67: 1191–1203.

Huang, Y., Zhang, X. F., Gao, G., et al. (2017). 3D bioprinting and the current applications in tissue engineering. *Biotechnology Journal*, 12.

Hull, C. (1988). StereoLithography: Plastic prototypes from CAD data without tooling. *Modern Casting* 78: 38.

Hüseyin, A., Güzel, F. D., Salim, E., et al. (2017). Recent advances in organ-on-a-chip technologies and future challenges: A review. *Turkish Journal of Chemistry* 42: 587–610.

Khoshnevis, B. (2004). Automated construction by contour crafting – related robotics and information technologies. *Automation in Construction* 13: 5–19.

Khoshnevis, B., Hwang, D., Yao, K.-T., et al. (2006). Mega-scale fabrication by contour crafting. *International Journal of Industrial and Systems Engineering* 1: 301–320.

Khosravani, M. R., and Reinicke, T. (2020). 3D-printed sensors: Current progress and future challenges. *Sensors and Actuators A: Physical* 305: 111916.

Knowlton, S., Onal, S., Yu, C. H., et al. (2015). Bioprinting for cancer research. *Trends in Biotechnology* 33: 504–513.

Kodama, H. (1981). Automatic method for fabricating a three-dimensional plastic model with photo-hardening polymer. *Review of Scientific Instruments* 52: 1770–1773.

Kruth, J.-P. (1991). Material incress manufacturing by rapid prototyping techniques. *CIRP Annals* 40: 603–614.

Kruth, J.-P., Froyen, L., Van Vaerenbergh, J., et al. (2004). Selective laser melting of iron-based powder. *Journal of Materials Processing Technology* 149: 616–622.

Kruth, J.-P., Leu, M.-C., and Nakagawa, T. (1998). Progress in additive manufacturing and rapid prototyping. *CIRP Annals* 47: 525–540.

Kruth, J.-P., Mercelis, P., Van Vaerenbergh, J., et al. (2005). Binding mechanisms in selective laser sintering and selective laser melting. *Rapid Prototyping Journal* 11 (1).

Krznar, M., and Dolinsek, S. (2010). Selective laser sintering of composite materials technologies. *Annals of DAAAM & Proceedings*, Annual 2010.

Kuang, X., Roach, D. J., Wu, J., et al. (2019). Advances in 4D printing: Materials and applications. *Advanced Functional Materials* 29: 1805290.

Le, H. P. (1998). Progress and trends in ink-jet printing technology. *Journal of Imaging Science and Technology* 42: 49–62.

Lee, J., Lu, Y., Kashyap, S., et al. (2018). Liquid bridge microstereolithography. *Additive Manufacturing* 21: 76–83.

Lee, M., Dunn, J. C., and Wu, B. M. (2005). Scaffold fabrication by indirect three-dimensional printing. *Biomaterials* 26: 4281–4289.

Lewis, G. (1995). Direct laser metal deposition process fabricates near-net-shape components rapidly. *Materials Technology* 10: 51–54.

Lewis, G. K., and Schlienger, E. (2000). Practical considerations and capabilities for laser assisted direct metal deposition. *Materials & Design* 21: 417–423.

Li, J., Monaghan, T., Nguyen, T., et al. (2017). Multifunctional metal matrix composites with embedded printed electrical materials fabricated by ultrasonic additive manufacturing. *Composites Part B: Engineering* 113: 342–354.

Liao, Y., Li, H., and Chiu, Y. (2006). Study of laminated object manufacturing with separately applied heating and pressing. *The International Journal of Advanced Manufacturing Technology* 27: 703–707.

Lim, S., Buswell, R. A., Le, T. T., et al. (2012). Developments in construction-scale additive manufacturing processes. *Automation in Construction* 21: 262–268.

Lim, S., Le, T., Webster, J., et al. (2009). Fabricating construction components using layered manufacturing technology. Global Innovation in Construction Conference, Loughborough University, 512–520.

Lipke, D. W., Zhang, Y., Liu, Y., et al. (2010). Near net-shape/net-dimension ZrC/W-based composites with complex geometries via rapid prototyping and displacive compensation of porosity. *Journal of the European Ceramic Society* 30: 2265–2277.

Liu, J., and Li, L. (2004). In-time motion adjustment in laser cladding manufacturing process for improving dimensional accuracy and surface finish of the formed part. *Optics & Laser Technology* 36: 477–483.

Liu, Q., and Orme, M. (2001a). High precision solder droplet printing technology and the state-of-the-art. *Journal of Materials Processing Technology* 115: 271–283.

Liu, Q., and Orme, M. (2001b). On precision droplet-based net-form manufacturing technology. *Proceedings of the Institution of Mechanical Engineers, Part B: Journal of Engineering Manufacture* 215: 1333–1355.

Lou, A., and Grosvenor, C. (2012). *Selective Laser Sintering: Birth of an Industry*. Austin, TX: University of Austin Cockrell School of Engineering.

Lu, L., Fuh, J., Chen, Z., et al. 2000. In situ formation of TiC composite using selective laser melting. *Materials Research Bulletin* 35: 1555–1561.

Lu, L., Fuh, J., Nee, A., et al. (1995). Origin of shrinkage, distortion and fracture of photopolymerized material. *Materials Research Bulletin* 30: 1561–1569.

Lui, Y. S., Sow, W. T., Tan, L. P., et al. (2019). 4D printing and stimuli-responsive materials in biomedical aspects. *Acta Biomaterialia* 92: 19–36.

Ma, S., Zhang, Y., Wang, M., et al. (2019). Recent progress in 4D printing of stimuli-responsive polymeric materials. *Science China Technological Sciences* 63: 532–544.

Maleksaeedi, S., Eng, H., Wiria, F., et al. (2014). Property enhancement of 3D-printed alumina ceramics using vacuum infiltration. *Journal of Materials Processing Technology* 214: 1301–1306.

Mandrycky, C., Wang, Z., Kim, K., et al. (2016). 3D bioprinting for engineering complex tissues. *Biotechnology Advances* 34: 422–434.

Markforged. (2020). https://markforged.com/materials.

Matthews, M. J., Guss, G., Drachenberg, D. R., et al. (2017). Diode-based additive manufacturing of metals using an optically-addressable light valve. *Optics Express* 25: 11788–11800.

Melican, M. C., Zimmerman, M. C., Dhillon, M. S., et al. (2001). Three-dimensional printing and porous metallic surfaces: A new orthopedic application. *Journal of Biomedical Materials Research: An Official Journal of the Society for Biomaterials, the Japanese Society for Biomaterials, and the Australian Society for Biomaterials and the Korean Society for Biomaterials* 55: 194–202.

Mohamed, O. A., Masood, S. H., and Bhowmik, J. L. (2015). Optimization of fused deposition modeling process parameters: A review of current research and future prospects. *Advances in Manufacturing* 3: 42–53.

Momeni, F., Hassani, S. M., Liu, X., et al. (2017). A review of 4D printing. *Materials & Design* 122: 42–79.

Mueller, B., and Kochan, D. (1999). Laminated object manufacturing for rapid tooling and pattern-making in foundry industry. *Computers in Industry* 39: 47–53.

Murphy, S. V., and Atala, A. (2014). 3D bioprinting of tissues and organs. *Nature Biotechnology* 32: 773–85.

Nadal, A., Pavón, J., and Liébana, O. (2017). 3D printing for construction: A procedural and material-based approach. *Informes de la Construcción* 69: e193.

Naghieh, S., and Chen, D. (2021). Printability – a key issue in extrusion-based bioprinting. *Journal of Pharmaceutical Analysis* (February).

Nano Dimension USA. (2021). The evolution of 3D printing. [Blog post]. Sunrise, FL: Nano Dimension USA. Available at https://www.nano-di.com/blog/the-evolution-of-3d-printing.

Ng, W. L., Lee, J. M., Zhou, M., et al. (2020). Vat polymerization-based bioprinting – process, materials, applications and regulatory challenges. *Biofabrication* 12, 022001.

Ngo, T. D., Kashani, A., Imbalzano, G., et al. (2018). Additive manufacturing (3D printing): A review of materials, methods, applications and challenges. *Composites Part B: Engineering* 143: 172–196.

Nguyen, T. T., and Kim, J. (2020). 4D-printing – Fused deposition modeling printing and PolyJet printing with shape memory polymers composite. *Fibers and Polymers* 21: 2364–2372.

Nie, B., Yang, L., Huang, H., et al. (2015). Femtosecond laser additive manufacturing of iron and tungsten parts. *Applied Physics A* 119: 1075–1080.

Niu, X., Peng, S., Liu, L., et al. (2007). Characterizing and patterning of PDMS-based conducting composites. *Advanced Materials* 19: 2682–2686.

Noorani, R. (2006). *Rapid Prototyping: Principles and Applications*. Hoboken, NJ: Wiley.

O'Neill, P. F., Ben Azouz, A., Vazquez, M., et al. (2014). Advances in three-dimensional rapid prototyping of microfluidic devices for biological applications. *Biomicrofluidics* 8: 052112.

On3DPrinting.com. (September 17, 2017). Inventor of 3D printing Scott Crump: "My Dreams Started in a Garage." Available from http://on3dprinting.com/2013/09/17/inventor-of-3d-printing-scott-crump-my-dreams-started-in-a-garage/.

Orme, M., and Huang, C. (1997). Phase change manipulation for droplet-based solid freeform fabrication. *Journal of Heat Transfer* 119 (4).

Osakada, K., and Shiomi, M. (2006). Flexible manufacturing of metallic products by selective laser melting of powder. *International Journal of Machine Tools and Manufacture* 46: 1188–1193.

Ozbolat, I. T. (2016). *3D Bioprinting: Fundamentals, Principles and Applications*. Academic Press.

Panda, B., Tay, Y. W. D., Paul, S. C., et al. (2018). Current challenges and future potential of 3D concrete printing. *Materialwissenschaft und Werkstofftechnik* 49: 666–673.

Park, J., Tari, M. J., and Hahn, H. T. (2000). Characterization of the laminated object manufacturing (LOM) process. *Rapid Prototyping Journal*.

Pasandideh-Fard, M., Qiao, Y., Chandra, S., et al. (1996). Capillary effects during droplet impact on a solid surface. *Physics of Fluids* 8: 650–659.

Pati, F., Jang, J., Lee, J. W., et al. (2015). Extrusion bioprinting. *Essentials of 3D Biofabrication and Translation*. Elsevier.

Petrovic, V., Vicente Haro Gonzalez, J., Jordá Ferrando, O., et al. (2011). Additive layered manufacturing: Sectors of industrial application shown through case studies. *International Journal of Production Research* 49: 1061–1079.

Pham, D. T., and Gault, R. S. (1998). A comparison of rapid prototyping technologies. *International Journal of Machine Tools and Manufacture* 38: 1257–1287.

ProMetal RCT. (2010). ProMetal RCT rapid prototyping and digital sand casting services. Available at http://www.youtube.com/watch?v=Z8MaVaqNr3U.

Rambo, C., Travitzky, N., Zimmermann, K., et al. (2005). Synthesis of TiC/Ti–Cu composites by pressureless reactive infiltration of TiCu alloy into carbon preforms fabricated by 3D-printing. *Materials Letters* 59: 1028–1031.

Rangarajan, S., Qi, G., Venkataraman, N., et al. (2000). Powder processing, rheology, and mechanical properties of feedstock for fused deposition of Si3N4 ceramics. *Journal of the American Ceramic Society* 83: 1663–1669.

Rännar, L. E., Glad, A., and Gustafson, C. G. (2007). Efficient cooling with tool inserts manufactured by electron beam melting. *Rapid Prototyping Journal*.

Reis, N., Seerden, K., Derby, B., et al. (1998). Direct inkjet deposition of ceramic green bodies: II–jet behaviour and deposit formation. *MRS Online Proceedings Library (OPL)*, 542.

Sachs, E., Cima, M., and Cornie, J. (1990). Three-dimensional printing: Rapid tooling and prototypes directly from a CAD model. *CIRP Annals* 39: 201–204.

Sachs, E., Cima, M., Cornie, J., et al. (1993). Three-dimensional printing: The physics and implications of additive manufacturing. *CIRP Annals* 42: 257–260.

Salmoria, G. V., Paggi, R. A., Lago, A., et al. (2011). Microstructural and mechanical characterization of PA12/MWCNTs nanocomposite manufactured by selective laser sintering. *Polymer Testing* 30: 611–615.

Salvo, P., Raedt, R., Carrette, E., et al. (2012). A 3D printed dry electrode for ECG/EEG recording. *Sensors and Actuators A: Physical* 174: 96–102.

Sanders, R. C. Jr., Forsyth, J. L., and Philbrook, K. F. (1996). 3-D model maker. Google Patents.

Sandia National Laboratories. (December 4, 1997). Creating a complex metal part in a day is a goal of commercial consortium [News Release]. Albuquerque, NM: Sandia National Laboratories. Available at http://www.sandia.gov/media/lens.htm.

Schiaffino, S., and Sonin, A. A. (1997). Molten droplet deposition and solidification at low Weber numbers. *Physics of Fluids* 9: 3172–3187.

Seitz, H., Rieder, W., Irsen, S., et al. (2005). Three-dimensional printing of porous ceramic scaffolds for bone tissue engineering. *Journal of Biomedical Materials Research Part B: Applied Biomaterials: An Official Journal of the Society for Biomaterials, the Japanese Society for Biomaterials, and the Australian Society for Biomaterials and the Korean Society for Biomaterials* 74: 782–788.

Shahzad, K., Deckers, J., Zhang, Z., et al. (2014). Additive manufacturing of zirconia parts by indirect selective laser sintering. *Journal of the European Ceramic Society* 34: 81–89.

Sheydaeian, E., and Toyserkani, E. (2018). A new approach for fabrication of titanium-titanium boride periodic composite via additive manufacturing and pressure-less sintering. *Composites Part B: Engineering* 138: 140–148.

Singh, G., Missiaen, J. M., Bouvard, D., et al. (2021). Copper extrusion 3D printing using metal injection moulding feedstock: Analysis of process parameters for green density and surface roughness optimization. *Additive Manufacturing* 38: 101778.

Singh, R. (2011). Process capability study of polyjet printing for plastic components. *Journal of Mechanical Science and Technology* 25: 1011–1015.

Sirringhaus, H., Kawase, T., Friend, R. H., et al. (2000). High-resolution inkjet printing of all-polymer transistor circuits. *Science* 290: 2123–2126.

Stansbury, J. W., and Idacavage, M. J. (2016). 3D printing with polymers: Challenges among expanding options and opportunities. *Dental Materials* 32: 54–64.

Stoof, D., and Pickering, K. (2018). Sustainable composite fused deposition modelling filament using recycled pre-consumer polypropylene. *Composites Part B: Engineering* 135: 110–118.

Sun, W., Dcosta, D., Lin, F., et al. (2002). Freeform fabrication of Ti3SiC2 powder-based structures: Part I – Integrated fabrication process. *Journal of Materials Processing Technology* 127: 343–351.

Suwanprateeb, J., Sanngam, R., and Panyathanmaporn, T. (2010). Influence of raw powder preparation routes on properties of hydroxyapatite fabricated by 3D printing technique. *Materials Science and Engineering: C* 30: 610–617.

Suwanprateeb, J., Sanngam, R., Suvannapruk, W., et al. (2009). Mechanical and in vitro performance of apatite–wollastonite glass ceramic reinforced hydroxyapatite composite fabricated by 3D-printing. *Journal of Materials Science: Materials in Medicine* 20: 1281–1289.

Swainson, W. K. (1977). Method, medium and apparatus for producing three-dimensional figure product. U.S. Patent 4,041,476. Google Patents.

Tang, D., Zhao, R., Wang, S., et al. (2017). The simulation and experimental research of particulate matter sensor on diesel engine with diesel particulate filter. *Sensors and Actuators A: Physical* 259: 160–170.

Tang, H.-H., Chiu, M.-L., and Yen, H.-C. (2011). Slurry-based selective laser sintering of polymer-coated ceramic powders to fabricate high strength alumina parts. *Journal of the European Ceramic Society* 31: 1383–1388.

Tarafder, S., Balla, V. K., Davies, N. M., et al. (2013). Microwave-sintered 3D printed tricalcium phosphate scaffolds for bone tissue engineering. *Journal of Tissue Engineering and Regenerative Medicine* 7: 631–641.

Tay, B., and Edirisinghe, M. (2001). Investigation of some phenomena occurring during continuous ink-jet printing of ceramics. *Journal of Materials Research* 16: 373–384.

Tay, B., Evans, J., and Edirisinghe, M. (2003). Solid freeform fabrication of ceramics. *International Materials Reviews* 48: 341–370.

Thayer, J. S., Almquist, T. A., Merot, C. M., et al. (2001). Selective deposition modeling system and method. Google Patents.

Tibbits, S. (2014). 4D printing: Multi-material shape change. *Architectural Design* 84: 116–121.

Tibbits, S., McKnelly, C., Olguin, C., Dikovsky, D., and Hirsch, S. (2014). 4D printing and universal transformation. Proceedings of the 34th Annual Conference of the Association for Computer Aided Design in Architecture (ACADIA), Los Angeles, October 23–25, 539–548.

Tofail, S. A., Koumoulos, E. P., Bandyopadhyay, A., et al. (2018). Additive manufacturing: Scientific and technological challenges, market uptake and opportunities. *Materials Today* 21: 22–37.

Travitzky, N., Bonet, A., Dermeik, B., et al. (2014). Additive manufacturing of ceramic-based materials. *Advanced Engineering Materials* 16: 729–754.

Turner, B. N., Strong, R., and Gold, S. A. (2014). A review of melt extrusion additive manufacturing processes: I. Process design and modeling. *Rapid Prototyping Journal.*

Uhland, S. A., Holman, R. K., Morissette, S., et al. (2001). Strength of green ceramics with low binder content. *Journal of the American Ceramic Society* 84: 2809–2818.

Utela, B., Storti, D., Anderson, R., et al. (2008). A review of process development steps for new material systems in three dimensional printing (3DP). *Journal of Manufacturing Processes* 10: 96–104.

Vaezi, M., Seitz, H., and Yang, S. (2013). A review on 3D micro-additive manufacturing technologies. *The International Journal of Advanced Manufacturing Technology* 67: 1721–1754.

Vanaei, S., Parizi, M. S., Vanaei, S., et al. (2021). An overview on materials and techniques in 3D bioprinting toward biomedical application. *Engineered Regeneration* 2: 1–18.

Vaupotič, B., Brezočnik, M., and Balič, J. (2006). Use of PolyJet technology in manufacture of new product. *Journal of Achievements in Materials and Manufacturing Engineering* 18: 319–322.

Vilardell, A. M., Takezawa, A., Du Plessis, A., et al. (2019). Topology optimization and characterization of Ti6Al4V ELI cellular lattice structures by laser powder bed fusion for biomedical applications. *Materials Science and Engineering: A* 766: 138330.

voxeljet AG, 2020. www.voxeljet.com.

Wang, X., Jiang, M., Zhou, Z., et al. (2017). 3D printing of polymer matrix composites: A review and prospective. *Composites Part B: Engineering* 110: 442–458.

Warszawski, A., and Navon, R. (1998). Implementation of robotics in building: Current status and future prospects. *Journal of Construction Engineering and Management* 124: 31–41.

Weisensel, L., Travitzky, N., Sieber, H., et al. (2004). Laminated object manufacturing (LOM) of SiSiC composites. *Advanced Engineering Materials* 6: 899–903.

Williams, J. D., and Deckard, C. R. (1998). Advances in modeling the effects of selected parameters on the SLS process. *Rapid Prototyping Journal.*

Wohlers, T. (1991). *Rapid pPototyping: An Update on RP Applications, Technology Improvements, and Developments in the Industry.* Fort Collins, CO: Wohlers Associates.

Wohlers, T. (2004). *Wohlers Report 2004: Rapid Prototyping, Tooling & Manufacturing State of the Industry Annual Worldwide Progress Report.* Fort Collins, CO: Wohlers Associates.

Wohlers, T. (2010). *Wohlers' Report: Additive Manufacturing State of the Industry, 7.* Fort Collins, CO: Wohlers Associates.

Wohlers, T. (2017). *Wohlers' Report: 3D Printing and Additive Manufacturing State of the Industry, 7.* Fort Collins, CO: Wohlers Associates.

Wohlers, T., and Gornet, T. (2014). History of additive manufacturing. *Wohlers Report* 24: 118.

Wong, K. V., and Hernandez, A. (2012). A review of additive manufacturing. *International Scholarly Research Notices.*

Wu, P., Wang, J., and Wang, X. (2016). A critical review of the use of 3-D printing in the construction industry. *Automation in Construction* 68: 21–31.

Xiong, Y. (2009). Investigation of the Laser Engineered Net Shaping Process for Nanostructured Cermets. University of California, Davis.

Yamaguchi, K. (2003). Generation of 3-dimensional microstructure by metal jet. *Microsystem Technologies* 9: 215–219.

Yamaguchi, K., Sakai, K., Yamanaka, T., et al. (2000). Generation of three-dimensional micro structure using metal jet. *Precision Engineering* 24: 2–8.

Zhang, K., Liu, W., and Shang, X. (2007). Research on the processing experiments of laser metal deposition shaping. *Optics & Laser Technology* 39: 549–557.

Zhang, Y. F., Zhang, N., Hingorani, H., et al. (2019). Fast-response, stiffness-tunable soft actuator by hybrid multimaterial 3D printing. *Advanced Functional Materials* 29: 1806698.

Zhao, J., Liu, M., Liang, L., et al. (2016). Airborne particulate matter classification and concentration detection based on 3D printed virtual impactor and quartz crystal microbalance sensor. *Sensors and Actuators A: Physical* 238: 379–388.

Zhong, W., Li, F., Zhang, Z., et al. (2001). Short fiber reinforced composites for fused deposition modeling. *Materials Science and Engineering: A* 301: 125–130.

Zhou, W., Loney, D., Degertekin, F. L., et al. (2013). What controls dynamics of droplet shape evolution upon impingement on a solid surface? *AIChE Journal* 59: 3071–3082.

Zusman, M., Schumacher, C. S., Gassett, A. J., et al. (2020). Calibration of low-cost particulate matter sensors: Model development for a multi-city epidemiological study. *Environment International* 134: 105329.

CHAPTER 13

Robotics

Thomas Paral, PhD

Introduction

The history of robotics is heavily connected to the history of gripping, sensing, and control. Whether a robot needs to do material or any other handling, an end-of-arm tool is necessary.

In 1946, Del Harder started his job as technical director at Ford; he started the first discussion about full automated production lines. His focus was to analyze the value chain in the production to increase efficiency and productivity. The first industrial robot dates back to the early 1950s. In 1954, George Devol applied for a US patent with his idea of Programmed Article Transfer. In 1956, George Devol and Joe Engelberger started their work in developing the first robot. In early 1960, they founded the company Unimate and launched their first product, called Unimate-Robot. This prototype of modern robots had a weight of two tons, and the program was stored on magnetic drums.

General Motors was the first customer; they installed a robot in their production at New Jersey in 1961. The first automated application with a robot was in the foundry of the plant. They bought the robot, which cost Unimate-Robot $65,000 to produce, for $18,000. In the same year, George Devol finally got his robot patent filed.

Industrial Robots

The International Organization for Standardization (ISO) defines industrial robots in its ISO 8373:2021 standard as "an automatically controlled, reprogrammable, multipurpose manipulator that is programmable in three or more axes, which can either be fixed in place or mobile for use in industrial automation applications."[1] By this definition, *industrial robot* includes:

- Manipulators
- Actuators
- Controllers, including a communication interface (hardware and software) as well as power supply

Industrial robots are equipped with end effectors, specifically designed to be attached to the wrist of an industrial robot or tool mounting plate, which enables the robot to perform its intended tasks. End effectors are also known as robotic peripherals, robotic accessories, robot tools, or robotic tools, end-of-arm tooling (EoAT), or end-of-arm devices.

Exhibit 13.1 shows a simple industrial robot system.

EXHIBIT 13.1   Simple industrial robot system

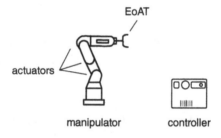

Manipulator

The manipulator is used to move and position parts or tools within a work envelope. Manipulators are formed from joints connected by large links. Manipulators are categorized into two-axis manipulators, three-axis manipulators, and so on, depending on the numbers of axes. The manipulator is responsible for moving an object from one defined location to another defined location for positioning or sorting applications.

For process applications like welding, cutting, or graving, the manipulator moves the process tool to its defined position. Equipped with measuring tools, the manipulator guides the tools to their measurement applications. To accomplish this, it is necessary to position the tool center point for the current tool in the work environment according to its relative position and orientation. For this purpose, each manipulator has several rotational or thrust axes, which are superimposed by combining the individual movements to form an overall movement. The elements of a manipulator are the:

- Base (foundation)
- Axes with the drive units (motors, gears, angle encoders)
- Connecting elements for the structural design and the mechanical coupling of the axes
- Cables or lines for power supply, control, and signal transmission
- Housing

The first manipulators were equipped with heavy hydraulic motors. Early electromechanically driven axes were introduced to improve both speed and costs in the 1970s. Electric-drive motors that are connected directly to the arms, eliminating the need for intermediate gears or chain systems, were introduced in the mid-1980s and delivered robust continuous movements while maintaining high accuracy.

Exhibit 13.2 shows a manipulator.[2]

EXHIBIT 13.2 A manipulator

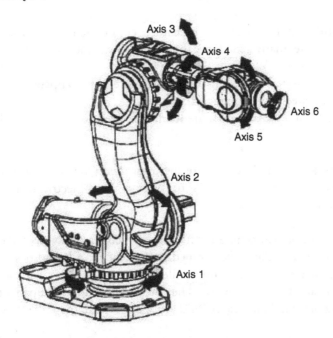

Source: ABB Robotics.

Actuators

Actuators are devices that convert energy into motion. There are two basic types of actuators:

1. **Linear actuators:** The shaft of the linear actuators will only move in a linear fashion.
2. **Rotary actuators:** The shaft of the rotary actuator will only rotate along an axis.

Linear and rotary actuators can be classified based on the energy they use to move the shaft of the motors, as described next.

Hydraulic actuators: The hydraulic actuators are used in robots that handle heavy loads. In hydraulic actuators, a hydraulic fluid in the cylinder moves the piston. These actuators produce a very high force compared to other actuators. They are deployed where higher speed, accuracy, and stability are required. These actuators have a cylinder and piston arrangement. The chamber is filled with hydraulic fluid. The pressure applied to the fluid will push the piston, and that will move the actuator output shaft. The hydraulic actuators can convert the piston movement into linear and rotary movements.

- **Advantages:** Hydraulic actuators are easy to control, accurate, and easier to maintain. They have constant torque or force regardless of speed changes. Leakages are easy to spot, and the actuators are not very noisy.
- **Disadvantages:** Hydraulic actuators require proper maintenance and are expensive. Leakages create environmental damage and using the wrong hydraulic fluid can damage the components.

Pneumatic actuators: In pneumatic actuators, compressed air moves the piston. Like hydraulic actuators, pneumatic actuators can produce linear and rotary movements.

- **Advantages:** Pneumatic actuators are inexpensive, safe, easy to operate, clean, and produce less pollution compared to hydraulic actuators.
- **Disadvantages:** Pneumatic actuators are loud, lack precision control, and are sensitive to vibrations.

Electric actuators: Electric actuators are the most used type of actuator in robotics. This actuator converts electric energy into linear or rotary motion. Electric actuators can be AC or DC actuators. Most robots use DC actuators.

- **Advantages:** Electric actuators have the highest precision, are easy to connect to a network, and are easy to program. They offer immediate feedback for diagnostics and maintenance, providing complete control on motion profiles. They can include an encoder to control the velocity, position, and torque of an axis. They produce less noise compared to hydraulic and pneumatic actuators and do not cause fluid leakages. Overall, they create fewer environmental hazards.
- **Disadvantages:** Electric actuators' initial costs are higher than the costs of pneumatic and hydraulic actuators. They are not suitable for all environments and can overheat, causing wear and tear issues when compared to pneumatic and hydraulic actuators. Their parameters are fixed, requiring replacement when making changes to torque or speed.

Controllers

A major factor influencing the development of robots was the development of control technology. In the mid-1990s, a central computer unit was used for the first time in robots. The connection to sensors, networks, screens, and other peripherals was simplified with the personal computer (PC) control and coupled to a mass market.

With a PC as the central control unit, it was easier to connect camera technology or additional sensors to the manipulator. Remote maintenance via the Internet and new operating options became more efficient. These are all important points that facilitated the use of robots in more complex tasks. The installation space for the controller itself also became more compact and disappeared into the robot base, leading to structural integration becoming more and more successful.

Operation became increasingly easier through the further development of teach pendants. Teach pendants are handheld devices to program the motion of a robot step by step, making it possible, for example, to program or precisely position the robots at the pickup point. In this way, the devices can be used not only during robot setup, but also during production as displays or human machine interfaces (HMIs).

The integrators and machine builders can thus use the high-quality hand-held operating devices to display error messages for the entire system or to enter commands that affect the peripherals. This makes the robot integration more cost-effective, as no additional display devices are needed in the applications.

End Effectors

An end effector is a device that is attached to the end or the wrist of a robot arm, which enables the robot to perform a specific task. ISO 14539 defines standards for grippers and clamps. Grippers, welding guns, and suction cups are some examples of end effectors that directly interact with the environment or object a robot needs to handle or treat.

More accessories like sensors, visions systems, and tool changers are installed, together with an end effector, to enable the robot to also perform complex tasks. A gripper enables the holding and manipulation of an object that performs motions such as holding, moving, tightening, and releasing. Most gripper fingers are made from hard materials, but soft grippers and gecko grippers are being increasingly adopted in handling sensitive objects. Collaborative end effectors are designed to be safe for coworkers.

Exhibit 13.3 shows end-of-arm tooling (EoAT).[3]

Types of Robots

Robots are classified by their mechanical structure, according to the International Federation of Robotics (IFR), into:[4]

- **Articulated robot:** A robot whose arm has at least three rotary joints.

    This robot design can range from simple two-joint structures to structures with 10 or more joints. The arm is connected to the base with a twisting joint. The links in the arm are connected by rotary joints. Each joint is called an axis and provides an additional degree of freedom or range of motion. Industrial robots commonly have four or six axes.
- **Cartesian (linear/gantry) robot:** A robot whose arm has three prismatic joints and whose axes are correlated with a cartesian coordinate system.

    These robots are also called rectilinear or gantry robots. Cartesian robots have three linear joints that use the Cartesian coordinate system (X, Y, and Z). They also may have an attached wrist to allow for rotational movement. The three prismatic joints deliver a linear motion along the axes.
- **Cylindrical robot:** A robot whose axes form a cylindrical coordinate system.

    The robot has at least one rotary joint at the base and at least one prismatic joint to connect the links. The rotary joint uses a rotational motion along the joint axis, while the prismatic joint moves in a linear motion. Cylindrical robots operate within a cylindrical-shaped work envelope.

EXHIBIT 13.3   End-of-arm tooling

VGC10 Gripper
For highly specific application needs and constrained environments

3FG15 Gripper
For gripping a wide range of cylindrical objects from 20 mm to 150 mm wide in machine-tending applications such as CNC lathe machines

Soft Gripper
For gripping a wide array of irregular shapes, sizes, and delicate items in pick-and-place food and beverage production, manufacturing, and packaging

Gecko SP Gripper
For small-footprint application handling objects with perforated or high shine surfaces

2FG7 Gripper
For tight spaces and demanding payloads, can withstand harsh environments, and ready for cleanroom use

VG10 Gripper
For large objects or applications where two parts must be moved separately

RG2-FT Gripper
For high-precision assembly type applications where F/T sensing in fingertips is required

RG6 Gripper
For movement of larger items, payload up to 6 kg, stroke 160 mm

RG2 Gripper
For movement of items, payload up to 2 kg, stroke 110 mm

Source: From OnRobot.

- **Parallel/Delta robot:** A robot whose arms have concurrent prismatic or rotary joints.

  These spider-like robots are built from jointed parallelograms connected to a common base. The parallelograms move a single actuator in a dome-shaped work area. These are heavily used in the food, pharmaceutical, and electronic industries, as this robot configuration is capable of delicate, precise movement.
- **SCARA robot:** A robot that has two parallel rotary joints to provide compliance in a plane. These robots are commonly used in assembly applications, as this selectively compliant arm for robotic assembly is primarily cylindrical in design. It features two parallel joints that provide compliance in one selected plane.

Exhibit 13.4 shows different types of robots.

Robotics Timeline: 1961 to 2011

The robotics timeline delivers basic knowledge of the innovations, driving technologies, and relevant companies in robotics.

1961. Unimation installed the first industrial robot at General Motors. The world's first industrial robot was used on a production line at the General Motors Ternstedt plant, which made door and window handles, gearshift knobs, light fixtures, and other hardware for automobile interiors. Obeying step-by-step commands stored on a magnetic drum, the Unimate robot's 4,000-pound arm sequenced and stacked hot pieces of diecast metal. The robot cost $65,000 to make, but Unimation sold it for only $18,000.

1962. The first cylindrical robot was introduced: the Versatran from American Machine and Foundry (AMF). Six Versatran robots were installed by AMF at a Ford Motor Co. factory in Canton, Ohio. It was named the Versatran from the words "versatile transfer."

1967. The first industrial robot in Europe, a Unimate, was installed at Metallverken, Uppsland Väsby, Sweden.

1968. The octopus-like Tentacle Arm was developed by Marvin Minsky.

1969. General Motors installed the first spot-welding robots at their Lordstown, Ohio, assembly plant. The Unimation robots boosted productivity and allowed more than 90% of body welding operations to be automated, versus only 20–40% at traditional plants where welding was a manual, dirty, and dangerous task dominated by large jigs and fixtures.

1969. Robot vision, for mobile robot guidance, was demonstrated at the Stanford Research Institute.

1969. The company Trallfa in Norway offered the first commercial painting robot. The robots were developed for in-house use in 1967 to spraypaint wheelbarrows during a Norwegian labor shortage.

1969. Unimate robots enter the Japanese market. Unimation signs a licensing agreement with Kawasaki Heavy Industries that allows them to manufacture and market Unimate robots for the Asian market. Kawasaki regarded the development and production of

EXHIBIT 13.4 Different types of robots

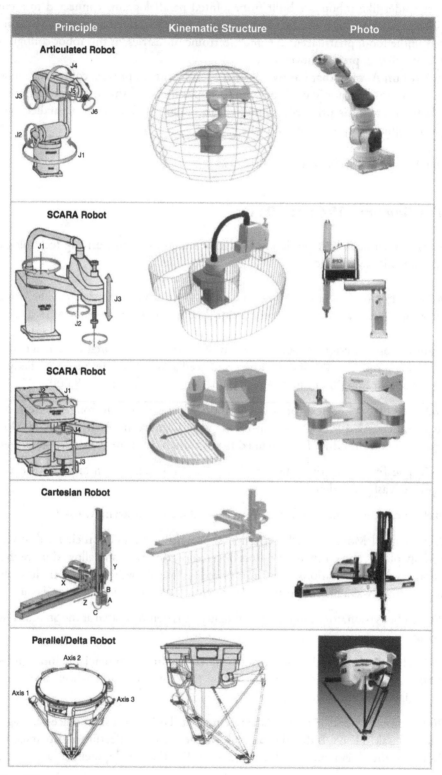

Source: From IFR.

labor-saving machines and systems as an important mission and became Japan's pioneer in the industrial robot field. Later that year, Kawasaki Heavy Industries succeeded in developing the Kawasaki-Unimate 2000, the first industrial robot ever produced in Japan.

1969. Hitachi in Japan developed the world's first vision-based fully automatic intelligent robot that assembles objects from plan drawings. This robot could build blocks based on information created from a direct visual image of assembly plan drawings.

1971. The company KUKA builds Europe's first welding transfer line with hydraulic actuated robots for Daimler-Benz Sindelfingen.

1972. Robot production lines were installed in Europe. FIAT in Italy and Nissan in Japan installed production lines of spot-welding robots.

1973. Kuka developed the first robot with six electromechanically driven axes. KUKA changed from using Unimate robots to developing their own robots. Their robot, the Famulus, was the first robot to have six electromechanically driven axes.

1973. Scheinman started production of the Vicarm/Stanford Arm at Vicarm Inc. The Stanford arm was a robotic arm that performed small-parts assembly using feedback from touch and pressure sensors. Professor Scheinman, the developer of the Stanford Arm, formed Vicarm Inc. to market a version of the arm for industrial applications. The new arm was controlled by a minicomputer.

1973. Hitachi developed the automatic bolting robot for the concrete pile and pole industry. This robot was the first industrial robot with dynamic vision sensors for moving objects. It recognized bolts on a moving mold and fastened/loosened the bolts in synchronization with the mold motion.

1974. The first minicomputer-controlled industrial robot came to market. The first commercially available minicomputer-controlled industrial robot was developed by Richard Hohn for Cincinnati Milacron Corporation. The robot was called T3, The Tomorrow Tool.

1974. The first arc welding robots went to work in Japan. In Japan, Kawasaki built on the Unimate design to create an arc-welding robot, used to fabricate motorcycle frames. Kawasaki also developed touch- and force-sensing capabilities in their Hi-T-Hand robot, enabling the robot to guide pins into holes at a rate of one second per pin.

1974. The first fully electric microprocessor-controlled industrial robot, the IRB 6 from the company ASEA, was introduced. With an anthropomorphic design, its arm movement mimicked that of a human arm, with a payload of 6 kg and five axes. The S1 controller was the first to use an Intel eight-bit microprocessor. The memory capacity was 16 KB. The controller had 16 digital I/O ports and was programmed through 16 keys and a four-digit LED display. The first model, IRB 6, was developed in 1972–1973 on assignment by ASEA's CEO, Curt Nicolin, and was shown for the first time at the end of August 1973. It was acquired by Magnussons in Genarp to wax and polish stainless steel tubes bent at 90° angles.

1974. Hitachi developed the first precision insertion control robot, HI-T-HAND Expert. This robot had a flexible wrist mechanism and a force feedback control system. This meant that it could insert mechanical parts with a clearance of about 10 microns.

1975. The Olivetti SIGMA, a Cartesian-coordinate robot, was one of the first robots used in assembly applications. The Olivetti SIGMA robot was used in Italy for assembly operations with two hands.

1975. The company ABB developed an industrial robot with a payload of up to 60 kg. This met automotive industry's demand for more payload and more flexibility. The robot, called the IRB60, was first delivered to Saab in Sweden for welding car bodies.

1975. Hitachi in Japan developed the first sensor-based arc welding robot, Mr. AROS. The robot is equipped with microprocessors and gap sensors to correct arc welding paths by detecting precise location of workpieces.

1976. The first robots went into space. Robot arms are used on Viking 1 and 2 space probes.

1977. Hitachi developed an assembly cell with eight TV cameras and two robot arms to assemble vacuum cleaners.

1978. The Programmable Universal Machine for Assembly (PUMA) was developed by Unimation/Vicarm, with support from General Motors. GM had concluded that 90% of all parts handled during assembly weighed five pounds or less. The PUMA was adapted to GM specifications for a small-parts-handling line robot that maintained the same space intrusion as a human operator.

1978. Hiroshi Makino, from the University of Yamanashi, Japan, developed the SCARA (Selective Compliance Assembly Robot Arm) robot. By virtue of the SCARA's parallel-axis joint layout, the arm was slightly compliant in the X-Y direction but rigid in the Z direction, hence the term "selective compliant." This is advantageous for many types of assembly operations, such as inserting a round pin in a round hole without binding. The second attribute of the SCARA is the jointed two-link arm layout, which is like our human arms, hence the often-used term articulated. This feature allows the arm to extend into confined areas and then retract, or fold up, out of the way. This was advantageous for transferring parts from one cell to another or for the loading/unloading process stations that are enclosed. In 1981, SCARA robots were launched by the companies Sankyo Seiki and Hirata in Japan.

1979. The first six-axis robot with own control system, the RE 15, was introduced by the company Reis in Obernburg, Germany.

1979. Nachi in Japan developed the first motor-driven robots. The spot-welding robots ushered in a new era of electric robots, replacing the previous era of hydraulic drive.

1980. The first use of machine vision demonstrated. At the University of Rhode Island, a bin-picking robotics system demonstrated the picking of parts in random orientation and positions out of a bin.

1981. General Motors installed the CONSIGHT, a machine vision system. The first production implementation of the system at their foundry in St. Catherines, Ontario, successfully sorted up to six different castings at up to 1,400 castings an hour from a belt conveyor using three industrial robots in a harsh manufacturing environment.

1981. PaR Systems in the United States introduced its first industrial gantry robot. Gantry robots provided a much larger range of motion than pedestal robots and could replace several robots.

1982. IBM in the United States introduced A Manufacturing Language (AML), a powerful, easily used programming language specifically for robotic applications. Using an IBM PC, manufacturing engineers could quickly and easily create application programs.

1983. The first flexible automated assembly lines were introduced. Westinghouse issued a research report on APAS (adaptable-programming assembly systems), a pilot project for using robots in a more flexible automated assembly line environment. The approach uses machine vision in the positioning, orienting, and inspecting of the component parts.

1984. Adept in the United States introduced the AdeptOne, which was the first direct-drive SCARA robot. Electric-drive motors connected directly to the arms, eliminating the need for intermediate gear or chain systems. The simplicity of the mechanism made AdeptOne robots very robust in continuous industrial automation applications, while maintaining high accuracy.

1984. ABB in Sweden produced the fastest assembly robot (IRB 1000). It was equipped with a vertical arm, which made it a sort of hanging pendulum robot. The robot could work quickly across a large area without the need to traverse. It was as much as 50% faster than conventional arm robots.[5]

1992. Wittmann in Austria introduced the CAN bus control for robots. The newly developed CAN bus CNC controls were able to run all the robot programs and related subroutines simultaneously on different microprocessors instead of sequentially on one microprocessor.

1992. ABB in Sweden launched an open control system (S4). The S4 controller was designed to improve two areas of critical importance: the human machine interface and the robot's technical performance.

1992. Demaurex in Switzerland sold its first Delta robot packaging application to Roland. The first application was a landmark installation of six robots loading pretzels into blister trays. It was based on the Delta robot developed by Reymond Clavel at the Federal Institute of Technology of Lausanne (EPFL).

1994. Motoman introduced the first robot control system (MRC), which provided synchronized control of two robots. MRC also made it possible to edit tasks from an ordinary PC. MRC offered the ability to control up to 21 axes, and could also synchronize the motions of two robots.

1996. KUKA in Germany launched the first PC-based robot control system. For the first time, it was possible to move robots in real time using a 6D mouse on an operator control device. This teach pendant featured a Windows user interface for control and programming tasks.

1998. Reis Robotics launched the fifth robot control generation ROBOTstar V, with one of the shortest interpolation cycle times for robot controls.

1998. ABB in Sweden developed the FlexPicker, the world's fastest picking robot, based on the Delta robot, again developed by Reymond Clavel at EPFL. It was able to pick 120 objects a minute or pick and release at a speed of 10 meters per second, using image technology.

1998. Güdel in Switzerland launched the roboLoop system, the only curved-track gantry and transfer system. The roboLoop concept enables one or more robocarriers to track curves and to circulate in a closed system, thereby creating new possibilities for factory automation.

1999. The first remote diagnosis for robots via the Internet was launched by KUKA in Germany.

1999. Reis Robotics introduced integrated laser beam guiding within the robot arm. Reis Robotics receives a patent on this technology and launches the RV6L-CO2 laser robot model. This technology replaces the need of an external beam guiding device, thus allowing the use of lasers in combination with a robot at high dynamics without collision contours.

2003. Robots were sent to Mars. The Mars Exploration Rover Mission is an ongoing robotic space mission involving two rovers, *Spirit* and *Opportunity*, exploring Mars. The mission began in 2003 with rovers to explore Mars's surface and geology.

2003. KULA in Germany created Robocoaster, the first entertainment robot based on an articulated robot. KUKA was the first robot manufacturer to bring people and robots into close contact: in the Robocoaster, the robot whirls passengers around in the air – an extraordinary ride for amusement parks and entertainment events.

2004. Motoman in Japan introduced the improved robot control system (NX100), which provided the synchronized control of four robots and up to 38 axes. The NX100 programming pendant has a touchscreen display and is based on the WindowsCE operating system.

2006. Comau in Italy introduced the first wireless teach pendant (WiTP). All the traditional data communication/robot programming activities can be carried out with absolute safety and without the restrictions caused by the cable connected to the control unit.

2006. KUKA in Germany presented the first lightweight robot. Developed in cooperation with DLR (Institute of Robotics and Mechatronics) in Germany, the outer structure of the KUKA lightweight robot is made of aluminum. With its 16 kg weight, the robot is energy-efficient and portable and can perform a wide range of different tasks. It has a payload capacity of 7 kg and, thanks to its integrated sensors, is highly sensitive. This makes it ideally suited for handling and assembly tasks.

2006. Motoman in Japan launched a human-sized single-armed (7 axis) and dual-armed (13 axis) robot with all the supply cables hidden in the robot arm, which significantly increases the robot's freedom of movement. Robots with dual robot arms, offering human-like flexibility in their movements, are ideal for machine tending and assembly – even beverage serving. The Robot Bar grabs national attention.

2007. Motoman, Japan, launched superspeed arc welding robots that reduced cycle times by 15%, the fastest welding robots in existence at that time. This speed is achieved with a 40% axis movement increase. Their design reduces air-cut time by 30%.

2007. KUKA in Germany launched the first long-range and heavy-duty robot with a payload of 1,000 kg. It expands the application possibilities of industrial robots and creates a new class of reach and payload combinations.

2008. FANUC in Japan launched a new heavy-duty robot with a payload of almost 1,200 kg. The M-2000iA was the world's largest and strongest six-axis robot. It had the longest reach and the strongest wrist as compared to other six-axis robots. The wrist strength set a record, but, more importantly, it moved large heavy parts a great distance with maximum stability.

2009. Yaskawa Motoman in Japan introduced a control system that synced up to eight robots and introduced the improved robot control system (DX100), which provided the fully synchronized control of eight robots and up to 72 axes using I/O devices and communication protocols. Dynamic interference zones protected robot arms and provided advanced collision avoidance.

2009. ABB in Sweden launched the smallest multipurpose industrial robot, IRB120, which weighed just 25 kg and could handle a payload of 3 kg (4 kg for vertical wrist) with a reach of 580 mm.

2010. KUKA in Germany launched a new series of shelf-mounted robots (Quantec) with a new controller, the KR C4. The Quantec robots have an extremely low base, allowing for a greater lower reach for unloading applications. The KR C4 controller generation was the first to combine a complete safety controller in a single control system. This allowed all tasks to be carried out simultaneously.

2011. A humanoid robot launched in space for the first time. A Robonaut 2 (R2) was launched to the International Space Station. Initially R2 was deployed on a fixed pedestal inside the ISS. Next steps included a leg for climbing through the corridors of the ISS and upgrades that would allow R2 to go into the vacuum of space.

Collaborative Robots

In 2008, a new era of industrial robots entered the market and had the highest growth rates in robotics. Collaborative robots are defined by ISO 8373:2021 as a robot designed for direct interaction with a human in compliance with ISO 10218-1 – "Robots and robotic devices" and intended for collaborative use.[6]

Collaborative robots, often called *cobots*, offer opportunities for companies of almost any size to cost-effectively automate processes that were previously out of reach for smaller companies. Cobots are available in a wide range of sizes, payloads, and capabilities, but they all share attributes that define the term *collaborative*.

The aspect that most people think of first is safety – the ability for cobots to work alongside human workers without bulky, expensive guarding. Cobots include several built-in safety mechanisms, including rounded edges and softer materials, as well as power- and force-limiting technologies. These innovations avoid injury and minimize discomfort in case of contact and sensors that automatically slow or stop the robot arm if a human enters its workspace.

However, beyond safety, there are many other important factors that define the concept of collaborative robots. These include flexibility, easy programming, and much lower costs than traditional industrial robots, all of which leads to fast return on investment. These characteristics make cobots desirable, especially for small and mid-sized enterprises (SMEs) that were historically not suitable prospects for large, complex, and expensive industrial robots.

It is important to remember, however, that a collaborative application is more than just a cobot. The robot does not complete any tasks without end-of-arm tooling. The innovation that is occurring in the field of end effectors is the next frontier for collaborative robotic innovation. As collaborative robotic arms move toward commoditization, the innovation occurring around the end-of-arm tooling will enable organizations to deploy robotic technology in new ways.

There are key benefits in using end-of-arm tooling (EoAT) for collaborative automation such as grippers, sensors, and tool changers. Some of the key benefits in collaborative automation using cobots are as follows.

Safe collaboration with human workers. Safe interaction between humans and robots is one of the primary advantages of collaborative automation, but it is important to note that no matter how cobots are designed and marketed, their use may not be free from danger in every application. For example, a cobot arm operating a welding torch or moving heavy objects can still injure nearby people. Therefore, a risk assessment must always be performed before implementation, and all elements of the application must be considered, including the end-of-arm tooling as well as the workpiece itself and the presence of other robots or equipment in the workspace.

End-of-arm tools for collaborative applications must be designed with certification in mind for the international standard ISO 10218, which defines safety requirements for robots and robotic devices.[7] For example, grippers that are rated for 10 kg or lower payloads are considered "collaborative" in nature, and thoughtful design elements such as rounded edges, soft materials, and force-limiting and sensor technologies also contribute to worker safety. Another standard, ISO/CS 15066:2016,[8] describes how a risk assessment for a collaborative application should be carried out, including specific ranges for acceptable amounts of force and pressure on human workers. Collaborative grippers are built with this standard in mind, with appropriate maximum gripping force and available safety shields to minimize the risk of pinch-point injuries.

Easy use and programming. Cobots are known for their simple programming that allows even inexperienced users to "teach" a robotic process by simply moving the robot arm to the desired waypoints and using a touchscreen teach pendant to set the required actions. Collaborative EoAT extends this benefit with easy-to-use programming capabilities for tools that are accessed directly from the robot's teach pendant. This is one way that collaborative EoAT helps maximize production with faster uptime.

Other ease-of-use attributes include plug-and-produce EoAT implementation with limited cables and connections to manage. Electric vacuum grippers are an ideal choice for collaborative applications, as they eliminate the need to manage air lines or clear space for pumps in smaller production cells. For environments that have implemented both collaborative and light industrial robots, tooling that can be used on both platforms interchangeably makes it easier for employees to learn the tools and move smoothly from one process to the other.

Flexibility for easy redeployment. The ability to quickly and easily redeploy cobots provides numerous advantages for companies with changeable product lines, seasonal demands, or multiple processes that need to be automated using single robots that are moved between tasks. Cobots can save multiple programs on the teach pendant and with collaborative EoAT; changing jobs can occur in minutes by simply plugging in the new tool and pulling up the right program on the teach pendant.

The addition of a collaborative quick changer allows for fast and easy tool changes for maximum uptime and productivity. Collaborative design elements of the tool changer, such as low weight and height, minimize its impact on the application's payload limits. Rounded edges help make safe interaction with human workers, and reliable, easy-to-use locking mechanisms allow tool changes within seconds.

A line of collaborative application tools that work seamlessly together can have a dramatic impact on long-term ease of use and flexibility, allowing tools to be easily moved between multiple robots and processes with no additional training. New capabilities can also be easily added, such as force/torque sensing on a finger gripper. This is ideal for sensitive applications such as placing parts in a blister pack, or for precision applications such as inserting a part into a chuck with a defined amount of force or polishing over a contoured surface.

Force/torque sensing is also ideal for processes where the robot and humans work collaboratively side by side, such as when a cobot hands off a part to a human worker for a secondary operation. A collaborative gripper with built-in force/torque capabilities senses the human's grasp of the part and automatically releases it, just as another human worker would.

Cost effectiveness for a fast return on investment (ROI). Low upfront costs and fast ROI are attractive benefits of collaborative applications, especially for small and mid-sized manufacturers. Because these tools are such a critical aspect of the success of the application, collaborative tools must be simple enough to offer quick success even for operators with no previous robotics experience. These tools minimize the number of cables that could interfere with the robot's actions or cause downtime and must be robust enough for ongoing productivity in an industrial environment running 24/7.

While collaborative automation often pays for itself within months, a line of collaborative tools can continue to build ROI over time. A manufacturer can start with a simple automation process and, once workers are comfortable and ROI has been provided, the company can extend the application with more capabilities for greater productivity enhancements. For instance, productivity can be increased over time by a change from a single gripper to a dual-finger gripper, which maximizes cycle time but is as simple to use and program as a single gripper and requires no reengineering. Similarly, a full line of interoperable collaborative finger and vacuum grippers can be exchanged or combined as needs evolve, or sensors can be added for new applications or to enhance quality. Implementing a quick changer nearly eliminates downtime between process changes for additional cost savings.

Easy decisions for common collaborative applications. Collaborative applications are designed for easy implementation, even for those who are not robotics experts. But to maximize success and ROI, it is worth considering what you want to accomplish and what technical capabilities and experience you have in-house. Some applications are more complex than others and may require outside integration help. For most companies, "walk before you run" is good advice.

Look at all the processes you are considering automating and start with one of the easiest applications. Even if this is not the highest-volume or most costly process, a business will likely see significant savings more quickly with a fast success. Choose a tool that can work across multiple production cells and processes and save programs for quick changes between tasks. Once employees gain confidence and familiarity with the robot and tool, the business can move on to more complex applications. Common applications to start with include machine tending and packaging, which can offer immediate results in terms of increased productivity, agility, and ROI. Each has its own considerations for EoAT decisions.

Machine tending applications. Machine tending requires repetitive motion and a high level of consistency, even after hours of work. By automating these processes, businesses can improve worker safety and ergonomics and make employees available for more valuable tasks. Automation can also increase output and allow production to continue after normal work hours. For example, companies may be able to automate processes that can run overnight to have material ready for workers in the morning, which allows for expansion to multiple shifts without an addition of workers. A tool with customizable fingertips allows the robot to handle components with diverse geometries and high surface finish.

With the right collaborative applications tool, businesses can also dramatically increase production with a single robot by moving from a single to a dual gripper. A dual gripper decreases cycle times and boosts CNC machine utilization as it can handle two objects and actions simultaneously, even when working with highly variable materials.

Packaging application. Packaging and palletizing applications can be easily implemented with one cobot and then replicated across similar lines to increase productivity and profitability. With versatile, collaborative EoAT, the robotic application can be easily redeployed if products or logistics contract change. With a line of plug-and-produce collaborative grippers, a single robot can also be used for multiple processes. For instance, a finger gripper can be used to pack a box with parts, then can be changed for a vacuum gripper to palletize. Dual grippers can precisely palletize two different shapes and sizes of boxes at the same time, and new electric vacuum grippers eliminate the need for an external air supply and hoses. Stacking and packaging tasks that require a human sense of touch can be automated quickly and easily with the addition of force/torque sensing on a finger gripper, which supports even delicate packaging applications, such as placing parts in a blister pack.

Collaborative EoAT drives innovation and results. The innovation occurring around collaborative applications enabling EoAT is empowering manufacturers to deploy robotic technology in new ways and gain new advantages. Collaborative EoAT such as grippers, sensors, and tool changers have become critical elements in the success of collaborative applications such as machine tending and packaging. The more manufacturers understand their EoAT options, the greater the increase they will see in output, quality, and profitability.

The Outlook

In the future, robots will no longer be used for executing a specific single task. Robots will be faced with thousands of different tasks that will rarely be repeated in ever-changing environments. It is not feasible to preprogram all these possible tasks and scenarios. Robots will need to learn and to adapt by themselves or with the help of humans. They will need to automatically adjust to stochastic, dynamic, and nonstationary environments and compensate for hardware degradation due to wear and tear. Machine learning and related AI techniques hold the promise of achieving this high degree of autonomy and reliability in analyzing data of sensors and actuators at any time.

Sample Questions

1. Who invented the first robot?
   a. Richard Hohn, Cincinnati Milacron Corporation
   b. Del Harder, Ford
   c. George Devol and Joe Engelberger, Unimate
2. What was the reason behind inventing robots?
   a. To increase efficiency
   b. To increase productivity
   c. To increase efficiency and productivity
3. What are the three major components of an industrial robot?
   a. Manipulator, actuator, controller
   b. Axes, end-of-arm-tools, PC
4. Classify robot actuators and name the different transformation energy to create the movements.
   a. Hydraulic (compressed air), pneumatic (electric energy), and electric actuators (hydraulic fluid)
   b. Hydraulic (electric energy), pneumatic (hydraulic fluid), and electric actuators (compressed air)
   c. Hydraulic (hydraulic fluid), pneumatic (compressed air), and electric actuators (electric energy)
5. What was the main driver of the mid-1990s growth of robots?
   a. PCs as central control unit
   b. HMIs as central control unit
   c. Cameras as central control unit
6. Explain the five different robot classes.
   a. Articulated robot with two parallel rotary joints, Cartesian robot with a minimum of three rotary joints, cylindrical robot with prismatic joints and a Cartesian coordinate system, delta robot with a cylindrical coordinate system, SCARA robot with prismatic or rotary joints
   b. Articulated robot with a minimum of three rotary joints, Cartesian robot with prismatic joints and a Cartesian coordinate system, cylindrical robot with a cylindrical coordinate system, delta robot with prismatic or rotary joints, SCARA robot with two parallel rotary joints
   c. Articulated robot with prismatic joints and a Cartesian coordinate system, Cartesian robot with prismatic or rotary joints, cylindrical robot with a minimum three rotary joints, delta robot with two parallel rotary joints, SCARA robot with a cylindrical coordinate system
7. What are the differences between industrial and collaborative robots?
   a. Industrial robots are robots designed for direct interaction with a human in compliance and cobots are designed for fast and precise repetitive tasks.
   b. Industrial robots are robots designed for fast and precise repetitive tasks and cobots are designed for direct interaction with a human in compliance

8. Name the four key benefits of collaborative applications.
    a. Safe collaboration with humans, easy use and programming, flexibility and easy rede-
        ployment, fast return on investment
    b. Fast execution, able to perform complex tasks, able to perform precise repetitive tasks,
        fenced and protected to prevent harm to humans
9. What are the drivers of robotics to help robots learn and adjust by themselves in the future?
    a. Sensors
    b. Vision systems
    c. Artificial intelligence
    d. Machine learning
    e. Integrated and connected devices

Bibliography

Christensen, H., et al. (September 9, 2020). A roadmap for US robotics: From Internet to robotics.
    https://www.hichristensen.com/pdf/roadmap-2020.pdf
Coronavirus Survey. (May 7, 2020). *IndustryWeek*. https://www.industryweek.com/resources/iw-best-
    practices-reports/whitepaper/21130815/download-industryweek-coronavirus-sur-vey-report?utm_
    source=UM_IWEmail051520&utm_medium=email%3Fcode%3DUM_IWEmail051520/
    download-industryweek-coronavi-rus-survey-report?utm_source=UM_IWEmail051520&utm_
    medium=email%3Fcode%3DUM_IWEmail051520.
COVID accelerates automation & job losses. *Asian Robotics Review*. https://asianroboticsreview.com/
    home344.html.
Global collaborative robot market (2020 to 2025) – Growth, trends and forecast. (April 16, 2020). *Business
    Wire*. https://www.businesswire.com/news/home/20200416005732/en/Global-Collaborative-
    Robot-Market-2020-2025--.
How to rebound stronger from COVID-19: Resilience in manufacturing and supply system. (April
    2020). World Economic Forum. http://www3.weforum.org/docs/WEF_GVC_the_impact_of_
    COVID_19_Report.pdf.
IFR Press Room. (May 14, 2020a). Post-COVID-19 economy: "Robots create jobs." International Fede-
    ration of Robots. https://ifr.org/ifr-press-releases/news/post-covid-19-economy-robots-create-jobs.
IFR Press Room. (September 24, 2020b). IFR presents World Robotics Report. International
    Federation of *Robotics*. https://ifr.org/ifr-press-releases/news/record-2.7-million-robots-work-
    in-factories-around-the-globe.
ISO 8373:2012 – Robots and robotic devices. (2012). International Organization for Standardization.
    https://www.iso.org/obp/ui/#iso:std:iso:8373:ed-2:v1:en.
Joseph, L. (April 13, 2020). How to choose an actuator for your robot? *RoboAcademy*. https://
    robocademy.com/2020/04/13/how-to-choose-an-actuator-for-your-robot/.
Kroupenev, A. (April 21, 2020). What will manufacturing's new normal be after COVID-19?
    IndustryWeek. https://www.industryweek.com/technology-and-iiot/article/21129334/what-will-
    manufacturings-new-normal-be-after-covid19.
Marin, D. (April 20, 2020). How COVID-19 is transforming manufacturing. *Financial Mirror*. https://
    www.financialmirror.com/2020/04/20/how-covid-19-is-transforming-manufacturing/.
Marlin, D. (August 3, 2020). How Covid-19 is transforming manufacturing. *Project Syndicate*. https://
    www.project-syndicate.org/commentary/covid19-and-robots-drive-manufacturing-reshoring-by-
    dalia-marin-2020-04.

NAM Coronavirus Outbreak Special Survey. (February/March 2020). National Association of Manufacturers. https://www.nam.org/wp-content/uploads/2020/03/NAM-SPECIAL-CORONA-SURVEY.pdf.

Robot history: Timeline. (2021). International Federation of Robotics. https://ifr.org/robot-history.

Sneader, K., and Singhal, S. (March 23, 2020). Beyond coronavirus: The path to the next normal. McKinsey & Company. https://www.mckinsey.com/industries/healthcare-systems-and-services/our-insights/beyond-coronavirus-the-path-to-the-next-normal

Steinmann, R. (August 31, 2004). *Greifer in Bewegung* (in German.) München: Hanser.

What are the main types of robots? RobotWorx. https://www.robots.com/faq/what-are-the-main-types-of-robots.

Why manufacturers turn to collaborative automation for business resilience. (2021). *White Paper.* OnRobot. www.onrobot.com.

World Robotics Report 2016 announces collaborative robots as market driver. (September 29, 2016). *Business Wire.* https://www.businesswire.com/news/home/20160929006150/en/World-Robotics-Report-2016-Announces-Collaborative-Robots.

Notes

1. ISO 10218-1:2011(en). (2021). Robots – Vocabulary. International Organization for Standardization. https://www.iso.org/obp/ui/#iso:std:iso:8373:ed-3:v1:en.
2. ABB Robotics. https://new.abb.com/products/robotics.
3. OnRobot. www.onrobot.com.
4. Industrial Robots, definition. International Federation of Robotics (IFR). https://ifr.org/industrial-robots (accessed July 1, 2021.)
5. Westerlund, L. (2000). *The Extended Arm of Man.* Stockholm: Informationsförlaget. https://www.amazon.com/Extended-Arm-Man-History-Industrial/dp/9177364678.
6. ISO 10218-1:2011(en). (2011). Robots and robotic devices – Safety requirements for industrial robots. International Organization for Standardization. https://www.iso.org/obp/ui/#iso:std:iso:10218:-1:ed-2:v1:en.
7. Ibid.
8. ISO/TS 15066:2016. (2016). Robots and robotic devices – Collaborative robots. International Organization for Standardization. https://www.iso.org/standard/62996.html.

CHAPTER 14

Improving Life on the Factory Floor with Smart Technology

Miles Schofield and Aaron Pompey PhD

Introduction

In this chapter we discuss the history, current state, and future trends of manufacturing and how it will impact the life of factory workers. The factories of the future will be a far cry from their historical counterparts, requiring high skill levels but offering rewarding careers. Smart Manufacturing will bring automation on a number of technological fronts and to all types of manufacturing. We also discuss the key challenges to be overcome, and the responsibilities of the factory workers of the future.

There are several definitions of what is entailed in Smart Manufacturing. Simply put, Smart Manufacturing is the future state of manufacturing that is creating new and exciting career paths. Smart Manufacturing is also the current state of manufacturing, especially in countries where labor expenses continue to rise rapidly. For this discussion, we define Smart Manufacturing as follows: *the automation of a modern dynamic manufacturing process such that only three primary job types are needed: service, in which the employee must perform scheduled and unscheduled maintenance on tools; application, in which employees must adapt the current structure to new product and tool types; and operator, where skilled labor that cannot be automated effectively must be performed.*

Life on the Factory Floor from 1700 to Today

Life on the factory floor has changed dramatically since the early days of the Industrial Revolution, when unhealthy and dangerous working conditions were the norm. Equipment was prone to breakdowns and fires with few safety protocols. Factory owners had no obligation to project their workers. With the shift from craftsmen working in rural areas to factory workers working in urban factories, skilled labor was no longer needed. Factory owners turned to women and children for cheap labor. If a worker was injured or killed, it was easy to replace them. Workdays ranged from 12 to 16 hours, six days per week, with no paid holidays or vacations. Factories were dirty, dusty, and poorly lit, as the only light was from sunlight. In many factories, smoke was everywhere since the machines ran on steam from fires.[1]

Exhibit 14.1 shows women working in a nineteenth-century factory.[2]

EXHIBIT 14.1   Nineteenth-century factory workers.

Source: From *Industrial Revolution.*

Working conditions on the factory floor did not improve significantly in the early twentieth century with 35,000 workers killed and 500,000 injured in workplace accidents in 1900. Life did improve by the mid-twentieth century, especially for union employees. Major legal protections were enacted in 1971 with the creation of the Occupational Safety and Health Administration (OSHA) and the enactment of workplace safety standards.[3]

Prior to the outsourcing and offshoring that started in the 1970s, manufacturing jobs were typically portrayed as providing a good income and benefits. In the past 40 years the portrayal has become more negative. The negative portrayal is understandable given that only about 9% of the American and European labor force works in manufacturing and few people have ever visited a production floor.[4]

In the past 20 years, Smart Manufacturing technology gained a negative reputation as a jobs killer, eliminating workers through automation. Critics cite a survey by the Center for Business and Economic Research at Ball State University that found that 85% of the 5.6 million manufacturing jobs lost from 2000 to 2010 were due to automation.[5] In fairness, Smart Manufacturing technologies did eliminate some worker activities, but these were often highly repetitive and boring tasks that few workers wanted to do in American and European economies. As Alex Owen-Hill discusses in Case Study 8, "Five Dangerous Jobs Robots Can Do Safely", there are several jobs that need to automated because they are simply too dangerous.

While Smart Manufacturing technology has eliminated some jobs, it has also made life on the factory floor safer and more enjoyable. For example, Industrial Internet of Things (IIoT) sensors are able to detect problems in equipment, avoiding line stoppages and workers losing pay. IIoT sensors and smart cameras using computer vision can flag safety violations, preventing accidents and injuries. Artificial intelligence (AI) through industrial sensors, computer vision, and robotics can also go a long way in improving quality. No one wants to apply their labor to products of poor quality. Finally, large touchscreens throughout the factory floor keep everyone on the same page as to production levels, department schedules, individual assignments, and so on. This type of real-time transparency presents only one version of reality, avoiding confusion, reducing errors, and preventing line stoppages.

Historically, factories with safe working conditions and making quality products in efficiently run facilities enjoy higher worker morale and less turnover. Employees are happier because they have more pride in their work and the place they work. It is hard to imagine a successful manufacturing organization in our hypercompetitive global economy without skilled workers who feel rewarded in their jobs and careers.

Exhibit 14.2 shows BMW's modern assembly line with good lighting and safe working conditions.[6]

EXHIBIT 14.2 BMW's modern assembly line

Source: From *Wired.*

The Smart Manufacturing Factory Floor

Innovation, like any resource, has the potential to impact manufacturing output and its effects on society. Traditionally, and often unfairly, the news media have portrayed life on the factory floor as boring and dismal, not a place where an educated and talented person would want to work. Nonetheless, factory workers have remained central to industrial and technological development over the past 40 years. With Smart Technologies eliminating most labor-intensive, dangerous, and boring tasks, there will be fewer factory workers in the future, but they will be more versatile and adaptable because they will continue to learn new skills as new technologies emerge. Smart Manufacturing will only be successful with skilled and dedicated workers on the factory and distribution floor.

How AI Is Powering Smart Manufacturing

Arguably, fictional AI dominates the way the general public perceives the possibilities and potential pitfalls of machine learning applications. The spectrum of questions, from the innocuous "What can AI do?" to the ominous "When will I lose my job?" represent gaps in general knowledge about a complex and relatively early-stage technology. For the layperson, it is difficult to comprehend the connection and differences between AlphaZero, a computer chess algorithm, and daily modern technologies such as Google that have been available since the mid-1990s. The slow rate of adoption has been expectedly shaped by the pace of commodities they support. These messy, inconsistent developments are hardly the pristine, well-defined, if certainly ominous, developments of the HAL supercomputer from Stanley Kubrick's 1970 science fiction classic movie *2001*. In the movie HAL suffers a mental breakdown and attempts to kill the entire crew during a mission to Jupiter. A more terrifying image of AI came in 1984 with James Cameron's sci-fi classic *The Terminator*, in which AI is used to create the ultimate killing machine.

Unlike fictional narratives, the reality lies in how modern AI systems work. One critical advancement in modern AI technology has been the optimization of neural networks. In a nutshell, neural networks permit a node-based data pattern to develop through training that allows them to identify key patterns in a data type. Most people are familiar with helping train these systems, as often as they are asked to "type the characters you see in the picture" or "select all images with a boat."

Generally speaking, neural networks provide estimations on how closely a new event or data matches an existing expected or trained data set. In chess, one of the neural networks analyzes all the games ever played previously once standard modern notation was created. In any given position, it can then nominate likely moves to be considered by another network that executes a tree or depth analysis. The process continues until an "intelligent" result is discovered.

More familiar AI examples predominant in daily life are wayfinding and logistics.

Wayfinding. Wayfinding uses data – like historical speeds, current speeds, throughput, emergency/planned restrictions, and distance – to ensure (and optimize) your journey and arrival time. In manufacturing, AI-based GPS navigation systems play a critical role by optimizing delivery routes in busy urban traffic. Timely deliveries eliminate production line shutdowns and help to meet customer commitment dates.

Logistics. Logistics applies to myriad products; the most problematic logistics issue being routing. While mail packages may be handled by internal AI, for example, external logistics in the age of global supply chains can be very challenging, especially with fluctuations of throughput and availability. Using AI, logistics providers are able to reroute shipments to avoid major storms, fires, and other traffic disruptions. The value of logistics modeling came into focus during the COVID-19 pandemic. Many workers lost their jobs because of global commodity shortages in manufacturing and distribution for organizations that did not invest in Smart Technologies to help develop alternative suppliers and logistic providers.

Smart Manufacturing Is Optimizing Factory Processes

The first step to optimizing any manufacturing process is to collect data that detects where the largest gains can be made. As with any data-driven solution, Smart Manufacturing data must be clean in order to be valuable. Neither humans nor AI can deliver "smart" execution without good data. Manufacturing execution and material planning and programmable logic control (PLC) systems have been in widespread use for decades. They collect massive amounts of data but fail to create a digital twin of physical operations. Only by using AI to watch the interaction among equipment, materials, and people is a true digital twin possible. The digital twin improves life on the factory floor by flagging potential safety hazards, eliminating wasted motion, and preventing frustrating bottlenecks.

For a variety of reasons, the technology most available and developed is visual analytics, or the technology of knowing what is in a picture. Video or photo data is quite versatile and can produce multiple data types or automate a number of manufacturing steps. A description of some of the more popular automation applications follows.

- **Quality Assurance.** The most common automated application is quality assurance (QA), which simultaneously delivers data about the state of a product and identifies key defect types. Automated visual inspection using AI and deep learning algorithms may be the most impactful automation of QA. High product rejection rates, customer returns, and product recalls hurt life on the factory floor, demoralizing and frustrating workers. Maintaining excellent quality levels brings prestige to organizations, helps recruit top talent, and improves worker job satisfaction.
- **Worker Tracking.** Another increasingly common camera application is worker tracking, in which a smart camera using AI can monitor multiple workers and derive potential differences in actions they take. This can detect, for example, line workers hand-tightening screws at different levels and the length of time required to complete the same task. Comparing various operators performing the same tasks exposes areas for additional training and may detect best practices that can be incorporated for other workers.
- **Camera-Based Tool Tracking.** Camera-based tool tracking monitors usage and throughput of a tool when the aforementioned metadata is not present. The value of AI camera technology lies in the consistency in monitoring results; however, its effectiveness in monitoring human results and actions may also be its potential drawback.

Although Smart Manufacturing has offered automation solutions to help address major operational inefficiencies for decades, it is natural for some workers to resent the constant monitoring and performance measurements that come with it. For this reason it is critical that supervisors not use Smart Technologies for punitive purposes. It is optimal to use it to improve worker training and operating processes.

Hurdles Faced in Implementing Smart Technologies

Following are three of the major limitations holding companies back from implementing Smart Technologies to transform their organizations.

1. **The Handling or Product Transfer Between Tools.** This is the first challenge in the transition to AI. Traditional manufacturing enabled elaborate handling that seamlessly connected each process step because product changes were rare. With shorter and shorter product life cycles, manufacturing has become more dynamic requiring frequent tooling changes. As Thomas Paral notes in Chapter 13, "Robotics," we remain years away from a generic, easy-to-train robotic handler that grips a variety of shapes and sizes at appropriate pressures, although, thanks to human dexterity, it may not be necessary to solve for complex tool and grip manipulation as a highest priority.

2. **The Manufacturing Environment Itself.** This is the second challenge. Companies that have created a successful Smart Manufacturing environment have effectively separated human movement from robotic movement. The difficulties faced when training a robot are exceeded by those faced when training one with the intelligence required to work alongside a human. To achieve the latter, the product can be transferred between stations either on or over the ceiling or under the floor to avoid collision with factory workers. This requires integration of rail-based ceiling systems. These systems run on a lift, which picks up the product, then has the intelligence to navigate the rail system and deliver the product to its next station. Since no humans are involved, the primary intelligence here is wayfinding, the process of determining one's location and position, then planning and following a route to avoid collisions. This comes with limitations: the rail must cross over any potential tool placement, so the ceiling pattern normally forms rows where tools can be placed. This technology requires a complete building overhaul, which makes it generally cost-prohibitive and impractical for most industries.

3. **Data and Communication.** This is the third challenge. Tools and operators in Smart Manufacturing environments all report on products in the same format so they can be processed by an overarching controller. Product data must not only include information from each step in the manufacturing process, but end-to-end data from all QA stations for triggering rework, scrap, or additional QA. To achieve this, the primary controller must be able to trigger automated calls across all tools (e.g., call lot #5 to tool 305 and perform preprogrammed action 124). The nonlinear layout allows for automatic short- and long-loop reworks and analyses.

Unfortunately, no software or standardization exists that addresses these issues comprehensively. Most product tracking software must be customized to fit the type of manufacturing; further, many tools are not yet designed to be fully automated and deliver status to a primary controller.

The manufacturing process itself cannot be standardized in job tracking software for individual companies since that contains much of the intellectual property (IP). Companies like Google and Microsoft are addressing this problem by creating task-based controllers that operate generically, similar to CRM systems like Salesforce. The most convenient and cost-effective method is with Cloud-based data-sharing. However, as Craig Martin notes in Chapter 5, "Improving Cybersecurity Using Smart Technology," manufacturing trails behind other industries in its adoption of new technology.

Solving for the command system is a simpler task, since it requires only the tool being able to read a remote text file on the controller. Even if every tool wrote some sort of Internet/intranet API, no controller can write thousands of changing calls; plain text calls to the tool itself is typically the standard.

Data coming off the tools is more complicated, since tools provide a variety of functions. A test head, which reports on the function of hundreds of products and components, wants to output a large data table; however, a process tool may simply report the status of the individual product. Currently, the best way to solve this problem is to use plain text and a standard data format, for example a comma-separated values (CSV) file with standardized headers, regardless of exactly the type of data coming off the tool. This puts the onus on the tools to have the flexibility to change their reporting structure to something that is configurable and easy to read.

Holistically, these three points may be the most significant since they address the cost factors influencing business decisions around manufacturing innovation. Notably, every technology brings unique challenges: consider a basic technology like a lathe machine tool. Although it appears to be a straightforward tool in its purpose and design, operating a lathe requires a measure of skill and intelligence to align the piece regardless of irregularities, test the alignment, then lock down the piece (ensuring that an irregular material like wood will not jam the tool on a knot) before clearing the area completely for the next piece. Companies must measure automation costs to produce and manage a Smart tool against employing an operator.

Three Essential Job Types in Smart Manufacturing

It is important to understand the three major roles mentioned earlier that will be needed in Smart Manufacturing: *service*, *application*, and *operator*. Although manufacturing facilities employ thousands of workers to execute hundreds of different job types, such as installation, technology, IT, security, materials, and maintenance, we focus on those essential to Smart Manufacturing in direct production.

The Service Worker

The primary role of a service worker is to maintain the operation of tools and equipment by performing scheduled and unscheduled maintenance. Scheduled maintenance may be caused by predicted slides, where mechanical factors may cause calibration of a tool to naturally shift over time. Scheduled maintenance may also be caused by consumables.

Consumables are the part of a tool that must be changed out after a certain amount of time due to wear. Typical examples of consumables include blades, dies, bulbs, motors, and end effectors in industrial manufacturing; solder, flux, and cleaning solvent are examples of consumables in electronics manufacturing. Unscheduled maintenance occurs when something breaks and can be among the most stressful and challenging jobs in manufacturing. Depending on the line design, an unexpected down may cause major throughput losses, adding manufacturing costs that likely flow through to the customer. As any car mechanic can attest, there are a number of factors and possibilities that contribute to a breakdown in mechanical systems. Identifying and eliminating potential causes, and optimizing troubleshooting work using fishbone techniques, are critical to this role.

Historically, planned maintenance was a matter of trial and error, but Smart Technologies provide the means to model maintenance historical data in order to avoid breakdowns and production line disruptions. As discussed in Chapter 2, "Lean Six Sigma in the Age of Smart Manufacturing," reducing setup and maintenance times has been critical in achieving smaller lot sizes and shorter production cycles.

Exhibit 14.3 shows examples of consumables used in electronics assembly such as solder wire, solder paste, flux, and flux cleaner.[7]

EXHIBIT 14.3   Examples of consumables used in electronics assembly

Source: From *Electronics and You.*

Smart Technologies are even being used to help workers by providing convenient vending machines to dispense the most popular consumable items. These systems use point-of-use technology to record and transmit the transaction back to the supplier so they will know when to replenish the consumed items. Besides eliminating production delays and shortages, many of these providers charge for items only when they are issued, reducing inventory carrying costs.

Exhibit 14.4 shows an example of an advanced vending machine for consumables.[8]

EXHIBIT 14.4   Advanced vending machine

Source: From *Process Engineering Control & Manufacturing.*

The Application Worker

The primary role of an application worker is to ensure that tools output is both consistent and expected. This often involves developing qualification and in-line calibration procedures that better control tools output. Smart Manufacturing is a dynamic process: the primary challenge for application engineers is determining how best to program the tool for each individual product or job type. Here are a few examples.

- A painting station uses a *process tool*; depending on the product, the paint chemistry could be different, and the tool may perform differently when using different finishes or color types.
- A camera system is a *process control tool* if it is designed to monitor the color accuracy for the painted product; the application engineer would determine which presets (camera, lighting, other configurations) to change, based on how the lighting and finish affects the color measurement.
- An application worker uses all possible tool settings to define best practices for individual customers or products.

Another role for application workers is to use statistics-based modeling to control and predict tool performance. In the example of the painting station, we can assess why, if a customer uses five tools, one of the tools creates 2% more defects than the other four tools. In our example, the tool has passed all factory qualifications and an application worker must now take steps to determine the root cause of the issue. The first step is to monitor calibration data. Thousands of local settings per tool can be compared between tools and between the entire fleet of tools. The second step is to monitor output or control data. Are the 2% defects being found by only a certain type of tool? Are there metrics that are in spec, but that exhibit a different type of data behavior? A value that is too stable may be more suspicious than one that has a natural noise level associated with it.

Modern systems govern not only by monitoring historical thresholds, but by seeing if the data becomes stagnant or the baseline shifts. Understanding the role of the application worker and analyzing how the entire process works around the tool is key to troubleshooting performance issues.

The Operator

The primary role of today's skilled professional operator is to produce work that is too expensive to automate. Until recently, the weakness in AI systems caused most types of processes to require operators, rendering automation unachievable. Physical welding, for example, is a type of art that requires that the weld meet physical strength metrics and that a variety of welds reflect nuanced differences and techniques that must be executed by a skilled craftsman, since these are not yet achievable by robots.

Predicting how and when a process may be optimized with automation requires a measure of human acumen, which may outperform the dexterity of robots. In the example of the camera system designed to monitor the color accuracy for the painted product, detecting defects like the flaking of layered transparent coatings would require a variable shifting low-angle source that was designed to catch light reflected off the edge caused by the missing transparent layer. On the other hand, without magnification, humans can discern defects down to ~20 microns. Humans can also discern difficult-to-detect defect types (like the flaking of layered transparent coatings).

Smart Manufacturing technologies create powerful tools for innovating century-old systems and methods; however, they are not without a wide range of challenging positions as technologies continue to be integrated. There will be growing roles for application engineers and service engineers in highly technical roles as products, components, and procedures continue to evolve. Like so many roles in Smart Manufacturing, success depends on the ability to adapt, diagnose, process data, and make meaningful decisions about products and customers. Further, the tools themselves are quite different. Whereas manufacturing roles were once defined by pump, hydraulic, and vacuum systems, these now include products, like camera systems, that require more computation and imaging.

Three Types of Tools Needed in Smart Manufacturing

Besides understanding the three job types that will be needed in Smart Manufacturing, it is also imperative to define the tool types used by service workers, application workers, and operators.

- **Process tools.** These are the tools used to perform an action that physically changes the product (e.g. painting, cutting, or gluing).
- **Process control tools.** Sometimes called metrology, these tools measure something about the product (e.g. finding defects, color, size, dimensions).
- **Handling tools.** These are tools or robots that are designed to hold or move the product either between tools or intratool. Intratool handlers are used where a tool type is not feasible or compatible with the automated type of transfer.

Exhibit 14.5 shows a variety of cutting tools, one of the most widely used process tools, that can be found any one of the 18,000 machine shops operating the United States.[9]

EXHIBIT 14.5 A variety of cutting tools

Source: From *Machine Tools World.*

Smart Manufacturing is making tools safer and more ergonomically friendly, reducing fatigue and work-related injuries such as carpal tunnel syndrome.

Exhibit 14.6 shows a zero-gravity mechanical arm to carry the weight of an ergonomically designed grinding tool.[10]

EXHIBIT 14.6 A zero-gravity mechanical arm

Source: From Ergonomic Partners.

Smart Manufacturing Design Choices

Manufacturing organizations make key decisions when deciding whether and where to integrate new technology into a process. Having already looked at the challenges of individually advancing technology types and how engineers work in the field to advance and control them, what are the considerations when integrating tools? The most straightforward answer is financial: the tool that works best for the bottom line. Making that determination, however, is complicated.

If an inline required camera QA station processes 10 units per hour, with an individual worker's goal to produce 10 units an hour, how many camera stations are required for the line? While the math seems simple, the answer is not necessarily "one." Consider the various factors mentioned already within this chapter, as well as the following questions: *What is the uptime for this tool? What is the maintenance schedule? What is the average utilization of my current line? What is the target average utilization of my line? How many product types use this tool? What is the potential that this tool could find additional uses through application engineering?*

Manufacturing organizations face the difficult task of balancing the cost and usage benefit of a single tool against many other tools and processes. A basic line is only as fast as its slowest tool; even then, that only considers product-based actions. If the slowest tool is the most reliable and consistent, then it may also be the most efficient.

Process tools and process control tools can be incorporated into nonlinear processing, which makes throughput calculations more difficult. Nonlinear processing occurs when the manufacturing process is more than a checklist of steps, each occurring once per product. Short control loops are basic examples of a nonlinear process. Short control loops occur when a product fails a metrology or QA check and is sent to other types of possibly in-line metrology for further study.

Additional processes introduce additional costs; however, optimizing the line requires analyzing outliers further to root-cause them. The final process should not require additional QA, although this is governed by achievability to yield x% out of the system. A rework loop occurs when a product fails a test and determines that process steps must be redone. A rework requires that the process be completed twice, effectively doubling the tool time for the product, so calculations must be done on tool throughput instead of scrapping that product and losing the yield.

While some manufacturers suffer from underutilization, facing challenges of consolidation and optimization, others suffer from bottlenecks that cause unneeded stress and output concerns. Available data is critical and solves some of these complex problems. For example, some manufacturers use optical character recognition (OCR) to pull data from legacy systems that only display their status on a single screen. The gathering and management of the data is almost an industry by itself and will likely offer generic commercial solutions in the future.

Summary

Ensuring efficiency through integration is a significant task since each process has its own unique challenge. New types of robots, new types of cutting, and new advanced modern materials will change existing tool types; the resulting potential for advancement is tremendous. With innovation comes the next chapter of problem-solving: as Smart Manufacturing lines advance, new technologies will create new challenges to be solved for the modern workforce.

Life on the production floor of the future will bear little resemblance to what workers experienced in the twentieth century. With most labor-intensive and repetitive tasks automated, the workforce of the future will be highly skilled and educated. With Smart Manufacturing continuing to introduce new automation and standardization opportunities on an ever more frequent basis, factory workers will need to commit to lifelong learning. The payoff will be higher-paying and more rewarding careers.

Sample Questions

1. Why were women and children selected to work in factories during the Industrial Revolution?
   a. They were more highly skilled than men.
   b. They worked longer hours than men.
   c. They worked cheaper than men.
   d. There was a lack of male workers.
2. What were working conditions like during the Industrial Revolution?
   a. 12- to 16-hour workdays
   b. Large numbers of injuries
   c. Unhealthy working conditions
   d. Low pay
   e. All of the above
3. What are three tool types used in Smart Manufacturing?
   a. Process tools, process control tools, handling tools
   b. Setup tools, process tools, maintenance tools
   c. Disposable tools, process tools, perishable tools
   d. None of the above
4. What are three types of workers used in Smart Manufacturing?
   a. Temporary, permanent, contract
   b. Skilled, unskilled, trainee
   c. Service, application, and operator
   d. Appliance, service, operator
5. What type of worker would maintain equipment?
   a. Operator
   b. Application
   c. Service
   d. Repair
6. The primary role of a skilled professional operator is to produce
   a. High-quality work
   b. Work that is too expensive to automate
   c. Small lots of work
   d. All of the above
7. What are the major differences between manufacturing of the past and Smart Manufacturing of the future?
   a. Workers will be highly skilled in the future.
   b. Workers will be committed to continued education.
   c. Workers will do jobs too complex for robots.
   d. All of the above

8. What government agency was created to improve worker safety in the 1970s?
   a. The Accident and Injury Prevention Administration
   b. The Occupational Safety and Health Administration
   c. The National Safety Institute
   d. The National Standards Institute
9. A smart camera system designed to monitor color accuracy in a paint line is a
   a. Quality control tool
   b. Process control tool
   c. Process tool
   d. Operator's tool
10. Traditional manufacturing was able to live with elaborate and time-consuming tooling changes because of
    a. Frequent product changes
    b. Infrequent product changes
    c. Automate tooling changes
    d. Small production quantities

Notes

1. Industrial Revolution working conditions. (October 29, 2021). *History on the Net.* https://www.historyonthenet.com/industrial-revolution-working-conditions.
2. Factories during the Industrial Revolution. *Industrial Revolution.* https://industrialrevolution.org.uk/factories-industrial-revolution/ (accessed September 4, 2021).
3. Tomyn, R. (25 June 25, 2018). What were the work conditions in American factories in 1900? *Classroom.* https://classroom.synonym.com/were-work-conditions-american-factories-1900-23383.html.
4. Employment by industry, 1910 and 2015. (March 3, 2016). US Bureau of Labor Statistics. https://www.bls.gov/opub/ted/2016/employment-by-industry-1910-and-2015.htm.
5. Savić, A. (2017). Using smart technology to aid factory workers. *The Atlantic.* https://www.theatlantic.com/sponsored/vmware-2017/human-ai-collaboration/1721/.
6. Knight, W. (April 12, 2021). BMW's virtual factory uses AI to hone the assembly line. *Wired.* https://www.wired.com/story/bmw-virtual-factory-ai-hone-assembly-line/.
7. *Electronics and You* (accessed September 5, 2021).
8. PECMN214. (February 23, 2014). Consumables Solutions launch revolutionary free* vending machines to help you control costs. Process Engineering Control & Manufacturing. https://pecm.co.uk/consumables-solutions-launch-revolutionary-free-vending-machines-to-help-you-control-costs/.
9. ToolKart's cutting tools e-commerce marketplace. *Machine Tools World.* https://www.mtwmag.com/toolkart-launches-online-e-commerce-marketplace-for-cutting-tools/ (accessed September 5, 2021).
10. Zero gravity tool balancers & part manipulators. Ergonomic Partners. https://www.ergonomicpartners.com/zero-gravity-tool-balancer (accessed September 9, 2021).

CHAPTER 15

Growing the Roles for Women in Smart Manufacturing

Maria Villamil and Deborah Walkup

Maria Villamil is vice president of WET Design, the company that designs and builds those famous Dubai and Las Vegas fountain shows; she has 26 years of experience in several areas of complex manufacturing. Deborah Walkup, a degreed mechanical engineer, is senior director of technical operations with InforNexus, with 30 years of experience providing technology solutions to manufacturing and distribution organizations.

The last section of this chapter relates the personal story of the two coauthors. Maria Villamil immigrated to the United States from the Philippines as a girl, and used her can-do attitude acquired growing up on a farm to enjoy a rewarding career in manufacturing. In the process she has learned virtually every aspect of complex manufacturing and taken on roles traditionally reserved for men, and rarely by a woman of color who speaks English as a second language. Maria shares her story of the steps she took to progress from receptionist to vice president of manufacturing.

Deborah Walkup also grew up on a farm learning a wide variety of mechanical and engineering skills from her father. Coming from a family of engineers and her love of fixing and designing things, it was natural for her to pursue an engineering education. She faced bias against women in STEM starting in middle school and continuing through high school, the formative early years shaping career paths. After receiving her degree in mechanical engineering, she benefited from government programs creating opportunities for women. She shares her personal story of the opportunities for women in technical and engineering areas traditionally dominated by men.

In this chapter they explore the history of women in traditional manufacturing, including the many obstacles they were forced to overcome and the many initiatives underway to encourage women to enter STEM education and training, which are essential for the success of manufacturing and distribution organizations.

Introduction

The history of women in manufacturing dates back to the early years of the Industrial Revolution. Back then, women were not always included or welcome in industrial workplaces. Factories that fabricated items in mass quantities have historically been male-dominated. Manufacturing jobs were considered inappropriate for women, and women weren't typically educated or trained for manufacturing jobs.

During the Industrial Revolution there was a migration of farm workers to cities to become factory workers. In some cases there were more farmers than land, in other cases men traded the uncertainty of farm life for set hours and a weekly paycheck, and more women were relegated to domestic roles. Most of the women in the workforce were low income and women of color.

In the United States, during World War II, the shortage of men opened the doors for female workers in factories and they remain open today. Factory jobs provide some of the highest wages for entry-level positions and, through the efforts of unions, offer more job protection than other careers. By working in factories, women became wage earners who could support a family, with or without a male partner.

A good description of the role women played in manufacturing during World War II is documented on the History.com website.[1] "Women were critical to the war effort: Between 1940 and 1945, the age of 'Rosie the Riveter,' the female percentage of the U.S. workforce increased from 27 percent to nearly 37 percent, and by 1945, nearly one out of every four married women worked outside the home. World War II opened the door for women to work in more types of jobs than ever before, but with the return of male soldiers at war's end, women, especially married women, were once again pressured to return to a life at home, a prospect that, for thousands of American women, had shifted thanks to their wartime service."[2]

While women had been participating in the workforce in greater numbers since the Great Depression, the United States' entry into World War II transformed the jobs open to women. Prewar, most women worked in traditionally female roles such as teaching and nursing. After Pearl Harbor, women were asked to take on roles formerly reserved for men. The aviation industry saw the greatest use of female workers, with over 300,000 women helping the war effort in 1943, which was 65% of total workforce. Prewar, only 1% of the aviation industry workforce was female. The munitions industry also became heavily dependent on female workers.

The iconic image of Rosie the Riveter was a great propaganda tool to recruit women to the war effort. Rosie was based in part on a real-life munitions worker but was primarily a fictitious character. The bandana-clad Rosie with a muscular arm became one of the most successful recruitment tools in US history. Although female workers were critical to the war effort, they were paid far less than their male counterparts, with most women earning about 50% of male wages.[3]

With men pulled into military service, mothers were faced with the burden of balancing work and childcare. The federal government acknowledged the issue by passing legislation providing childcare services for industries supporting the war effort.

Not all women were treated equally in the workplace. African American women found that white female workers did not accept working next to a person of color. African American women were paid less than their white peers. Japanese American women fared even worse, being interned in camps for the duration of the war.

Exhibit 15.1 is the iconic poster created by J. Howard Miller. While it was only shown briefly in Westinghouse factories, the poster in later years has become one of the most famous icons of World War II.[4]

EXHIBIT 15.1   J. Howard Miller's iconic poster

Source: American History Museum.

Women as Innovators

Women in households are typically the decision makers for most purchases, and for many items they are also the end user. They wash dishes, clean floors, and do the laundry. They also operate most of the appliances. Women have a deep understanding of how products are used and know how they could improve the lives of anyone who uses these products to perform household chores.

Their input for product design is valuable in the development and manufacturing environment. Unfortunately, most product designers are men, and they design products for men; occasionally they design smaller versions in pink for the ladies. Even the design of products like women's underwear or feminine hygiene products is dominated by male designers. Most of these aren't very comfortable and can be hard to use. Why do we expect men to have a better idea of how to design these products that they almost never use?

Exhibit 15.2 shows a notorious example of patronizing product design with the BiC pen "for Her."[5] This product was roundly skewered in Amazon reviews for being unnecessary and patronizing.

An important and obvious reason why women are great contributors in the world of manufacturing is the fact that they have a unique perspective on normal day-to-day life experiences. These daily life experiences should be strongly considered when adopting manufacturing as a career choice. How can things be better and what can be done to make life easier are questions

EXHIBIT 15.2   BiC pen "for Her"

Source: From *Business Insider.*

that should be asked. This is what allows us to be creative and to freely innovate and make our vision come to life. This type of mindset was instilled in both Maria and Deborah by their fathers growing up and has been a valuable asset in their careers.

Women Hold the Answers (Skills Where Women Excel)

Women make great leaders. Women have many traits that organically lend themselves to being great leaders such as being great listeners, experts in work–life balance, being able to multitask, being empathetic and nurturing, and being able to manage a crisis well. Anyone running a household and raising children had better be good at crisis management.

Jack Zengler and Joseph Folkman in the *Harvard Business Review* investigated the management skills that women and men possess and how they rank in effectiveness. "But the women's advantages were not at all confined to traditionally women's strengths. In fact at every level, more women were rated by their peers, their bosses, their direct reports, and their other associates as better overall leaders than their male counterparts – and the higher the level, the wider that gap grows."[6]

Exhibit 15.3 shows research by Jack Zenger and Joseph Folkman comparing male and female leadership effectiveness.[7]

EXHIBIT 15.3 Overall Leadership Effectiveness by Gender by Position (Percentile Scores)

	Male	Female
Top Management, Executive, Senior Team Members	57.7	67.7
Reports to Top Management, Supervises Middle Managers	48.9	56.2
Middle Manager	49.9	52.7
Supervisor, Front Line Manager, Foreman	52.5	52.6
Individual Contributor	52.7	53.9
Other	50.7	52.0
Total	51.3	55.1

Source: Zenger Folkman Inc., 2011.

Zenger and Folkman maintain that women are superior to men in 12 of 16 leadership competencies they measured. Interestingly, women had the highest scores in two traits thought traditionally to be male strengths, taking initiative and driving for results.

Exhibit 15.4 is from Zengler and Folkman's research and shows the comparison across 16 leadership qualities.[8]

EXHIBIT 15.4 The Top 16 Competencies Top Leaders Exemplify Most

	Male	Female	
	Mean Percentile	Mean Percentile	T value
Takes Initiative	48	56	−11.58
Practices Self-Development	48	55	−9.45
Displays High Integrity and Honesty	48	55	−9.28
Drives for Results	48	54	−8.84
Develops Others	48	54	−7.94
Inspires and Motivates Others	49	54	−7.53
Builds Relationships	49	54	−7.15
Collaboration and Teamwork	49	53	−6.14
Establishes Stretch Goals	49	53	−5.41
Champions Change	49	53	−4.48
Solves Problems and Analyzes Issues	50	52	−2.53
Communicates Powerfully and Prolifically	50	52	−2.47
Connects the Group to the Outside World	50	51	−0.78
Innovates	50	51	−0.76
Technical or Professional Expertise	50	51	−0.11
Develops Strategic Perspective	51	49	2.79

Source: Zenger Folkman Inc., 2011.

Zengler and Folkman ask why organizations are not employing exemplary women as leaders and believe that the answer is because of blatant discrimination. When they shared their findings with a group of women about why they believe their colleagues were rated so highly on taking initiative and self-development, their answers were around the still-tenuous positions they felt themselves to be in the workplace with such comments as "We need to work harder than men to prove ourselves," and "We feel the constant pressure to never make a mistake, and to continually prove our value to the organization."[9]

Women's Inspiration

Women's inspiration to enter the manufacturing industry has evolved over the years. It is important to learn how these women were inspired and how they got into the field of manufacturing. Inspiration comes from different sources; from supporting a household to becoming the substitute workforce while men were out fighting the war, inspiration for women to enter the manufacturing industry is quite diverse.

Women as inventors and those in leadership roles serve as perfect inspirations for girls to pursue careers in manufacturing. Women innovators such as Hedy Lamarr have certainly been an inspiration. Hedy Lamarr was considered one of the most beautiful actresses of all times who was also a brilliant engineer, inventing a radio guidance system using frequency-hopping spread spectrum technology that was incorporated in Bluetooth. She epitomized glamour, which could be easily romanticized by young girls.[10]

While there is data that shows that girls are good at math and science in grade school and middle school, their interest declines in high school and college. As Carly Berwick writes in Edutopia, there are three main barriers and solutions as we discuss next. [11]

Barrier One: Building a Math Identity

One explanation as to why the gender differences in STEM participation may be the commonly held beliefs that mostly men are scientists or mathematicians. When girls become aware of cultural messages about the male superiority in math, their encounters with math and technology become more fraught, fostering self-doubt in even studious young girls.

Building a math identity begins by showing images of famous female mathematicians and scientists throughout classroom materials and including the achievements of women in the class curriculum.

Barrier Two: The Question of Race and Class

Our perceptions as to who does well in math and science may impact performance more than raw aptitude does. This has especially profound implications for females who are Black, Latino, and low-income. As a group they are less likely to take advanced STEM courses and to pursue STEM professions after school. Nicole Joseph, a professor at Vanderbilt University, argues that Black girls and their teachers view themselves as outsiders in mathematics.[12]

Joseph suggests several measures to build math identity and interest among Black girls, such as making fundamental changes in how math is taught. One idea is to eliminate accelerated math in middle school, allowing all students to take Algebra I in the ninth grade. This worked in San Francisco's public schools to raise test scores in algebra and to lower the class repeat rates for all students, including Latino and Black students.

Another suggested change comes from educators Norman Alston and Patricia Brown. They believe that interest in mathematics can be fostered through after-school STEM programs. Alston requires her middle school program graduates to teach younger students and brings in women of color as inspirational speakers for female students.

Barrier Three: It's Not Just Content; It's Context, Too

Format matters when it comes to learning and teaching STEM, research shows. A recent analysis of admission tests at an elite New York City high school found that girls had better grades in higher-level math, but tended to score slightly lower on admission tests, leading to lower admission rates. According to various studies this may be because girls tend to guess less, which can be a disadvantage on multiple-choice tests.

One suggestion is to move away from multiple-choice tests and place a greater emphasis on open-ended assessments that allow students to demonstrate proficiency through word problems or writing, where they feel more confident. Another suggestion is to use project-based instruction because it addresses problems of girls rejecting engineering careers. A study by the National Academy of Engineering in 2008 asked students if they wanted to be engineers. Girls were twice as likely to say no as boys. But when the girls were asked if they would like to design a safe water system, save the rainforest, or use DNA to solve violent crimes, the girls answered yes.

Companies Working to Overcome Barriers to Women's Entry

There is a lot of work to be done to bridge the gap between women as a minority group in manufacturing and equity in numbers. The first step is to break down historic barriers of entry for women into these industries. Organizations that are involved in getting women to enter the field of manufacturing are a step in the right direction: they can be used to encourage more and more women to not only consider a role in manufacturing and engineering, but to embark on lifelong learning to improve their standing in a male-dominated field.

RippleMatch has compiled a list of 28 companies that have initiatives to close the gap for women in manufacturing and leadership roles. These are the manufacturing companies that made the list.[13]

> **Abbott.** Since the late 1800s, Abbott has always had women leaders. Therefore, it is not surprising that Abbott is now a leading workplace for women. Several of Abbott's female leaders rose through the ranks and benefited from Abbott's one-year mentorship program and commitment to women's career development.
>
> **Aramark.** Aramark has taken major strides to be more inclusive and supportive of women. Their corporate board is now made up of 30% women, up from none five years ago. Aramark also offers a business resource network for women employees.
>
> **Carbon.** Carbon is a manufacturer of 3D printing and led by a female CEO. Carbon's leadership is 45% female, with a goal of reaching 50%. In a traditionally male-dominated field, Carbon is leading the way to increasing representation with many majority-female teams.

Honeywell. Honeywell offers women an great environment to tap into, including women-only networking events, longer maternity leaves, and a seamless transition back into the workplace after having a child.

Juniper Networks. Juniper Networks is committed to enabling women to fill representation gaps in their workforce. One program to achieve this is their WeTech Scholarship program, which grants four $10,000 scholarships to female STEM majors. Juniper Networks also partners with Anita B.Org and Grace Hopper to ensure they are continuing to engage and build a more fully representative workforce.

Kimberly-Clark. The company has increased the number of women in their senior management by 66% over the past decade by a combination of on-the-job leadership training, professional development opportunities, and flexible work schedules that help women with various personal obligations.

Lilly. After a 2015 analysis revealed the poor representation of women in leadership roles, Lilly took quick action. The company launched a program to improve women's trajectory, including inclusivity and bias trainings, and concrete goals to increase the number of women in management positions.

Medtronic. Over the past several years Medtronic has been proactive in closing the gender gap. In 2020 they achieved a 99% gender pay equity globally. Medtronic is now working on evolving development programs to create more career opportunities and learning styles.

Nestlé. Nestlé is committed to increasing female career opportunities and is also working quickly to achieve equal pay for all employees.

Pfizer. Pfizer has an outstanding portfolio of programs by their global women's network that has served to unlock the full potential of female employees. Programs and initiatives include development, robust mentorship, and very visible support from Pfizer's senior executives.

Unilever. In March 2020, Unilever announced that it had achieved gender balance across its management team globally. Their management level employees are now 50% female and their nonexecutive board members are 45% female. Significantly, they closed the gap in their historically underrepresented supply chain, finance, operations, and technical engineering teams.

Programs to Develop STEM Skills for Women

WorldWideLearn has compiled a list of 15 initiatives to encourage girls in STEM careers.[14] They are:

1. **National Girls Collaborative Project.** A group of organizations working together to encourage girls to enter STEM fields when they start their careers.
2. **Association for Career and Technical Education.** The largest association that addresses pedagogical issues in STEM education. The association is involved in advocacy work,

lobbying policymakers on issues that affect technical education in schools. They also offer resources to help educators meet the needs of technology students.

3. **Association for Women in Science.** A professional organization for women who work in STEM jobs, they offer mentoring and coaching services, as well as advice on work–life balance issues. They have resources to educate women in the field, including workshops, books, educational videos, and journals. They also support advocacy for female STEM workers to address issues impacting their careers.

4. **Change the Equation.** This organization works to increase STEM literacy to encourage young students to become more interested in science and math and then receive quality STEM education. They offer online resources for educators and discuss forums to discuss STEM issues and advocate for improved STEM education.

5. **National Math and Science Initiative.** This organization works to address the candidate shortfall of qualified women to work in STEM-related jobs. Their goal is improve student performance in engineering, science, math, and technology courses by shaping the way these classes are taught and improve the quality of educators teaching them.

6. **National Center for Women & Information Technology.** The nonprofit works to increase the number of women in technology and computing fields. The organization includes companies, government organizations, and universities.

7. **National Council of Teachers of Mathematics.** Their goal is to support mathematics teachers by helping them teach at a higher standard for all their students. Their vision is that everyone sees the value in mathematics and is enthusiastic about math subjects.

8. **Million Women Mentors.** This is a campaign that brings together higher education, government agencies, and nonprofits to mentor young female students who are interested in pursuing STEM careers.

9. **American Association of University Women.** Founded in 1881, this association work to empower women by addressing the economic, educational, political, and social issues that impact them. They also created a math and science summer camp and a STEM conference for female middle school students.

10. **Society of Women Engineers.** The society aims to inspire female students to pursue engineering careers and to support women entering the field. This includes helping female engineering college students. The society also runs outreach events to encourage female elementary school students to consider engineering careers.

11. **Mathematical Association of America.** The MAA works to ensure the accessibility of mathematics to all students. They have an interest in improving math education, research, and public policy.

12. **Girls Who Code.** This organization works to address the major gender gap in the software industry by inspiring young girls to enter these fields and by giving them a solid technical education.

13. **NASA.** NASA has many resources for students and teachers designed to encourage an interest in technology and science. NASA offers a number of educational resources and publications for teachers to keep them current in the agency's scientific work, and to help them teach science.

14. **Anita Borg Institute.** The institute celebrates women who are successful in technology careers and inspires those interested in following in their footsteps. Anita Borg was a pioneer who championed the advance of women in technology.

15. **TechWomen.** This organization is dedicated to supporting women from the Middle East and Africa who are interested in entering STEM careers. The organization runs mentoring programs introducing these women to their successful counterparts from the United States, allowing them to share ideas and connect with one another personally and professionally.

Growing the Role of Women in Smart Manufacturing

Today, women are making a tremendous impact to help industries make gains in every facet of manufacturing. Women have been instrumental in developing cost-saving technologies, leading the way in the adoption of Industry 3.0, mentoring young talent, implementing diversity, and implementing equity and inclusion policies. Industry 4.0 or Smart Manufacturing is the automation of traditional manufacturing practices using Smart Technology. Smart Manufacturing introduced the integration of machine-to-machine communications, and the Internet of Things (IoT) is implemented for the automation, analysis, and diagnosis of problems without the need for human intervention.

These achievements come, but the number of women in manufacturing continue to lag behind the number of men. In manufacturing, diversity, equality, and inclusion (DEI) is often focused on women because the industry has historically been male-dominated. Today, fewer than one in three manufacturing professionals are women, despite representing nearly half of the overall workforce in the United States, according to the "2021 Manufacturing Talent Study" from Deloitte and Manufacturing Institute (MI).[15] In fact, many of the largest, most successful corporations, such as Procter & Gamble, Cisco, and HP, are focused on transforming their internal culture by prioritizing DEI, often with a woman at the helm of these initiatives. They've recognized that in order to effectively serve a diverse, multicultural consumer base, the makeup of their employees must reflect that diversity. By focusing on inclusion, they're also able to lower the attrition rate for women and those of color. DEI, although often seen as a political stance, is simply good business.

Marketability

Once women are in the door, staying relevant requires a love of lifelong learning. It is unrealistic to expect that a person can keep doing the same job year after year with the same set of skills that they started their career with. Automation is inevitable in manufacturing and other industries. Artificial intelligence (AI) is evolving and improving. There are many tasks that AI and automation can do so much better than a human can. It is important to continuously learn new skills to stay relevant and marketable. Over the past few decades, we have all witnessed manufacturing and other related jobs being sent offshore. Taking courses offered at community colleges and trade schools is the best thing anyone in the workforce can do to stay knowledgeable and employable long term. Taking courses is a good way to advance in one's career and can lead to other opportunities for higher-level positions. Having technical know-how, expertise, and educational experience is empowering and can give a big confidence boost.

In America, there is an unhealthy obsession with college degrees and graduate schools. While higher education is important, it may not be for everyone. Sometimes we lose sight of what is most effective and achievable for a person. When a student considers the cost of a four-year college degree, how much of a student loan is required, and the time it takes to pay off the debt, they need to have confidence that they will be a marketable commodity at the end of the process and be able to land a good-paying job. Taking vocational courses or starting at a community college can be a very cost-effective way to get on track for a good-paying career. Even if you haven't graduated from high school, you can be accepted at community college, finish that requirement, and take more classes. There is no wrong way to get started.

Pay Inequality

In spite of the gains over the past decades, income inequality between men and women remains a chronic problem. Exhibit 15.5 shows the current breakdown of pay inequality by gender and race.[16]

EXHIBIT 15.5 Earnings gap by race (median weekly earnings in 2020)

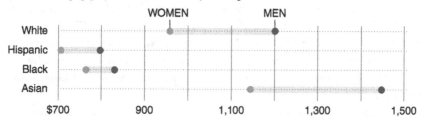

In a nutshell, women make 82 cents for every dollar a man makes. The disparity is even larger for women of color. Traditionally, women were discounted in the workplace because it was assumed they wouldn't be there long, especially if they were unmarried; to be fair, many women do leave the workplace because they become primary caregivers for children and need more flexible hours and more supporting infrastructure like affordable daycare.

Maria Villamil's Story

How I Got into Manufacturing

My path to a manufacturing career has been a journey of excitement. It was unexpected, as I felt pressure from my family and the community I grew up in (a small town in South Philippines) to be in medicine. I was fairly certain that I would be the first US-educated cardiologist coming from my hometown in Digos. Growing up as a little girl, my family maintained a farm in a town named Santa Maria, located south of Digos. I spent my first formative years growing up on that farm, where I lived a self-sufficient lifestyle with my family. Seeing my dad do everything and make things out of a block of wood, or pieces of metal, stone, and rubber, among other raw materials, made me think that was normal, or how life was supposed to be. If you needed something, you could make it. My family did not believe in the phrase "that cannot be done."

The family-centric values and self-sufficient lifestyle of living on a farm began my interest and passion in manufacturing. The idea of fabricating something you need out of necessity has always been embedded in my mindset. If a tool or an object that is needed to accomplish a task does not exist, it can always be designed and fabricated. I vividly remember the strong women in my family. My mother, grandmothers, aunts, and other women helpers made such an imprint on my mind. I learned early on that there was no substitute for hard work if you wanted higher yields or better harvests or thriving livestock and other ways to earn income from a farm. This was also the reason I saw innovation around me – from improving how we harvested our crops, to managing our irrigation, even to dealing with having clean chicken cages that have a clever way of catching and recycling dung.

Immigrating to the United States as a 16-year-old high school graduate, I struggled with acclimating to the environment in what seemed to be a super-fast-paced world, Los Angeles. Comparing Los Angeles to my hometown in the Philippines was like night and day – there are still no traffic lights in Santa Maria. In attempting to start my long journey as a healthcare professional, I had to take an anatomy class as a prerequisite, and I was immediately discouraged. Working on those cadavers made me feel uneasy and it was then that I made the decision to pursue a different path. In the early 1990s, personal computing was gaining popularity, and I was always interested in tinkering and making something out of parts and more parts and more parts! I frequented Radio Shack. I took a computer certification program to learn computer programming and how to build a network infrastructure.

Not having been born in this country, as a woman of color, with English as a second language, I had my work cut out for me. I was obsessed with learning and knowing everything I possibly could to advance in my career and to get acclimated and be worthy of entering the American workforce. I encountered many instances where being discouraged was so easy, but on the path to the technology and manufacturing world I embarked on, there were just so many positive aspects that kept me going and wanting to embrace the industry.

Conquering the common fear that is associated with entering the unknown and unfamiliar is empowering for me. Learning something difficult always becomes easy once it is tackled head-on. Taking welding, soldering, and woodworking classes sounds hard, but it does not have to be, and these classes provided a great foundation for learning other valuable skills. Taking trade school and community college courses helped tremendously. Learning basic computer skills allowed me to navigate more intermediate and advanced digital tools. Learning Excel and the office suite at a minimum and mastering this at an intermediate level was very helpful in making me marketable in the manufacturing environment.

I am a mother of two girls, a 16-year-old and a 12-year-old, who are similar in many ways, but also different in a lot of ways. I made sure they had a science fair entry every year during their days in elementary school. As their Girl Scout troop leader, I made sure that their activity involved technology, science, and manufacturing projects. When they were younger and into dolls I taught them about plastic injection molding, which is one of the skills I learned at WET, where I have worked most of my professional life. Introducing them to and keeping them interested in STEM is a big undertaking and a tough endeavor, because there are always other influences that distract them and draw them away, but it has been a joyful experience.

My Career at WET Design

WET Design is a water feature design firm based in Los Angeles, California. Founded in 1983 by former Disney Imagineers Mark Fuller, Melanie Simon, and Alan Robinson, the company has designed over 200 fountains and water features using water, fire, ice, fog, and lights. It is known for creating the Dubai Fountain, the world's largest performing fountain, along with the eight-acre (3.2 ha) Fountains of Bellagio in Las Vegas. WET has designed features in over 20 countries around the world, in North America, Europe, Asia and the Middle East.[17]

WET strives to stay be ahead of the game when it comes to educating its staff. WET has world-renowned experienced instructors to teach and provide training to our staff members. I am and have been surrounded by people who taught me generously. WET's culture is designed to nurture an individual's desire to learn meaningful and useful things, and the culture's emphasis on continuous and lifelong learning is evident – from a fully functional woodworking shop that is available to staff members who want to learn woodworking or make their own projects to an entire building dedicated to learning that is complete with a university-style amphitheater, a large convertible computer lab/classroom, and a large lab area for manual tinkering with every device that WET has designed and created. Having had my own personal set of challenges, this environment just made me more driven for success, knowing that I can take little steps and improve upon them by making them a part of my foundational knowledge. WET provides us with opportunities to grow. These challenges require us to research various engineering and manufacturing methods and processes, which brings endless possibilities when it comes to solving problems. There is nothing more gratifying than seeing something fabricated out of a raw material and see it come to life in a project. At WET we pride ourselves on providing the tools and the environment necessary for our talent to take a concept from ideation to life.

Engaging with our clients starts with understanding their vision. Inspiration to create comes from various elements we see all around us. Engineering is key to making these elements perform the way we envision them. Standard approaches, templates, and equipment that are readily available have proven inadequate in our processes and what we need to accomplish in our projects. We are constantly reinventing what we do on almost every project, and that is by design. We have ambitious requirements for our water features. Each project is a new idea and doesn't exist until we create it.

It is difficult to hire the talent we need, and WET has adopted the philosophy of hiring talented engineers who can grow into the position. They have to be able to adapt to a changing environment and learn easily. We also expect them to challenge how we design, manufacture, and build projects. They should always be asking "why?" In our industry we are balancing budgets, deadlines, resource scarcity, and sometimes political upheaval, civil unrest, high-profile scandals, and other nontraditional and complex issues.

WET's Use of Smart Technologies

WET has become heavily reliant on Smart Technologies because our projects are so unique and not mass produced. It allows us to be more fluid and flexible in operations and not be tied down by what we have done in the past. Our research and development process allows us to push the boundaries of water, fire, ice, and fog to create long-term works of art. We have state-of-the-art manufacturing facilities where we conceive and manufacture the technologies and equipment used in our features to achieve ultimate precision and quality. We pride ourselves on not relying on outsourcing. One of the most gratifying parts of my job is finding a solution to a problem by coming up with new ways to manufacture and then bringing the previously nonexistent manufacturing capability in-house. This includes studying and learning a particular manufacturing method, acquiring any necessary machines and building infrastructure, and implementing these new manufacturing processes.

Deborah Walkup's Story

Like Maria, I grew up on a family farm; ours was near State Center, Iowa, a very small town in the middle of the state. I loved working with my dad, and my first job was to hold the light while he fixed something. Both of my parents worked as professionals, my dad as an industrial engineer for a local valve manufacturer, and my mother as a registered nurse. My dad started as a machinist with the company and really understood everything about manufacturing.

My father's family is also chock-full of engineers. It wasn't inevitable that I would become one, but it was likely. Two of my dad's brothers were electrical and chemical engineers, two of his sisters married mechanical and electrical engineers, and in my generation, there is one mechanical, one computer, one aerospace, one chemical, and two industrial engineers. One of my father's sisters would have been a great engineer but she wasn't allowed in the program because she wasn't a man. Family reunions usually started with something like a softball game and then devolved into a working session to redesign the bases.

Eventually I graduated from holding the light to learning the different tools and handing them to my dad like a surgical assistant. My favorite was a crescent wrench – it had a great name, and it replaced a whole socket set. On a farm a crescent wrench and a Phillips screwdriver were essential for most repairs and every tractor toolbox had at least these two items.

The farm I grew up on had been my father's parents' home before we moved in and we had all the equipment my grandfather owned, even the harnesses from when he used mules instead of tractors. I learned to drive an 8N Ford Tractor when I was four years old, and I could work the brakes if I stood on them with both feet.

We also did several construction projects while I was growing up. We built an addition to the 100-year-old farmhouse that added a family room and walkout basement. We also built a large machine shed and a garage. My job was usually fabricating the rafters. My dad would make one and I would copy it to make the full set. I also learned basic framing, drywall, cement work, tile work and grouting, and doors and windows. I feel very fortunate to have had this giant workshop available to learn all of these skills.

Life on a farm is also a series of production lines. When the chickens get big enough you set up a line to slaughter them, pluck them, cut them up, and package them for the freezer. Once a year we would also butcher a steer. We would corn feed them and dry age the meat for some of the best beef I've ever eaten. Sweet corn was another production that would consume our dining room for a week and fill our freezer for the year.

Because of this hands-on background I would always select shop class instead of home economics in middle school and high school. Every year I would sign up for carpentry, auto shop, welding, sheet metal work, or drafting. Every year these classes wouldn't show up on my schedule and I'd have to go into the principal's office and ask why. Every year they would say, "We thought you made a mistake." That's also when I perfected my eye roll when school administrators in a small school continued to ignore my efforts to enroll in classes historically reserved for boys. I could understand that in a larger school, but my graduating class was fewer than 100 people, and that was consolidating kids from five towns.

I had the opportunity to do contract drafting in high school. A local construction company needed someone to put together perspective drawings of grain bin installations to bid on jobs. The owner visited our drafting class and asked who was the best drafter and everyone said it was me, and it was, so I was hired. My dad brought home an unwanted T-square, and I worked on the drawings in the evening, tracked my own hours, and got checks my parents didn't think I deserved.

College was another adventure. I worked part-time for minimum wage after school and summers for four years to earn enough money for my first year of school. Luckily, I was offered a co-op job with the company my dad worked for that paid five times more than minimum wage and that carried me through. It was also a great apprenticeship. I worked in the machine shop running drill presses, engine lathes, milling machines, and big CNC machines. The CNC machines were the best. There was a hoist, so moving big parts around was a piece of cake and the machines included programs, so if you set up your part correctly, zeroed the machine and hit start, it did all the work. I also worked in the assembly area, creating finished goods for customers. The last area I worked in was the office for the facilities team, the industrial engineering team, and the design team.

My focus at Iowa State was mechanical design and my first job, with Texas Instruments, was electronics packaging, which is the design of circuit cards and electronics housings. My next job was electronics packaging at Boeing, and then at Ford Aerospace. Part of the reason my early career was based on military and space applications is that these companies worked on government contracts, and they had quota requirements for diversity. So I benefited from affirmative action, and I am a strong supporter of correcting imbalances in gender and ethnicity. Before these programs, women might be able to study engineering, but even if they graduated, they probably couldn't get a job. These were jobs that gave me great experience and paid well.

While I was working at Ford Aerospace, the company purchased mechanical CAD workstations and software and I learned how to design parts in virtual space. The company we bought the hardware and software from was expanding and they asked me if I would work for them and help sell the software. This was my jump into the technical sales world and enterprise software. I went on to work for i2 technologies, FreeMarkets, Ariba, Sockeye Supply Chain, Docusign, E2Open, Atollogy, and Infor, where I am currently a senior director of technical operations.

Solution engineering is a wonderful blend of technology, personnel skills, magic, and applications. You respond to requests for information, meet with customers to discover their issues and goals, put together a solution and present it to them, and, if you are lucky, they select you and buy your software. I have always been amazed at the opportunities I've had and have enjoyed having the ability to follow my interests.

Summary

Women have played pivotal roles in manufacturing since the beginning of the Industrial Revolution (Industry 1.0). World War II dramatically increased their participation in manufacturing when they took on several roles that had been reserved for men. In the past decade, major corporate, university, and government initiatives have been launched to encourage more girls and women to pursue STEM majors and careers in Smart Manufacturing. The future for women in Smart Manufacturing is very bright, with growing opportunities in all sectors of manufacturing and distribution, from small companies to global enterprises.

Sample Questions

1. The aviation industry employed around _____ women during World War II.
   a. 50,000
   b. 300,000
   c. 800,000
2. True or false? Women are the primary decision makers for household purchases.
   a. True
   b. False
3. True or false? The Industrial Revolution had no effect on women in the workplace.
   a. True
   b. False
4. True or false? During World War II all women who joined the workforce in the United States benefited equally.
   a. True
   b. False
5. The current Pay Inequality indicates that women make _____ cents for every dollar a man makes.
   a. 53
   b. 82
   c. 107
6. Select all of the competencies where women rank higher than men:
   a. Practices Self-Development
   b. Develops Strategic Perspectives
   c. Develops Others
   d. Takes Initiative
7. Within the context of this chapter, STEM stands for
   a. Space, Time, Energy, and Mass
   b. Systems Telecommunication Engineering Manager
   c. Six Through Eight Mathematics
   d. Science, Technology, Engineering, and Math
8. Select all examples of creating a math identity for women.
   a. Including the achievements of women mathematicians and scientists in the curriculum
   b. Reinforcing common stereotypes
   c. Posting images of women scientists in the classroom
9. Select the companies that are known for promoting women in the workplace:
   a. Abbott
   b. Phillip Morris
   c. Pfizer
   d. Juniper Networks
   e. Icahn Enterprises

10. Which of the following organizations have active initiatives to encourage women and girls to pursue STEM careers?
    a. The Anita Borg Institute
    b. Girls Who Code
    c. NASA
    d. Society of Women Engineers

Notes

1. History.com Editors. (August 27, 2021). American women in World War II. A&E Television Networks. https://www.history.com/topics/world-war-ii/american-women-in-world-war-ii-1.
2. Ibid.
3. Ibid.
4. "We Can Do It!" produced by J. Howard Miller. American History Museum. https://american history.si.edu/collections/search/object/nmah_538122 (accessed August 30, 2021).
5. Samantha Felix. (August 28, 2012). Here are the Bic pens for women that everyone is laughing at. *Business Insider*. https://www.businessinsider.com/the-bic-pens-for-women-that-everyone-is-laughing-at-2012-8.
6. Jack Zenger and Joseph Folkman. (March 15, 2012). Are women better leaders than men? *Harvard Business Review*. https://hbr.org/2012/03/a-study-in-leadership-women-do.
7. Ibid.
8. Ibid.
9. Ibid.
10. Wikipedia. Hedy Lamarr. https://en.wikipedia.org/wiki/Hedy_Lamarr#Frequency-hopping_spread_spectrum (accessed August 31, 2021).
11. Carly Berwick. (March 12, 2019). Keeping girls in STEM: 3 barriers, 3 solutions. Edutopia. https://www.edutopia.org/article/keeping-girls-stem-3-barriers-3-solutions.
12. TODOS podcast. (January 8, 2021). S02 E09: Dr. Nicole Joseph, advocate for Black women in STEM and beyond. https://www.podomatic.com/podcasts/todosmath/episodes/2021-01-08T14_50_20-08_00.
13. Abbey Gringas. (March 2021). 28 companies invested in the success of women at work. *RippleMatch*. https://ripplematch.com/journal/article/companies-invested-in-the-success-of-women-at-work-d819cb0b/.
14. WorldWideLearn. 15 innovative initiatives bringing women into STEM. https://www.worldwide learn.com/articles/15-innovative-initiatives-bringing-women-into-stem/.
15. Paul Wellener, Heather Ashton, Visitor Reye, and Chad Montray. (May 4, 2021). Creating pathways for tomorrow's workforce today. *Deloitte Insights*. https://www2.deloitte.com/us/en/insights/industry/manufacturing/manufacturing-industry-diversity.html.
16. Francesca Donner and Emma Goldberg. (March 25, 2021). In 25 years, the pay gap has shrunk by just 8 cents. *New York Times*. https://www.nytimes.com/2021/03/24/us/equal-pay-day-explainer.html.
17. Wikipedia. WET (company). https://en.wikipedia.org/wiki/WET_(company) (accessed August 31, 2021).

CASE STUDIES

Automating Visual Inspection Using Computer Vision

Anthony Tarantino, PhD

Introduction

Deloitte's 2018 manufacturing survey projects that 4.6 million new manufacturing jobs will be needed in the coming decade, but that only about 2.1 million of them will be filled by qualified candidates, leaving a shortage of 2.5 million positions.[1] The shortage will be even more critical for quality assurance professionals who conduct visual inspections. These positions require advanced skills and dedication beyond the usual abilities of many factory workers. Even with the most skilled inspectors, high error rates are a fact of life.

The Federal Aviation Administration (FAA) defines visual inspection as "the process of using the unaided eye, alone or in conjunction with various aids, as the sensing mechanism from which judgments may be made about the condition of a unit to be inspected."[2] Joseph Juran, a pioneer in the field of quality assurance, argues in his *Quality Handbook* that 100% manual visual inspection can be expected to yield no more than 87% accuracy and that it would require 300% manual visual inspection to yield 99.7% accuracy.[3] Other studies argue that visual inspection error rates can be as low as 3–10% under optimal conditions, such as those with skilled and experienced inspectors working in a comfortable environment and good lighting. However, error rates can be as high as 20–30% under suboptimal conditions.[4]

Ironically, it has been found that as defect rates go down, inspection accuracy suffers, and as defect rates go up, inspection accuracy improves. Therefore, if you enjoy high quality levels, it is less likely that you will find defects.[5]

There are several factors that impact physical inspection. Exhibit CS1.1 shows a list of factors impacting physical inspection.[6,7]

This list would have looked much the same if created by Dr. Juran after World War II when he became the evangelist for quality assurance. Of the 41 factors listed, at least 24 are human factors. In the age of Smart Manufacturing and Smart Technology, this list will be fundamentally simplified by eliminating the human factors.

EXHIBIT CS1.1 Factors impacting physical inspection

Task	Individual
• Defect Rate	• Gender
• Defect Type	• Age
• Defect Salience	• Visual Acuity
• Defect Location	• Intelligence
• Complexity	• Aptitude
• Standards	• Personality
• Pacing	• Time in Job
• Multiple Inspections	• Experience
• Overlays	• Visual Lobe
• Automation	• Scanning Strategy
	• Biases

Environmental	Organizational
• Lighting	• Management Support
• Noise	• Training
• Temperature	• Retraining
• Shift Duration	• Instructions
• Time of Day	• Feedforward Information
• Vigilance	• Feedback
• Workplace Design	• Feedback
	• Incentives
	• Job Rotation

Social	
• Pressure	• Isolation
• Consultation	• Communications

Source: From Judi E. See and Harish's Notebook.

Why is it so important to reduce human factors? The answer is simple. Each human factor presents its own unique risk. Eliminating human factors eliminates risk. Alex Owen-Hill in Case Study 8, "Five Highly Dangerous Jobs That Robots Can Do Safely," describes some great examples where human factors should be eliminated to prevent accidents and injuries.

An example of reducing human factors that I liked to use in my risk management classes at Santa Clara University is not manufacturing-based, but is simple to comprehend.[8] It is the risk of the first nonstop flight over the Atlantic Ocean. In May 1927, Charles Lindberg became the first pilot to fly nonstop from New York to Paris. He succeed after all other attempts failed, usually resulting in the death of the aviators. One reason Lindberg succeed is by eliminating the human factor of a copilot. His was the only attempt made without a copilot and without a larger, more complex plane.

Automated visual inspection using Smart Technologies eliminates all human factors and focuses on a much shorter and more easily managed list of factors. With many fewer factors and with the advantages of high-speed cameras, deep-learning AI, and Edge computing, the quality and consistency of inspections can approach 100%.

Exhibit CS1.2 is a list of the more typical factors found in automated visual inspections.

EXHIBIT CS1.2 Automated visual inspection factors

Lighting
Camera Location
Camera Specifications
Speed of the Object Being Inspected
Location of the Defect
Complexity of the Defect
Level of Accuracy Required

Some industries automated visual inspection because human inspection was not practical or too dangerous and the direct and indirect costs of defects were high. The direct costs include warranty, rework, and replacement. The indirect costs can be much higher, with lawsuits, government intervention, major quality issues threatening brand acceptance.

Exhibit CS1.3 shows a bottling line example where automated visual inspection is essential to operations.[9] No person or group of people can be expected to notice and remove cracked bottles running down a production line at a high rate of speed.

EXHIBIT CS1.3 Bottling line with automated visual inspection

Source: From Jamshed Khan, in *Nanonets*.

In the bottling line example, cameras used in computer vision take 100 or more frames per second, deep learning analyzes the imagery and flags defects in a second, and finally an application programming interface (API) sends a signal to a robotic arm to immediately remove the defective item. While Edge computing is used for all the immediate action, data is sent to the Cloud for analysis that is used for continuous improvement.

Exhibit CS1.4 shows the automated visual inspection adoption rates for various industries.[10] The chart shows that adoption rates in manufacturing factories are low when compared to other industries, at less than 10%.

EXHIBIT CS1.4   Automated visual inspection adoption rates

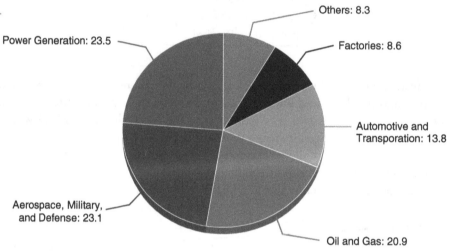

Source: From Jamshed Khan, *Nanonets.*

This begs the question, if artificial-intelligence-based computer vision has so many advantages over human visual inspection, why are adoption rates so low? After all, machine learning and machine vision have been in use for years. The answer lies in recent and rapid advances in deep learning technology that greatly reduce the effort required to accurately inspect objects.

Steven Herman, in Chapter 9, "Artificial Intelligence, Machine Learning, and Computer Vision," defines artificial intelligence as "software performing tasks traditionally requiring human intelligence to complete. Machine learning is a subset of artificial intelligence wherein software 'learns' or improves through data and/or experience. Deep learning is a subset of machine learning, usually distinguished by two characteristics: (1) the presence of three or more layers, and (2) automatic derivation of features."

Before deep learning came on the scene, computer vision typically used image processing algorithms and methods. This required extracting image features such as detecting edges, colors, and corners of objects. This in turn required human intervention and labor. As a result, model reliability and accuracy depended on the features extracted and the methods used in the feature extraction. Haritha Thilakarathne, writing in *NaadiSpeaks*, describes the problems this presents: "The difficulty with this approach of feature extraction in image classification is that you have to choose which features to look for in each given image. When the number of classes of the classification goes high or the image clarity goes down it's really hard to come up with traditional computer vision algorithms."[11]

Deep learning uses neural networks that contain thousands of layers that are good at mimicking human intelligence in order to distinguish between parts, anomalies, and characters while tolerating natural variations in complex patterns – a major advantage over earlier technologies. As a result, deep learning gets closer to merging the adaptability of humans conducting visual inspection with the speed and robustness of computerized systems conducting visual inspection.[12]

Next are examples of computer vision using deep learning algorithms for visual inspection in manufacturing. Exhibit CS1.5 shows how parts are classified on a printed board assembly (PCBA).[13]

EXHIBIT CS1.5 Parts classification on a printed board assembly

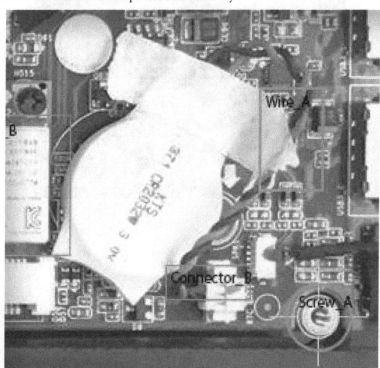

Source: From Radiant Vision Systems.

Exhibit CS1.6 shows the verification of model numbers on an automotive part.[14]
Exhibit CS1.7 shows the verification of a car door assembly.
Exhibit CS1.8 shows the discovery of fabric defects.
Exhibit CS1. 9 is courtesy of MobiDev showing the detection of plastic bottle cap defects.[15]

EXHIBIT CS1.6   Verification of model numbers

Source: From Machine Vision Experts.

EXHIBIT CS1.7   Verification of a car door assembly

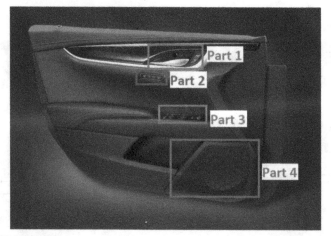

Source: From Machine Vision Experts.

EXHIBIT CS1.8   Discovery of fabric defects

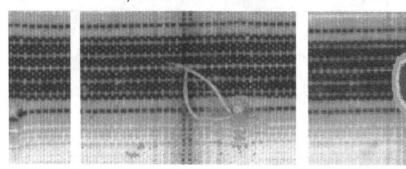

Source: From Machine Vision Experts.

EXHIBIT CS1.9 Detection of bottle cap defects

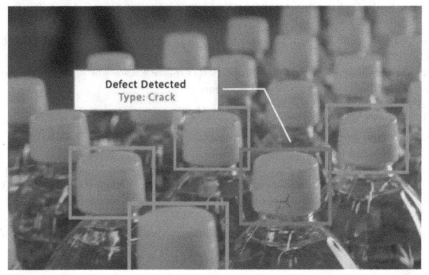

Defect Detected
Type: Crack

Source: From MobiDev.

Conclusion

Deep learning teaches machines to learn by example, something that comes naturally to people. With hardware and software costs continuing to drop, manufacturing is given amazing new abilities to distinguish trends, recognize images, and make intelligent decisions and predictions.

Automated visual inspection using deep learning has proven that it can overcome the limitations of human inspection and do so at lower costs and less time than traditional manual methods. Examples of successful applications can be found in virtually every industry and in all stages of manufacturing and distribution.

The number of computer vision solution providers has grown rapidly. *AI Startups* published a list in August 2021 of its top 99 computer vision startups.[16] The startups come from across the globe: Australia, Bangladesh, Belarus, Canada, Chile, China, France, Germany, India, Israel, the Netherlands, Russia, Singapore, Sweden, Switzerland, Taiwan, Turkey, the UK, Ukraine, and the United States. The list is far from being complete, as I know at least another dozen startups in Silicon Valley that are focused on computer vision. With so many organizations working to improve automated inspection technology, solution capabilities are guaranteed to grow at a rapid pace and make computer vision affordable for even the smallest of manufacturers.

Notes

1. Pajulla, S., Wellener, P., and Dollar, B. (2018). 2018 skills gap in manufacturing study. Deloitte. https://www2.deloitte.com/us/en/pages/manufacturing/articles/future-of-manufacturing-skills-gap-study.html.

2. Harish. (October 11, 2015). 100% visual inspection – being human. Harish's Notebook. https://harishsnotebook.wordpress.com/2015/10/11/100-visual-inspection-being-human/.

3. De Feo, J. (2017). *Juran's Quality Handbook*. New York: McGraw-Hill. https://www.amazon.com/Jurans-Quality-Handbook-Performance-Excellence/dp/1259643611?asin=1259643611&revisionId=&format=4&depth=1.

4. Solving for the limits of human visual inspection. *Creative Electron*. https://creativeelectron.com/solving-for-the-limits-of-human-visual-inspection/.

5. Harish, 100% visual inspection.

6. See, J., Drury, C., and Speed, A. (September 28, 2017). The role of visual inspection in the 21st century. *Sage Journals*. https://journals.sagepub.com/doi/10.1177/1541931213601548.

7. Harish, 100% visual inspection.

8. Tarantino, A. (2011). *The Essentials of Risk Management in Finance*. Hoboken, NJ: John Wiley & Sons. https://www.amazon.com/Essentials-Management-Finance-Anthony-Tarantino/dp/0470635282/ref=sr_1_3?dchild=1&keywords=essentials+of+risk+management+in+finance&qid=1629660991&sr=8-3.

9. Khan, J. (May 2021). Everything you need to know about visual inspection with AI. *Nanonets*. https://nanonets.com/blog/ai-visual-inspection/.

10. Ibid.

11. Thilakarathne, H. (August 12, 2018). Deep learning vs. traditional computer vision. *NaadiSpeaks*. https://naadispeaks.wordpress.com/2018/08/12/deep-learning-vs-traditional-computer-vision/.

12. Khan, Everything you need to know.

13. Radiant Vision Systems. https://www.radiantvisionsystems.com/applications?gclid=CjwKCAjw64eJBhAGEiwABr9o2EoLU6-pcL8M10lrIjZ-zHqrZ0tmugqTvI8syYyiukaVUE1IK1UqchoCHdgQAvD_BwE (accessed August 22, 2021).

14. Top 10 Deep Learning application types in industrial vision systems. Machine Vision Experts. https://www.mvexperts.com/top-10-deep-learning-application-types-in-industrial-vision-systems/August 22, 2021).

15. Krasnokutsky, E. (August 26, 2021). AI visual inspection for defect detection. MobiDev. https://mobidev.biz/blog/ai-visual-inspection-deep-learning-computer-vision-defect-detection.

16. Top 99 computer vision startups. AI Startups. https://www.ai-startups.org/top/computer_vision/.

Bar Coding, the Most Ubiquitous and Most Critical IIoT Technology

Anthony Tarantino, PhD

Introduction

Of all the smart technologies in use today, none is better known by both manufacturing and the general public than the barcode. It was invented in 1951 by Bernard Silver and Norman Joseph Woodland. They based their invention on a type of Morse code that they extended to thin and thick bars. It took another 20 years to make barcode technology commercially viable. One of the earliest use cases for barcodes was to identify individual railcars, but the technology proved unreliable and was abandoned.

The big breakthrough for barcodes came in supermarkets, to automate the checkout system. The system is now used universally in all types of retail and industrial operations based on George Laurer's Uniform Product Code (UPC), which uses a system of vertical bars.[1] In the mid-1990s a type of matrix, or 2D bar code, called a QR code was developed that has grown in popularity with smartphone users.[2]

Barcodes play a similar and vital role in just-in-time (JIT) Lean manufacturing by providing point-of-use inventory movements. As with retail, barcodes are scanned to reduce or decrement inventory levels when items move through the various stages of production and finally reach the shipping dock. Without barcodes Lean systems would not be practical, requiring workers to key in part numbers and inventory information for every item that moves into finish goods or onto loading docks for shipment.

The average error rates for data entry have been debated for years, with 1% usually offered. This may have been true in the past with dedicated data entry clerks, but the rates will be higher when data entry is handled by workers wearing multiple hats, and with higher labor turnover rates. Error rates will also vary depending on the part or item number schemes used. Before the age of data entry into computerized systems, it was common for engineers to use significant or intelligent part numbers (also called item numbers) containing meaning.

EXHIBIT CS2.1   Handheld wireless barcode scanner

These numbers were often quite long. With this type of scheme, a part number created for a capacitor might look like this: CAP-220-0005, in which "CAP" stands for capacitor, "220" is the capacitor's value, in this case in pico-Farads, and "0005" is a serialized suffix.

In the early days of implementing manufacturing systems, we struggled to convince manufacturing organizations to abandon their traditional significant part numbering schemes for typically shorter nonintelligent numbering systems. The reasons for the change were simple: the error rates for longer part numbers were unacceptably high. As a general rule, the longer the number, the higher the error rate.

While this may seem like ancient history given today's smart technology, it is doubtful we have gotten here without simple barcoding that eliminated data entry error issues while dramatically increasing the speed of data entry. How dramatic of an increase? Today's handheld barcode scanners can scan 60 to 120 images per second.[3]

You can imagine trying to do data entry in a high-speed packaging line with hundreds of packages moving by each station per hour. Since it is physically impossible to perform data entry this quickly, a modern distribution facility could not exist without barcoding.

Exhibit CS2.1 shows a handheld wireless barcode scanner designed for use by retail consumers.

Barcode Technology

At their most basic level, barcodes are a square or rectangle containing a combination of vertical black lines of varying height and thickness, numbers, and white space. Combined, they identify specific items and their relevant information. Scanners working with computers read barcodes, which are then used to track an amazing variety of items from retail stores and supermarkets, and checked luggage, rental cars, and hospital bands for newborn babies.[4]

EXHIBIT CS2.2 Components of a 1D barcode

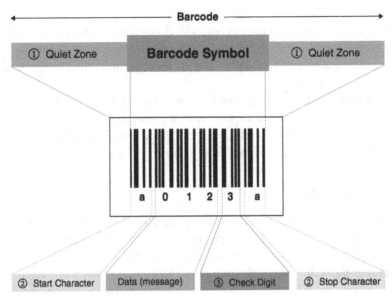

Source: From Oracle.

Two types of barcodes are in use:

1. 1D, or one-dimensional, codes used in manufacturing, distribution, and retail industries. These use either a European Article Number (EAN) or an American Universal Product Code (UPC).
2. 2D, or two-dimensional, QR codes used in advertising, directing users to a retailer's website.

For the 1D codes, the black bar's width usually represent the numbers 0 or 1, while the bar's sequence signifies a number between 0 and 9. Computers connected to scanners contain all the needed data as to what is associated with the unique combination of spaces and bars and may add, multiply, or divide the numbers to identify the correct product, which shows up on the screen. For manufacturing and distribution, barcodes would typically encode an item's color, size, country of origin, manufacturer, and other attributes. It may also include location information.[5] Exhibit CS2.2 shows the components of a 1D barcode.

These are the components of a 1D barcode:

- **Quiet zone.** The quiet zone is the empty white space on the barcode's edges and is needed for the scanner to be able to read the label.
- **Number system digit.** The first digit is the product category used in UPC codes. For example, coupons start with a 5, pharmaceuticals start with a 1, and retail with a 0.
- **Manufacturer code.** The first group of characters after the initial number identifies the name of the manufacturer.
- **Product code.** The next set of characters identifies specific product information created by the manufacturer.
- **Check digit.** Check digits confirm data accuracy tied to the barcode and help flag any potential errors.

There are three types of technology required to create a barcoding system:

1. **Scanner.** Scanners are needed to read barcodes. Laser scanners are popular because of their low cost and ability to read codes from as far as two feet away. Charge-coupled-device (CCD) scanners use several LED lights and are more accurate than laser scanners, but they operate at a shorter range.
2. **Printer.** There is a wide variety of wired and wireless printers used for barcoding.
3. **Central database.** A link to a computer with a central data source is needed to complete the barcoding system. Data is typically stored using a software application that supports point-of-sale (POS) systems and inventory control systems.

Popular Uses for Barcode Systems

Mail: As with inventory management, barcoding can track the status of mail and packages that are sent out. This includes scanning packages and letters prior to handing them off to couriers. If mail is returned, barcodes can quickly identify customers.

Asset Tracking: Barcodes are widely used to track physical assets. Connected to asset tracking software, barcodes help businesses monitor the location and condition of assets, including maintenance scheduling.

Inventory Tracking: Barcodes are the tool of choice to manage inventories on the production floor, in the warehouse, and in the distribution center. In larger-volume facilities, mounted scanners are able to scan items as they move along a conveyor.

Sales Order Fulfillment: A warehouse worker scans a sales order barcode that lists all the items needed to fulfill the order. They scan the items as they are picked in the warehouse, with instant visibility around the completeness of the sales order. Once the order is complete, the barcode is scanned in the shipping department, creating an invoice for the accounts receivable (AR) system.

Lean (Just-in-Time) Manufacturing: As mentioned earlier, barcodes are an essential component of Lean inventory and production management. Lean replaced traditional warehouses with point-of-use inventory on the production floor. Rather than issuing materials from a warehouse, a kanban system pulls orders through production, and reduces the various stages of inventory when their barcodes are scanned. This is a process known as backflushing (technically post-deduct issuing). Without the combination of barcoding and backflushing, Lean and its just-in-time inventory would not be practical.

Mobile Barcodes: Radio-Frequency Identification (RFID)

Standard short-range barcode scanners are designed to scan objects within a few feet. Long-range barcode scanners are accurate out to 50–60 feet with a clear line of sight. For greater distances and where a clear line of sight is not available, radio-frequency identification (RFID) systems can be used.

RFID is based on transponder technology developed in World War II to differentiate friendly from enemy aircraft but did not become commercially viable until the 1980s.[6] Today the global RFID market is valued at over $12 billion and expected to grow to over $16 billion by 2029.[7]

An RFID system consists of a small (typically passive) radio transponder, a radio receiver, and a transmitter. A passive tag is powered by energy from a RFID reader's interrogating radio waves. Active tags are powered by a battery and can therefore be read at a greater range by the RFID reader, up to hundreds of meters. RFID tags can be affixed to almost any object and used to track people, pets, tools, equipment, inventory, assets, or other objects. RFID tags can be miniaturized and placed on objects as small as poker chips.

Exhibit CS2.3 shows examples of RFID tags.[8]

Exhibit CS2.4 describes the RFID system and RFID tag circuit. "RFID tag is influenced by the correct frequency; it becomes energized via its antenna. The RFID tag will then transmit

EXHIBIT CS2.3 Examples of RFID tags

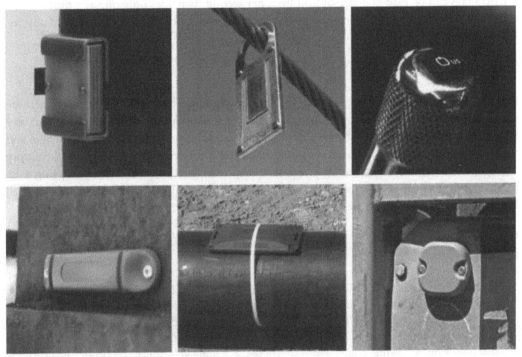

Source: From Glynn Warren in notes, EE Publishers.

EXHIBIT CS2.4 RFID system and RFID tag circuit

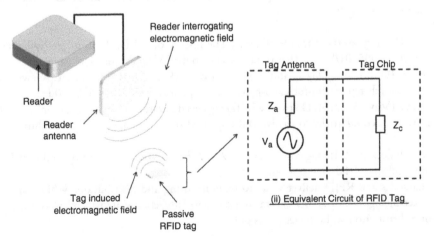

Source: From Glynn Warren in notes, EE Publishers.

its data (such as ID code) to the RFID reader. Read/Write RFID tags can also have their data updated by the RFID reader."[9]

In comparing the advantages and disadvantages of barcodes and RFID, the higher costs of RFID are often cited. Passive RFID tags costs average 7 to 15 cents, but active tags (with their own power source) can cost over $30. The cost of barcodes is only the paper they are printed on.[10] But RFID does have unique advantages over barcodes:[11]

- RFID tags can be read if passed near a reader, even if it is covered by the object or not visible.
- RFID tags can be read inside a closed container.
- RFID tags can be read at a rate of hundreds per second, barcodes only one at a time.
- RFID tags can be implanted in animals and attached to clothing and a wide variety of objects.

Summary

Barcodes are used in so many applications that it is easy to forget how valuable they are. There are benefits to all sizes and complexity of operations. Barcodes automate inventory transactions at very high rates of speed, ensuring accurate inventory and asset tracking. Barcodes are affordable and easily installed by smaller organizations, only requiring a printer, scanner, and basic inventory management software. Barcodes offer real-time data management, immediately flagging sales transactions, production order status changes, inventory movements, and asset changes. Of all Industrial Internet of Things (IIoT) devices, none has played a more vital role than barcoding.

Notes

1. Roberts, S. (December 11, 2019). George Laurer, who developed the bar code, is dead at 94. *New York Times.* https://www.nytimes.com/2019/12/11/technology/george-laurer-dead.html.
2. G. F. (November 2, 2017). Why QR codes are on the rise. *The Economist.* https://www.economist.com/the-economist-explains/2017/11/02/why-qr-codes-are-on-the-rise.
3. Barcode scanner buying guide: How to choose a handheld scanner. Camcode. https://www.camcode.com/barcode-scanner-buying-guide.html (accessed August 26, 2021).
4. McCue, I. (November 17, 2020). Barcodes defined – how they work, benefits & uses. Oracle NetSuite. https://www.netsuite.com/portal/resource/articles/inventory-management/barcode.shtml (accessed August 27, 2021).
5. Ibid.
6. Landt, J. (January 2001). Shrouds of Time: The history of RFID. AIM. https://web.archive.org/web/20090327005501/http://www.transcore.com/pdf/AIM%20shrouds_of_time.pdf.
7. Das, R. RFID forecasts, players and opportunities 2019–2029. IDTechEx. https://www.idtechex.com/en/research-report/rfid-forecasts-players-and-opportunities-2019-2029/700.
8. Warren, G. (May 16, 2019). Do passive RFID tags need hazardous area certification? EE Publishers. https://www.ee.co.za/article/do-passive-rfid-tags-need-hazardous-area-certification.html.
9. Ibid.
10. How much does an RFID tag cost today? (2021). *RFID Journal.* https://www.rfidjournal.com/faq/how-much-does-an-rfid-tag-cost-today.
11. How barcodes and RFID deliver value to manufacturing and distribution. White paper. (2016). Zebra Technologies. https://www.zebra.com/content/dam/zebra/white-papers/en-us/value-barcodes-rfid-mfg-distribution-white-paper-en-us.pdf.

Improving Safety with Computer Vision

Anthony Tarantino, PhD

Introduction

Navigating around industrial forklifts is a common experience for anyone who has worked in manufacturing or distribution. When I started working in manufacturing as a summer hire in the 1970s, workplace safety was often an afterthought. Typically, forklifts had the right of way, and the person on the floor was supposed to get out of the way. This was especially important because some drivers were paid bonuses for moving higher volumes of materials. It was also common to use forklifts as convenient extension ladders, and young summer hires were the first choice to be lifted on a forklift blades to fetch hard-to-reach items, sometimes 10 feet above the floor. I remember this well, as I had a fear of heights and always dreaded being volunteered to ride the blades.

Later in my career, I was in charge of supply chain operations for the world's largest lockset facility, and they had a 250,000-square-foot warehouse. I liked to stay late and walk the facility during the swing shift. We had some impressive forklift drivers operating on swing shift, almost all of whom were temporary contract workers. When I asked why we did not hire some of best drivers, the answer was simple: none of them could pass the drug tests. After that, I proceeded with more caution when walking the shop floor.

The Occupational Safety and Health Administration (OSHA) was formed in 1970 under the US Department of Labor to ensure workplace safety. OSHA added much-needed oversight for reporting injuries, but did little to analyze near-misses, which were very rarely reported in most organizations. Even under OSHA, after-the-fact accident reporting was viewed as a bureaucratic requirement with limited corrective action or root cause analysis.

Exhibit CS3.1 shows two ways of using a forklift to provide a work platform. On the left is an example of a very dangerous practice that was all too common in the past. On the right is an example of a much safer practice.[1]

EXHIBIT CS3.1 Using a forklift to provide a work platform

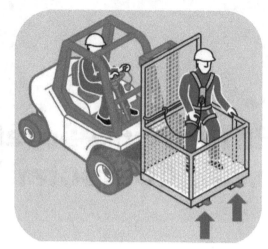

Source: From Safe Work Australia.

To put this in perspective, consider the annual costs from accidents in the workplace. According to the Bureau of Labor Statistics, American companies pay out over $60 billion per year to cover the costs of workplace injuries. An injury requiring medical attention costs an average of $39,000, but a workplace death costs an average of $1.15 million.[2]

According to statistics from the Australian government, manufacturing is overrepresented in workplace injuries, with 70% of workplace injuries and deaths caused by unwanted interactions between heavy vehicles and people. Of the 70%, over 30% of injuries are caused by unwanted interactions between people and forklifts.[3]

Of course, the real costs of workplace accidents and injuries are much higher when factoring in worker morale, increased insurance premiums, and the time and effort to respond to OSHA and insurance investigations. This is especially true for a workplace fatality, which inevitably leads to lawsuits and government investigations that can take several months, if not years, to complete. For the facilities I have supported, a workplace fatality created a dark cloud over the entire operation, regardless of management efforts to assure workers of their safety.

Before the introduction of computer vision that used deep learning algorithms and Edge computing, it was very challenging to prevent dangerous work practices, such as our previous forklift example. Even the most dedicated and diligent supervision cannot see all the action on busy manufacturing and distribution floors and in their yards.

Computer vision provides an automated means to watch all the action on the floor all the time. Unlike a manual supervisor, computer vision operates on a 24/7 basis, not in just one spot at a time, but in all critical areas of interest.

Computer vision has the unique ability to watch how people interact with the equipment they are operating, and the materials or products they are working on. There are many solutions that read machine and vehicle data, but only with computer vision can you be assured that safety regulations are being followed.

In order to flag safety violations in real time, advanced computer vision solutions use neural networks in their AI and locate Edge computers close to the action. Only with this combination is it possible to determine when an anomaly occurs and send an alert within a few seconds. None of this is simple, given the variations in lighting, obstructions that block the camera's views, and, most importantly, wide variations in the imagery being analyzed.

The following is a list of some of the safety violations that can be flagged with computer vision.

- A person steps in an area that has been deemed unsafe.
- A person stands behind a vehicle that is backing up.
- A truck driver in the yard is speeding.
- A forklift driver in the warehouse does not stop at blind corners.
- A forklift or other vehicle is parked in an unsafe area.
- A machine is operating with the safety door open.
- A person is not wearing safety equipment.
- A group of people are violating social distancing guidelines.
- A person is injured on the production floor.
- A person has a medical emergency on the floor.

After supporting computer vision startups for five years, it was interesting to see how many users discovered unexpected anomalies. These anomalies occurred in facilities run by very experienced and dedicated managers. The problem was that the managers could not have eyes on all the activities in their plants and yards. For example, while monitoring trucks loading cement to determine cycle times, they discovered trucks backing up illegally, trucks driving in wrong directions, and drivers leaving their vehicles in restricted areas. The lesson here is to keep an open mind as to what anomalies may be occurring in your facilities. While it was important to measure cycle times, the anomalies that presented the greatest risks were not part of the original project scope.

Besides the obvious benefits of flagging safety violations in near real time, computer vision generates valuable data sets for analysis. Traditional safety management programs took a reactive approach, analyzing accidents and injuries using tools we discuss in Chapter 3, such as root cause analysis and failure modes and effects analysis (FMEA). No matter how good the analysis was, it still suffered from being after the fact. Before computer vision, traditional analysis was compelled to rely on small sample sizes in a labor-intensive process.

With computer vision sending massive data sets to the Cloud, it is possible to capture every transaction for analysis, often resulting in gaining surprising insights and trends. Examples include discovering the days of the week or hour of the day with the highest accident rates. It is also now possible to compare shifts, departments, and other facilities to see the problems and the best practices.

The Deep Learning Revolution

Scientists first started experimenting with computer vision in the 1960s. The concept was very simple: to help computers see, much like a human does. A challenge was the major limitations of computer technology of the time. Labor-intensive manual training was required for building rules-based classifications and picking an object's relevant features. Deep learning dramatically improved the process, making it possible to commercialize the technology.[4]

Before deep learning, traditional machine learning (ML) required feeding image data into the system, which was then manually extracted for classification of its features. Next, each component was coded as a rule, which detects the elements within the image.[5]

Sandeep Pandya, writing for *Forbes*, describes how deep learning simplifies this process. "The deep neural network training process uses massive amounts of data sets and numerous training cycles to teach the machine what an object looks like. Instead of manually extracting features, the algorithm automatically extracts relevant parts. One of the most valuable capabilities is that the deep learning model can be applied to previously unseen images and still produce an accurate classification. In contrast, a traditional machine learning flow would fail."[6]

There are challenges in implementing computer vision solutions. Consider the problem with building models for workplace accidents – because you cannot ask volunteers to get run over by a forklift. One solution is to create 3D models that simulate the physical environment without endangering anyone. Another challenge is the requirement for large data sets. This is becoming less of a problem in that the amount of data required for the models continues to drop, from thousands of images down to hundreds of images.

Examples of Computer Vision's Role in Improving Safety

The following exhibits are examples of the many uses of computer vision to prevent accidents and improve safety.

Exhibit CS3.2 shows Australian-based Bigmate's computer vision solution that captures the distance between pedestrians and moving vehicles.[7]

Exhibit CS3.3 shows Australian-based Bigmate's computer vision solution that uses thermal imaging to detect if anyone is running a fever.[8]

Exhibit CS3.4 shows the use of computer vision to ensure compliance in wearing hard hats in designated areas. Notice the violator in the background.[9]

Exhibit CS3.5 shows the use of computer vision to detect a major safety violation. Company rules prevent drivers from leaving their trucks. Not only did this driver ignore the rule, but he is standing in a dangerous location, one in which rocks can break loose and hit him.

EXHIBIT CS3.2   Computer vision capturing the distance between pedestrians and moving vehicles

Source: From *Venture Beat.*

EXHIBIT CS3.3 Computer vision using thermal imaging to detect a fever

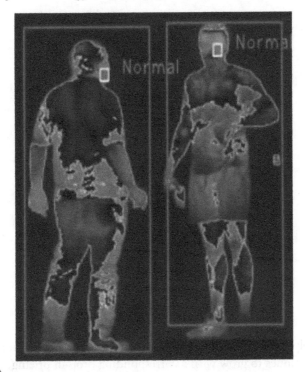

Source: From *Venture Beat.*

EXHIBIT CS3.4 Computer vision ensuring compliance in wearing hard hats

Source: From V-Soft Consulting.

EXHIBIT CS3.5   Computer vision detecting a major safety violation

Source: From *Atollogy*.

Conclusion

Computer vision helps reduce accidents both inside and outside of facilities. As the power of deep learning AI continues to grow with a corresponding drop in pricing, we can expect even small facilities to adopt computer vision solutions in the coming years. The costs of injuries and deaths are too high for most businesses to bear. Now they can implement proactive solutions to provide alerts about hazardous conditions in near real time and analyze the data computer vision provides to create a safer workplace environment.

Notes

1. General guide for industrial forklifts. (July 2014). Safe Work Australia. https://www.safework australia.gov.au/system/files/documents/1703/industrial-lift-trucks-general-guide.pdf.
2. Harter, A. (December 5, 2019). Cost of a lost work day: The true costs of workplace incidents. Anvl. https://anvl.com/blog/cost-of-a-lost-work-day-true-costs-of-workplace-incidents/.
3. General guide for industrial forklifts.
4. Pandya, S. (August 3, 2021). Seeing the future of AI: An introduction to computer vision for safety. *Forbes.* https://www.forbes.com/sites/forbestechcouncil/2021/08/03/seeing-the-future-of-ai-an-introduction-to-computer-vision-for-safety/?sh=88152693a1c9.
5. Ibid.
6. Ibid.
7. VB Staff. (September 29, 2020). Applying machine learning to keep employees safe and save lives. *Venture Beat.* https://venturebeat.com/2020/09/29/applying-machine-learning-to-keep-employees-safe-and-save-lives/.
8. Ibid.
9. V-Soft Consulting. https://blog.vsoftconsulting.com/blog/top-usecases-of-computer-vision-in-manufacturing (accessed August 21, 2021).

COVID-19 Accelerates the Adoption of 3D Printing

Anthony Tarantino, PhD

Introduction

In Chapter 12, "3D Printing and Additive Manufacturing," Professors Bahareh Tavousi Tabatabaei, Rui Huang, and Jae-Won Choi provide an excellent deep dive into the history, technology, processes, growing acceptance, and various types of 3D printing, also known as additive manufacturing (AM). This case study examines how the acceptance of 3D printing accelerated and became front-page news because of the COVID-19 pandemic.

Additive manufacturing, or 3D printing, is a fast-growing technology to manufacture items using computer-aided design (CAD) models and then to add materials in a layer-by-layer manner. 3D printing has the advantage of minimal waste of materials, the capability of printing parts with complex geometry, and the ability to create highly customized parts without the need for tooling.

Although there had been steady growth prior to COVID-19, the pandemic acted as a wakeup call to the value of additive manufacturing during periods of global supply chain disruptions. These disruptions created critical and sometimes life-threatening shortages of everything from simple nasal swabs to complex medical equipment.

3D Printing During the COVID-19 Pandemic

3D printing was pressed into service during the COVID-19 pandemic when the leading economies realized how fragile their supply chains were. The economic and military dangers of outsourcing and offshoring have been debated for years, usually around rare earth minerals, metals, computer chips, and so forth. Few analysts showed concerns over medical equipment items, especially simple medical consumable items such as face masks, swabs, personal protective equipment (PPE), and hypodermic needles. 3D printing used to produce not only basic medical items, but also continuous positive airways pressure devices (CPAP).[1] CPAP was critical in helping to alleviate a severe shortage of ventilator machines, saving thousands of lives over the course of the pandemic.

3D printing has also helped create training and visualization aids used by healthcare workers to address critical shortages of trained personnel. These aids included producing highly accurate models of the human body, sometimes with transparent materials showing internal structures of the mouth, throat, and nasal cavity.

3D printing was even used to produce temporary emergency dwellings where patients could be isolated. These structures were produced in less time and at lower labor costs than traditional construction methods.

Exhibit CS4.1 shows examples of 3D printing applications that were introduced during the pandemic.[2]

The Society of Manufacturing Engineers conducted a survey of 700 US manufacturing professionals in April 2020. In the survey, 25% of respondents said that they were compelled to change their supply chain in response to the pandemic, and seven of the surveyed industries ranked 3D printing as one of their top three technology priorities for investment once the COVID-19 pandemic ended.[3]

Exhibit CS4.2 depicts how engineers from various industries ranked additive manufacturing as an investment priority post-COVID-19.[4]

Thompson argues in his *Barclays Research* article that 3D printing represents a paradigm shift in design, manufacturing, and distribution. "By creating objects layer by layer, AM allows for greater design freedom with little or no added cost for greater complexity, and results in less waste overall. It can create lighter, better performing, greener and potentially cheaper industrial products, all with enhanced operational flexibility, speed-to-market, plant productivity and supply chain resiliency."[5]

EXHIBIT CS4.1   3D printing applications

3D-printed
Charlotte valve

Medical devices
• Ventilator valves
• Mask connectors for CPAP and BiPAP
• Emergency respiration device
• Non-invasive PEEP mask

3D-printed
respirator

Personal protective equipment (PPE)
• Face shield
• Respirators
• Metal respirator filters

3D-printed
NP swab

Testing devices
• Nasopharyngeal (NP) swabs

3D-printed
customizable mask

Personal accessories
• Face masks
• Mask fitters
• Mask adjusters
• Door openers

3D-printed
medical manikin

Training and visualization aids
• Medical manikins
• Bio-models

3D-printed
isolation wards

Emergency dwellings
• Isolation wards

Source: From Yu Ying Clarrisa Choong et al. in *Nature Reviews Materials.*

EXHIBIT CS4.2 Additive manufacturing as an investment priority

Source: From William Thompson in *Barclays Research*.

When comparing additive manufacturing (AM) to traditional manufacturing, Thompson predicts that AM will open the door to decentralized manufacturing, enabling on-location production in remote locations, including for the military, energy, and space exploration. AM also makes possible lighter weight, greener designs.

AM has proved its value in prototyping for some years, but using it to produce durable goods remains a challenge. While the global market for AM has grown at an annual rate of about 25% since 2015, it is still only a $15 billion market, or 0.10% of global manufacturing.[6] AM's biggest challenge may be in the area of mass production, where manufacturers achieve greater economies of scale with traditional methods and AM faces reliability issues. The reliability issues come from creating new materials for new parts simultaneously, which accelerates new product development but presents potential design and quality issues.

Advances in artificial intelligence (AI) may offer the breakthrough technology needed to move AM into more areas of mass production. Today's AM progresses primarily through trial and error. Computer vision using deep learning algorithms and neural networks offer the potential for 3D printers to see and the intelligence to optimize new designs.

3D Printing in a Post-COVID-19 World

Jabil, a leading contract manufacturer, conducts a customer survey every two years, starting in 2017. In 2021 Jabil surveyed 300 of their customers in companies with over $500 million in annual revenue. Virtually all decision-makers among the respondents expected their revenues to increase and that the role of 3D printing will grow and facilitate revenue growth.[7]

The Jabil survey found the following major trends in using 3D printing in manufacturing:

- Use of 3D printing for production parts continues to grow, with over 50% of respondents reporting that they use one quarter of their 3D printing capacity in producing functional or end-use parts.
- Prototyping, where 3D printing got its start, has stayed fairly flat.
- Manufacturers using 3D printers report that they have over 100 3D printers on average.
- Over 85% of respondents predict their use of 3D printing will double in the next five years.
- Expectations for growth of 3D printing for production remain high.
- Issues with 3D printing in production continue; compared to 2019, platform issues rose by 8% and ecosystem issues increased by 13%.
- Respondents are still having problems with 3D printing; scalability challenges rose 5% from 2019.

- Respondents predict a wide variety of 3D printing benefits, such as the ability to quickly deliver parts, lower production costs, and quickly respond to production line issues.
- More than half of top leadership views 3D printing as strategic to their organization.
- 95% face financial barriers in adopting 3D printing.[8]

The Jabil survey also covered materials used in 3D printing:

- A majority reported significant growth in the use of all types of materials.
- Plastics remain the most popular material, but more are now using metals.
- There is a major increase in post-processing of 3D parts, especially with machining and polishing.
- Challenges continue with materials used in 3D printing, with a 26% jump in frustration over the time taken to develop new material uses.

Finally, the Jabil survey covered outsourcing of 3D printing:

- 73% of respondents do 3D printing in-house, but also outsource when needed for additional capacity or for specialized applications.
- About 50% respondents do most of their 3D printing in-house.
- Design houses are more likely than manufacturers to do 3D printing in-house.
- About 50% of respondents expect to increase their outsourcing in the future.

Exhibit CS4.3 shows how Jabil's customers use 3D printing.
Exhibit CS4.4 shows at what stages of the product life cycle Jabil's customers use 3D printing.

EXHIBIT CS4.3   How 3D printing is used

In what ways is your company currently using 3D printing? *Choose all that apply.*

By number of 3D printers

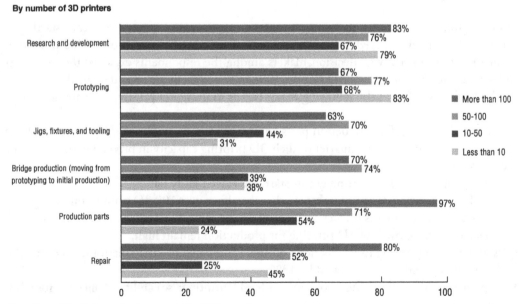

Source: From 3D printing technology trends (the Jabil survey).

EXHIBIT CS4.4   3D printing and product lifecycle stage

What part of the product lifecycle is the most positively impacted by 3D printing? *Choose up to two of the following.*

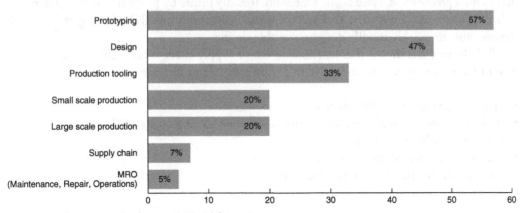

Source: From 3D printing technology trends (the Jabil survey).

EXHIBIT CS4.5   Expected increases in using 3D printing

How do you **expect your use of 3D printing for production parts or goods to change in the coming 3-5 years?** *Choose the answer that most closely applies.*

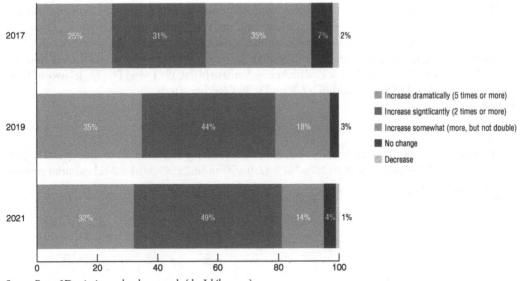

Source: From 3D printing technology trends (the Jabil survey).

Exhibit CS4.5 shows the changes Jabil's customers expect in their use of 3D printing. Notice that over 80% of customers expect their 3D printing use to increase dramatically or significantly, up from 56% in the 2017 survey.

Summary

The global supply chain crisis sparked by the COVID-19 pandemic created a major breakthrough for additive manufacturing, commonly known as 3D printing. It has proved its value in quickly responding to supply chain shocks that created critical shortages of vital medical equipment and supplies. The global crisis highlighted 3D printing's ability to become a foundation for a greener and more environmentally friendly future by producing parts on demand, reducing waste and inventory. Unlike traditional manufacturing, 3D printing is inherently flexible, able to quickly modify designs available online.

While history has shown that reactions to major supply chain shocks will fade over time, a host of factors are increasing the move to 3D printing. They include:

- The spread of COVID-19 variants
- Trade wars between major trading partners like the United States and China
- The rise of isolationist nationalism
- The continued shrinkage of product life cycles
- Robotics eliminating labor-intensive work

Notes

1. Choong, Y., Tan, H., Patel, D., et al. (August 12, 2020). The global rise of 3D printing during the COVID-19 pandemic. *Nature Reviews Materials* 5, 637–639. https://www.nature.com/articles/s41578-020-00234-3.
2. Ibid.
3. Thompson, W. (May 11, 2021). Additive manufacturing: Advancing the 4th Industrial Revolution. *Barclays Research.* https://www.investmentbank.barclays.com/our-insights/3-point-perspective/additive-manufacturing-advancing-the-fourth-industrial-revolution.html?cid=paidsearch-textads_google_google_themes_additive-manufacturing_us_research_additive-manufacturing_phrase_2125345119&gclid=CjwKCAjw95yJBhAgEiwAmRrutJ4PxCi9eTMwjTnqYd_h2uwmu9u3oiNV_9UFLZr68OHK5vLNlJmJthoCxZ0QAvD_BwE&gclsrc=aw.ds.
4. Ibid.
5. Ibid.
6. Ibid.
7. 3D printing technology trends: A survey of additive manufacturing decision-makers. (March 2021). Jabil. https://www.jabil.com/dam/jcr:82f12c7a-7475-42a0-a64f-0f4a625587d8/jabil-2021-3d-printing-tech-trends-report.pdf.
8. Ibid.

How Mobile Apps Benefit Small to Midsize Enterprises

Anthony Tarantino, PhD

Introduction

In Chapter 10, "Networking for Mobile Edge Computing," Jeff Little describes in great detail the history, technology, and applications for mobile computing, and how mobile computing powers Smart Manufacturing. Without mobile computing, Smart Manufacturing would not be practical. Mobile devices are the platforms by which manufacturing workers and managers can digitize physical operations and connect easily to the Cloud. The Industrial Internet of Things (IIoT) generates massive amounts of data with connected devices. By combining mobile's ability to provide networks with the IIoT's ability to generate operational data, manufacturers and distributors have powerful new sources of information to improve operations and eliminate labor-intensive paper-based practices.

This case study highlights how small to midsize enterprises (SMEs) can utilize mobile computing's power to transform their manufacturing and distribution operations. By using low-cost mobile apps (applications) on smartphones and tablet computers, SMEs can deploy software tools historically only available to larger organization using large, expensive, and complex computer systems.

For those wondering why they should make the jump to mobile apps, especially when they have invested in more traditional, nonmobile, software applications, there are three reasons to change. First, mobile apps are everywhere in our lives, making them easier for younger workers to embrace and to master. Second, mobile apps require no in-house technical expertise. Finally, mobile apps are typically much less expensive than traditional enterprise software systems.

We begin by describing mobile apps that are useful to any smaller organization and then describe mobile apps that are useful to manufacturers and distributors.

Mobile Apps for All SMEs

Mobile Payment Processing Apps

Apps to support point-of-sales (POS), also known as mobile money, typically refer to regulated payment services that can be performed using a mobile device. Mobile payments do not need to use cash, checks, or credit cards for a wide variety of products and services. Although the concept of using non-coin-based currency systems dates back to 2000 when the first patents were filed, improved technology has now made mobile payments popular and more affordable.

Before mobile payment apps a business would need to create, print, and mail an invoice in order to get paid. The process could take several days or weeks before a check arrived and was cashed at a bank. Mobile payment apps simplify and speed the process, helping to increase cash flow, lower processing costs, and eliminate the need for mail services or a merchant bank.

Mobile payments secure transactions using token technology. A token is used as a temporary representation of credit card data that can be created in a mobile app. The token permits a business to obtain a customer's credit or debit card details without processing or storing card sensitive data on a business's computer system.[1]

With an estimated half of the adult world's population lacking banking services, mobile payment apps are facilitating commerce in both developed and developing societies.[2] According to Boku, the global growth over the next five years will result in half the world using mobile payments by 2025.

Exhibit CS5.1 shows the regional growth in mobile payments (in millions).[3]

Mobile Accounting Apps

There are several mobile accounting apps that can provide the major accounting and bookkeeping functions normally found in major accounting systems. These are the features that you should expect from a mobile accounting app.[4]

EXHIBIT CS5.1   Regional growth in mobile payments (in millions)

	2020	2025	CAGR
North America	184.7	275.4	8.3%
Latin America	227.3	605.7	21.7%
West Europe	200.1	331.9	10.7%
Central and East Europe	76.3	248.9	26.7%
Asia Pacific	1,343.40	1,541.40	2.8%
Indian Subcontinent	269.2	550.4	15.4%
Rest of Asia Pacific	179.7	520.7	23.7%
Africa and Middle East	322.9	798.2	19.8%
Global	**2,803.70**	**4,872.70**	**11.7%**

Source: From Boku, Inc.

EXHIBIT CS5.2   Examples of mobile accounting app screens

Source: From Zoho Books.

- **Issue invoices.** Generate and email invoices to customers easily.
- **Receive payments.** Record customer payments and apply to outstanding invoices.
- **Pay bills.** Pay bills through the mobile app.
- **Track mileage.** Track business mileage automatically using your phone's GPS.
- **Scan receipts.** Scan receipts and extract the data to create new accounting transactions.
- **Categorize and reconcile bank transactions.** Transfer and categorize transactions automatically from your checking account.
- **Bank reconciliation.** View daily bank transactions and reconcile them.
- **Live cashflow dashboard.** Review cash flow with dashboards showing invoices, payments, revenue, expenditure, and past due invoices.
- **View reports.** View financial reports online without printing.

Exhibit CS5.2 shows examples of mobile accounting app screens.[5]
Exhibit CS5.3 shows a sample of QuickBooks' mobile accounting screens.[6]

Mobile Project Management Apps

The concept of project management dates back to the early twentieth century, when Henry Gantt introduced his Gantt chart to track projects.[7] Anyone who has managed complex projects has probably used one of the desktop project management tools that were introduced in the 1980s. I know of no users who have judged these tools as being user-friendly, primarily because projects have so many moving parts. For instance, a project needs to deal with task dependencies, resource allocations, compressed schedules to meet project completion dates, and other projects competing for your resources. The earliest project management applications such as Primavera and MS Project were indeed complex, requiring training to be used effectively.

EXHIBIT CS5.3 QuickBooks' mobile accounting screens

Source: From QuickBooks.

There are a variety of mobile project management applications that are easier to use than the cumbersome systems of the past. These are the features you should expect from a mobile project management app, but not all apps will offer all these features:

- Time tracking
- Job cost tracking
- Purchase order generating
- Project quotes
- Invoicing
- Document tracking
- Status tracking
- Custom fields
- Budget management
- File sharing
- Gantt charts
- Issues management
- Percent-complete tracking

Exhibit CS5.4 shows WorkflowMax's Leads dashboard for its mobile project management app.[8]

Exhibit CS5.5 shows WorkflowMax's project quote screen.

Exhibit CS5.6 shows a sample of Trello's collaborative Gantt chart.[9]

EXHIBIT CS5.4   WorkflowMax's Leads dashboard

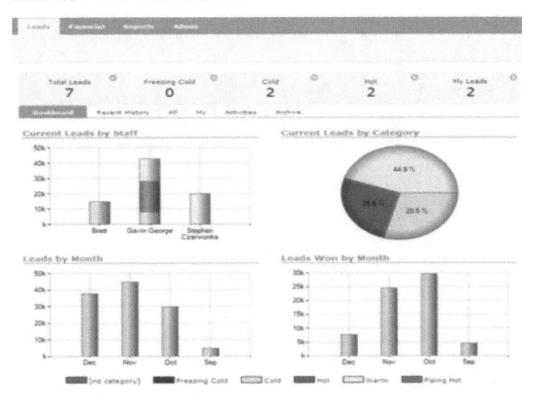

Source: From Software Advice.

Mobile Time Tracking Apps

Mobile apps are a good way to track the hours worked for specific tasks by remote employees and contract workers. The app should allow time logs to be exported to spreadsheets for analysis as to the profitability of each project.

Exhibit CS5.7 shows an individual timesheet.[10]

Exhibit CS5.8 shows an individual calendar used for tracking work time.[11]

Mobile Expense Reporting Apps

For those of you who spent hours stapling or taping receipts onto paper sheets to document your expense reports, mobile expense reporting apps will come as a great relief. Most of these apps allow the user to take a cellphone photo of receipts. The app reads the receipt, as long as the image is clear enough, and it transcribes receipt details, categorizes each receipt, and submits it for reimbursement. A tedious paper process is replaced with a painless paperless process. The mobile aspect is especially helpful for road warriors who are traveling with limited access to printers.

Exhibit CS5.9 shows an individual expense report.[12]

Exhibit CS5.10 shows a receipt and its transcription into an expense report.[13]

EXHIBIT CS5.5 WorkflowMax's project quote screen

Source: From Software Advice.

Mobile Apps for Manufacturing and Distribution

Mobile Inventory Management Apps

For many manufacturing and distribution businesses, inventory is their most valuable asset. Maintaining accurate inventory balances is critical in meeting customer demand while minimizing inventory costs. I learned as a long-time supply chain manager that not maintaining accurate inventory balances was the fastest way to get fired, especially if the imbalances led to inventory write-downs. Inventory write-downs typically mean that a company is making less money than reported in their financial statements. Maintaining accurate inventory balances requires timely transactions and physical and system controls. Making the process mobile helps to speed transactions, as there is no need to log on to a terminal personal computer away from where the transaction occurred.

There are several mobile apps to control inventory. These are the features you should expect from a mobile inventory management app, but not all apps will offer all these features:[14]

- Maintain perpetual inventory balances
- Scan barcodes
- Generate UPC and SKU numbers

EXHIBIT CS5.6 Collaborative Gantt chart

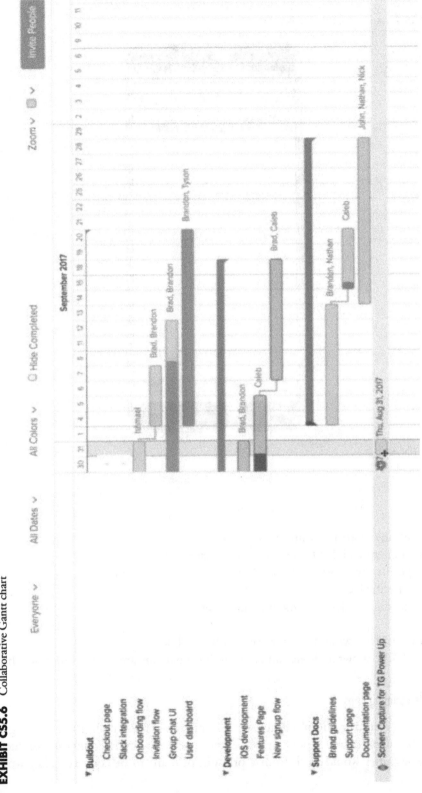

Source: From Trello.

EXHIBIT CS5.7　An individual timesheet

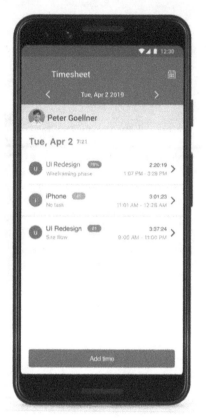

Source: From Hubstaff.

- Offer an offline mode for scanning inventory
- Generate an item pick list for kitting to meet sales orders
- Suggest specific inventory levels based on historical usage levels
- Maintain specific locations for all items
- Support customer order fulfillment management

Exhibit CS5.11 shows an inventory summary dashboard.[15]
Exhibit CS5.12 shows a phone camera scanning a barcode and a sales order pick list.[16]

Mobile Cycle Counting Apps

A robust cycle counting and physical inventory program is essential in maintaining accurate inventory balances. Before mobile apps, people conducting cycle counts or conducting full physical inventories would work from a printout of items to be counted. They would then

EXHIBIT CS5.8 An individual calendar for tracking work time

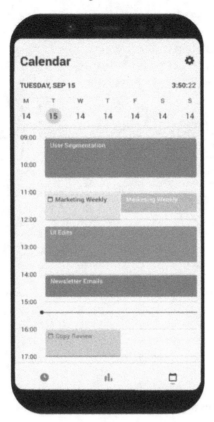

Source: From Toggl.

count items on the list and return to a computer terminal to input the inventory levels. With a mobile app, the paperless process is very simple. Inventory balances are updated on a mobile device, so all system users can see the new inventory levels in real time. As mentioned earlier, making the process mobile increases the speed of transactions, which helps improve accuracy.

These are the features you should expect from a mobile cycle counting app, but not all apps will offer all these features:

- Supports the ABC classification of inventory in which the most valuable or critical items are classified as "A" and the least valuable items are classed as "C"
- Allows multiple users to conduct counts simultaneously
- Compares minimum quantities with on-hand quantities
- Adjusts inventory levels in real time
- Provides visibility to other users updating inventory levels
- Enables both cycle counting and full physical inventories
- Recalculates cycle counting frequency based on usage
- Makes available barcode scanning
- Flags items that were missed in a cycle count

EXHIBIT CS5.9   An individual expense report

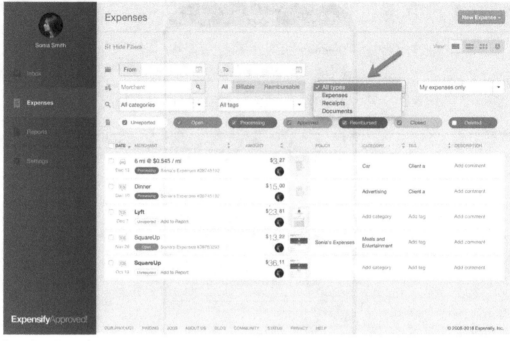

Source: From Expensify.

EXHIBIT CS5.10   A receipt and its transcription into an expense report

Source: From Xero.

EXHIBIT CS5.11 Inventory summary dashboard

Source: From Sortly.

Exhibit CS5.13 shows a screen to update inventory levels and a pick list to fulfill a customer order.[17]

Exhibit CS5.14 shows a screen to filter cycle counts by status and by location plus a cycle count input screen with barcode-scanning capability.[18]

Mobile Material Requirements Planning (MRP) Apps

Material requirements planning (MRP) systems date back to the 1970s and became widely implemented in the 1980s. They were complicated and typically took many months to implement, especially as they typically contained bugs (software glitches), which we jokingly called undocumented features. Viable PC-based MRP systems were introduced in the 1990s that provided the core features of larger software solutions. Most users hired system integrators to implement these systems due to their complexity and the specialized skills required to get them to go live. As MRP systems took on more functionalities including finance, sales orders, and distribution, they were rebranded, first as MRP II and later as enterprise resource planning (ERP) systems, but the core logic of the MRP remained the same. I discovered this firsthand after leading or supporting over a dozen such system implementations over a 25-year period. Regardless of the way screens were laid out, they all worked the same.

EXHIBIT CS5.12 Phone camera scanning a barcode and sales order

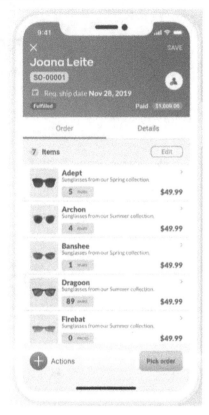

Source: From inFlow.

EXHIBIT CS5.13 Screens to update inventory levels and to fulfill orders

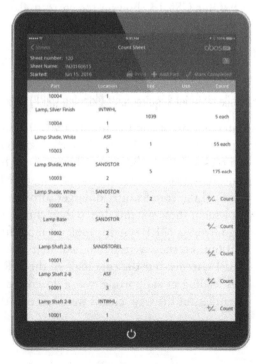

Source: From Abas MRP.

EXHIBIT CS5.14   Screens showing filters and stock

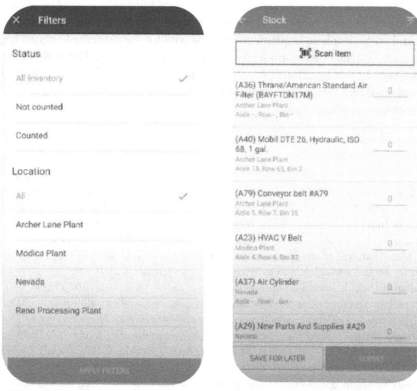

Source: From Fiix.

The interconnected nature of MRP systems will remain a challenge to any user as they demand 100% accuracy for bills of material, inventory, purchase orders, work orders, and sales orders, along with timely transactions. Today's mobile MRP apps facilitate timely transactions and bring the system to any user with a smartphone or tablet.

In selecting a mobile MRP system, you should insist that it offers all the functionality you use in your existing, desktop-version systems. This is an important distinction between selecting a system with several features you are unlikely to use and selecting one with the needed core functionality. With MRP systems, less is more, especially for SMEs.

While mobile functionality is valuable in updating work orders, inventory levels, and customer orders, complex tasks such as master production scheduling and forecasting lend themselves to larger monitors than a tablet or smartphone.

These are the features you should expect when selecting a MRP system that supports mobile apps:

- Update work order statuses at all stages of manufacturing
- View work order instructions
- Update inventory balances at all stages of manufacturing
- Update cycle counting transactions
- Update receiving and shipping transactions
- View and update sales order pick lists

Summary

In this case study we have provided some of the more popular mobile applications that are affordable and fairly simple to implement. Anyone who has implemented or used large enterprise systems understands their complexity and long learning curves. These mobile apps bring the great majority of this functionality to SMEs and at a fraction of the cost. Younger generations have adopted and embraced mobile apps for years, which will make them very open to accepting mobile operational apps.

Notes

1. Tokens. SecurionPay. https://securionpay.com/docs/api#tokens (accessed August 28, 2021).
2. Chaia, A., Dalal, A., Goland, T., et al. (November 2020). Half of the world is unbanked. Financial Access Initiative Research Brief. https://web.archive.org/web/20141222020037/http://www.financialaccess.org/sites/default/files/publications/Half-the-World-is-Unbanked.pdf.
3. Study: More than half of the world's population will use mobile wallets by 2025. (July 8, 2021). Boku Inc. https://www.globenewswire.com/en/news-release/2021/07/08/2259605/0/en/Study-More-than-half-of-the-world-s-population-will-use-mobile-wallets-by-2025.html.
4. 5 best mobile accounting apps. Fits Small Business. https://fitsmallbusiness.com/?url=https%3A%2F%2Ffitsmallbusiness.com%2Fmobile-accounting-apps%2F (accessed August 29, 2021).
5. Zoho Books. https://www.bing.com/images/search?view=detailV2&ccid=V%2BcW4dYZ&id=FEC7303D33AAB5FE6758F14F1B28FDC4DE8A1779&thid=OIP.V-cW4dYZMNSS_7pN_RBRTQHaFc&mediaurl=https%3A%2F%2Fblogs.zoho.com%2Fwp-content%2Fuploads%2F2014%2F01%2Fzoho-books-mobile-apps.png&exph=420&expw=571&q=mobile+accounting+apps+screens&simid=607990910530377622&form=IRPRST&ck=3348A8BD3D5F757365645B2F6586EBA0&selectedindex=14&ajaxhist=0&ajaxserp=0&vt=0&sim=11&cdnurl=https%3A%2F%2Fth.bing.com%2Fth%2Fid%2FR.57e716e1d61930d492ffba4dfd10514d%3Frik%3DReK3sT9KBtP8Q%26pid%3DImgRaw%26r%3D0.
6. Mobile app. QuickBooks. https://quickbooks.intuit.com/accounting/mobile/ (accessed August 29, 2021).
7. Henry Gantt. Wikipedia. https://en.wikipedia.org/wiki/Henry_Gantt (accessed August 29, 2021).
8. WorkflowMax software. Software Advice. https://www.softwareadvice.com/project-management/workflowmax-profile/ (accessed August 29, 2021).
9. Trello (accessed August 29, 2021).
10. Hubstaff. https://hubstaff.com/android_time_tracking_app_with_gps (accessed August 29, 2021).
11. Toggl. https://toggl.com/track/time-tracking-android/ (accessed August 29, 2021).
12. Trepanier, S. (January 2018, edited December 2018). A new way to view expenses, receipts and documents in Expensify. Expensify. https://community.expensify.com/discussion/752/a-new-way-to-view-expenses-receipts-and-documents-in-expensify.
13. Macaire, M. (2019). Two new Xero Expenses features to save you and your team even more time. Xero Blog. https://www.xero.com/blog/2020/04/new-xero-expenses-features/.
14. Stevens, C. (December 17, 2020). The best inventory management apps of 2021. Business.org. https://www.business.org/finance/inventory-management/best-inventory-apps/.
15. Sortly. https://www.sortly.com/ (accessed August 30, 2021).
16. inFlow. https://www.inflowinventory.com/ (accessed August 30, 2021).
17. Abas. https://abas-erp.com/en/products/mobile-cycle-count (accessed August 30, 2021).
18. Fiix. https://www.fiixsoftware.com/ (accessed August 30, 2021).

Using Factory-Floor Touch Screens to Improve Operations

Miles Schofield and Aaron Pompey

Introduction

Standardizing reporting and task tracking is crucial to Smart Manufacturing practices, although optimizing reporting is often a contentious discussion. If a worker is executing a manual reporting protocol (e.g. taking or sending notes), then they are not directly contributing to a production action. Conversely, if a worker is executing no reporting, manual or otherwise, then the end product becomes the single (lagging) metric, yielding little data to optimize or troubleshoot technical and efficiency issues. Ideally, notes and reporting would be automated through a number of intelligent monitoring systems; however, advances in reporting automation have not yet reached a level of practicality, which locates us in our current state: an age where boxes must be checked, and notes must be "handwritten."

Depending on the environment and the type of manufacturing line, and in the absence of fully automated systems, several input options are available for streamlining reporting protocols – from the simple post-shift action of transcribing paper notes (which have been taken during a shift) into a laptop, to more sophisticated solutions (like real-time controllers positioned at each station) for collecting dynamic real-time data. Many technical options exist in between, like a standard computer, acting as a controller, linked to a tracking database, or the use of mobile communication devices (mobile phones, tablets, pagers) by workers.

The following case study profiles a large-scale production environment that deployed large-format 22-inch touchscreen computers to solve this communication/tracking problem.

Problem Definition

One of the initial key challenges to large-format production is that the product cannot be moved, disallowing for traditional stations or controllers. Facilities that manufacture boats, trains, ships, and other large machines do so in specialized areas that are designed to accommodate the unique challenges posed by large-scale assembly. Stations, in the traditional sense, are impractical, since they potentially obstruct the production line, are misplaced, or require constant repositioning, depending on the stage of the production. This case study examines the reporting challenges facing large-scale production facilities and, in particular, how a scalable solution for digitizing process workflow in one facility delivered key data to interactive workstation displays and streamlined the reporting process.

This global, large-scale production manufacturer followed a traditional model. Tasks were tracked on a linear shift basis. Workers received daily task assignments. Time was allocated to perform these tasks on the product. At the end of each shift would be a passdown to report on progress and potential issues while performing the tasks. Workers then would be responsible for inputting tasks and details using a standard desktop computer.

By improving on this traditional model with better tools, this manufacturer targeted better results in the following areas: tracking worker progress, reducing overhead (limiting pre- and post-shift data entry), and handling dynamic and problem situations more efficiently.

Solution Description

To solve this problem effectively, this manufacturer developed a cart-mounted, battery-powered rolling workstation (5 feet high by 3.5 feet wide) for standing-position use that workers can move with them as they perform production line tasks. On the cart is an AOPEN 22-inch, all-in-one touch computer, keyboard, and mouse (primarily for log-in convenience, but also to be used to take notes to highlight procedural issues or send issue reports), and charging equipment, which allows the computer to be used in a standing position over a sustained period. Pre-shift, each worker picks up a fully charged cart, which will accompany them as they complete each task. After picking up the cart, the worker uses an individual log-in, which allows the station to identify them and their individual tasks. Not unlike a typical nursing cart found in hospital environments, the work cart served another functional role for this solution, to store tools and extra equipment.

To provide information for the workstation display, the entire process workflow needed to be digitized. Typically, worker tasks are stored in large lists or Gantt charts. For this solution, tasks were added to custom-written software that displays the status of several different contiguous actions simultaneously. The screen displays key information: the stage and status of overall production, what tasks are being done on the product, who is completing those tasks, and which tasks are assigned to an individual worker.

Real-time data enables management to control dependencies (especially important for dynamic environments), thereby improving process efficiencies. A dependency exists when one task must be completed before another can begin. Dependencies can cause considerable lag and worker procedural outcomes may not be known until completion (typically at the end of a shift). The manufacturer, in this case, looked for a solution to control dependencies by managing individual worker tasks, including estimated time to completion and any data resources surrounding the task. This effectively eliminates issues caused by knowledge gaps: if a worker is unsure about specific task information, or requires clarification, all documentation on a particular action is accessible via the mobile workstation. The individual task data is presented in a checklist format, enabling the worker/user to then instantly report on status and completion of tasks and further enabling the management (or controller) to orchestrate improved work efficiency.

Solution Choices

Beginning with a customized, web-based software application, designed to pair with a human computer interaction (HCI) hardware solution to best fit the operational requirements, the application's key component was a data and status display to keep workers updated. Although there are many web-compatible devices on the market (e.g. mobile phones, tablets, laptops, all-in-one computers, standard desktop computers), there are several factors that effectively narrow which hardware options are best suited for a dynamic, industrial environment. Consumer devices such as mobile phones may offer front-end cost savings but fail to meet two critical requirements: highlighting key, complex data in a comprehensive dashboard and standing up to a harsh, industrial environment.

The decision to drive the solution with a commercial all-in-one touchscreen instead of a laptop came down to three factors: quick interaction, an easy-to-clean surface, and reliability.

1. **Quick Interaction.** First, the performance of a hardware solution is fundamental to its quick interaction. In this case, workers used the touchscreen to report on basic tasks; the option (rather than the requirement) to use a keyboard and mouse further facilitated faster interaction. The commercial device is powered by an Intel Celeron processor, which reduces common failure modes, enables faster touch processing, and supports modern web design elements essential to a clean, navigable interface.
2. **Easy-to-Clean Surface.** Second, industrial environments, especially those involving large machinery, are notoriously difficult to keep clean. The ability to wipe down a commercial touchscreen surface routinely with a bleach solution ensured that the devices were easy to clean.
3. **Reliability.** Finally, hardware design plays a fundamental role in the reliability of a device. Like other consumer mobile devices, the touchscreens on consumer laptops are more delicate and designed for use in the more forgiving environments of daily life. Commercial touchscreens have much thicker, harder glass and stand up in harsh work environments.

The choice of a commercial work cart was based on several factors. First, similar nursing work carts in hospital environments are successful based on design and portability. These large work carts are ill-suited to uneven or outdoor sites with limited maneuverability (outdoor construction sites, construction vehicles, tight compartments); in these scenarios, commercial laptops or tablets are the hardware of choice. However, in this environment, the manufacturer has the common configuration of large concrete-floored warehouses, which allows the cart to be moved with ease.

Adjacent Applications

Commercial all-in-one touchscreens are used commonly in retail environments due to their reliability, robustness, and ease of workflow for an optimized touch-based interface. It is not uncommon to interface with touch devices to perform daily actions like banking (ATMs), shopping (retail kiosks), dining (ordering and POS kiosks), working (restaurant management and bump screen kiosks), and healthcare (check-in and intake kiosks).

The most commonly deployed screen sizes for touchscreen computing applications are 10, 15, 19, and 22 or 24 inches. A 10-inch device is used commonly when only one or two data points are required (e.g. bump screen, enterprise AV management). Several manufacturing tools use a 10-inch internal commercial screen to provide a basic interface and status screen for the user. A 15-inch device would deploy to applications that have size restrictions (e.g. point of sale). A 19-inch device would deploy to retail applications that have no specific size restriction (e.g. kiosks). A 22-inch or 24-inch device would be best suited to applications in which multiple data types are displayed to an individual user. These data types can be data and images, in a retail kiosk, or the types detailed in this case study.

Larger-format (over 24 inches) touch interfaces are less often deployed, due to their narrow range of applications and high cost. The most common use of large-format touchscreens are interactive whiteboard solutions for enterprise and education. The current limitation when developing touchscreens for applications is the touch response itself on large-format screens. Capacitive-based touch technology is slower as screen size becomes larger. Infrared (IR) touch technology is used for these larger format screens. Unlike capacitive touch technology (which measures the charge depleted through the conductive glass into the user's finger), IR touch uses sensors around the edge of the screen to pinpoint the location of the touch. It is possible to create responsive large-format touchscreens; however, these often-customized designs are expensive and the value they add to targeted applications remains in question.

Although screen technologies are driven primarily by the consumer television market, the screens themselves remain difficult to manufacture due to their large, delicate format and numerous breakable components (like glass). For capacitive touchscreens, manufacturing costs are further complicated by the need to integrate a transparent conductive layer, which makes the touch detection possible. Among the key advances in this market will be considerable reductions in manufacturing costs, especially as screens reach their practical resolution limit. The general practical limit of screens determines how many pixels the human eye can resolve at a given distance. As technology approaches this limit, many computers and television screens already deliver 4K resolution (a horizontal display resolution of 4,000 pixels), and few applications require anything beyond this (such as 8K). The future sits with the ubiquity of this technology (turning every pane of glass into an advertising canvas), rather than its resolution.

The Future and Conclusions

As humans continue to execute complex mechanical work, they are also likely to remain at the core of productivity and key operations. As a result, the largest gaps to fill with solutions like this are related to information and efficiency. Solutions like this fill those gaps, move industrial automation quickly and inevitably closer to becoming a standard approach, and highlight three theoretical technologies: *augmented reality* (AR), *automatic recording*, and *generalized proactive support*.

1. **Augmented Reality (AR).** Modern augmented reality tools already assist in a variety of both real and experimental work environments, with the capacity to guide even untrained

users through complex procedures with clear, step-by-step instructions and clarification. Using a transparent display (instead of a virtual-reality-enabled opaque display), users see the real world overlaid with useful and assistive information. This approach potentially lowers productivity costs by reducing (perhaps eliminating) training time and error rates. While this technology is not yet practical for commercial environments, due to the high cost of overlays and current methods of instruction, these systems continue to evolve; as innovation aligns to utility, the best versions of these systems (like direct projection methods onto user retinas) will become increasingly accessible.

2. **Automatic Reporting.** Automatic reporting (hands-free) introduces another opportunity to innovate while further reducing overhead costs, especially with out-of-the-box, generic solutions that eliminate individual command programming. Rudimentary tracking of products with cameras (e.g. entering and then leaving a station) exists already. Advanced reporting would leverage advanced AI camera systems (recording the progress automatically based on the job being performed) to further compare, contrast, and analyze these actions. Generic voice command systems are another way to solve this efficiently and dynamically ("Step 4, Section C is completed successfully; mark it down for me"). Both eliminate the need for a worker to disengage from the task, cease core productivity actions, and focus (even just temporarily) on the less productive task of documentation.

3. **Proactive Support.** The most common proactive support (or true intelligence) that a worker brings to the job is handling discrepancies and issues related to certain tasks. The value of AI is less in its ability to cull data from multiple public online sources, such as "the population of Lesotho is 2.125 million," but rather for its ability to draw specific, even proprietary, data that solves particular, individualized issues such as "resolved internal procedure 402, sub-section C." The value of AR lies in its ability to increase productivity across both trained and untrained users. The value, then, of proactive support is to complete mundane tasks, deliver relevant information, and generally remove obstacles to the information, time, and opportunities for higher-level decision making. It is likely that this technology will emerge first for things that are well-referenced, and then eventually for specific positions and technologies.

This case study already represents a notable development in smart workstations and controllers and, by extension, appreciable advancement in Smart Manufacturing technologies. Further advancements will likely focus on creating data and visibility for managers to make intelligent solutions and mitigate procedural bottlenecks and weak points. Positioning commercial interactive computers near workers introduces possibilities for future advancements (connecting cameras, connecting hand tools to Bluetooth, initiating video chats), making this the genesis of many smart manufacturing developments to come.

Edge Computing to Improve Operations

Allison Yrungaray

In Chapter 11, "Edge Computing," Vatsal Shah and Allison Yrungaray detail the technology and the growing applications supporting Smart Manufacturing. In this case study Allison highlights use cases for Edge computing.

Edge Computing Deployment Use Case: Food and Beverage

A bottled water manufacturer operates numerous bottling plants around the world. They sought an Edge platform for Smart Manufacturing that could collect and process data at the Edge for immediate visualizations and business value, then send the clean and valuable data to the Cloud for further processing.

Challenges

- Heterogeneous industrial systems on the shop floor
- Multiple data acquisition systems, creating data silos
- No manufacturing analytics
- Lack of insight into data for predictive maintenance
- Lack of shop floor visibility across the enterprise
- Insufficient maintenance practices due to missing data

Solution

The customer implemented the Litmus Edge-to-Azure platform by adding Litmus Edge to the shop floor. Litmus Edge connected to all of their disparate devices, collected, normalized, and analyzed the data, and then integrated with Azure IoT Hub for further processing in the Cloud. Models were then sent back to Litmus Edge for continuous improvement through the Azure Container or Litmus Edge Manager, a centralized management system for all Edge deployments.

Business Outcomes

- Connected and collected data from a myriad of industrial systems in a matter of days
- Eliminated data silos to easily share data across operational technology (OT) and information technology (IT) teams
- Enabled anomaly detection to see what is affecting quality
- Integrated the solution into their maintenance system to create automatic work orders
- Reduced downtime and optimized maintenance schedules to save time and money
- Created statistical models to predict machine failures

Architecture

- Litmus Edge on virtual machines in each plant data center
- Data sent to Azure Cloud and Oracle Maintenance System
- Litmus Edge Manager deployed in the Cloud
- Plant visualizations and local processing on Litmus Edge

Exhibit CS7.1 shows the Litmus architecture created for the food and beverage client.

EXHIBIT CS7.1   Litmus architecture created for food and beverage client

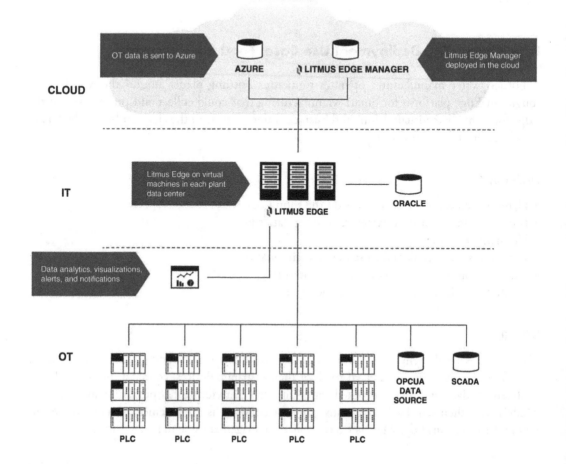

Edge Computing Deployment Use Case: Automotive

A multinational corporation that manufactures building and high-performance automotive materials set out to achieve a simple and centralized Industry 4.0 strategy – to transform all shop floors into a digital workplace and generalize a data-driven approach for use cases. The end goal was to provide the right data to the right person for the right purpose. Their plants were highly automated and the company aimed to complete their digital transformation for better visibility into assets and valuable data to make better business decisions.

Challenges

- No visibility into assets on the factory floor
- Data trapped in machines, Historian, and manufacturing execution system (MES)
- Several production lines with a high level of automation
- Machines from a wide variety of vendors, including brownfield assets
- Up to 20 different machines on each production line

Solution

The customer chose to install Litmus Edge on the shop floor for its ability to connect to the full breadth of OT and information technology IT assets with out-of-the-box support for any driver, protocol, or connection. They started small and then added data points and use cases as the solution showed a good return on investment (ROI). The solution could integrate the data to any destination and was easy to use, with little training required and a reasonable total cost of ownership.

Business Outcomes

- Rapid deployment to 20 plants in the first year
- Collected a multitude of data points that could be accessed by anyone in the plant
- Achieved a consolidated view of all data across all cells
- Improved operations and production
- Derived instant value from live analytics with dashboards, alerts, and process improvements
- Performed "offline" machine learning on historical data by collecting, storing, analyzing, and building an analytics model
- Enabled Edge application deployments and Edge analytics with complex dashboards

Architecture

- More than 100 Litmus Edge installations on HPE GL20 gateways
- Litmus Edge Manager hosted on-premises to manage devices at each location
- Send data to MES and Historian systems

Exhibit CS7.2 shows the Litmus architecture created for the automotive client.

EXHIBIT CS7.2 Litmus architecture created for automotive client

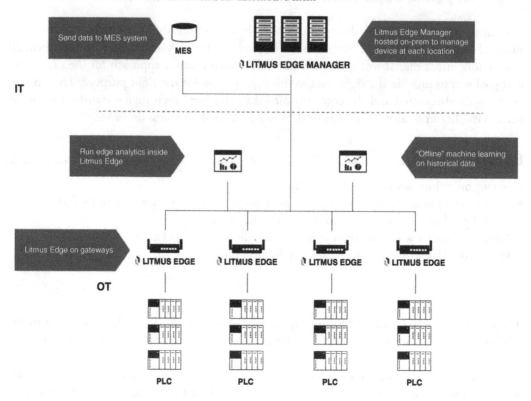

Five Highly Dangerous Jobs That Robots Can Do Safely

Alex Owen-Hill

Some jobs are just too dangerous for human workers to carry out safely. Robots are increasingly used to improve the safety of those workers who would usually perform these tough and dangerous tasks. Discussed here are five highly dangerous jobs that robots can do safely without unnecessarily risking humans' lives and health.

When selecting potential tasks for robotic automation, a common rule of thumb within industrial settings is to ask how well the task aligns with the criteria of the "4 D's of Industry" (this is an extension of the original "3 D's of Industry,"[1] revised to include cost).

These four D's are:

1. **Dull.** The task is not mentally stimulating for a human, such as loading dozens of identical parts onto a conveyor.
2. **Dirty.** The task is dirty or messy. This can be unpleasant for humans and the cleanup adds extra work.
3. **Dangerous.** The task is actively dangerous for a human to perform. What constitutes "danger" is the focus of this chapter.
4. **Dear.** The task is more expensive when performed by a human, usually unnecessarily expensive, such as paying a person for a whole day's work for them to only perform the mindless task of arranging products on trays.

Ideal applications for the use of robots will match all four criteria. However, even if a task meets only one of the criteria, it can be a good candidate for robotic automation.

The criterion that is probably most easily understood by industrial workers is "dull." Most people can identify boring tasks in their job that they would prefer to give to a robot. However, it's not always obvious what level of danger is required when the "dangerous" criterion is discussed.

While some robotic tasks present only a low level of danger, other jobs are so dangerous that a robot is absolutely the best way to achieve them. Before long, these highly dangerous jobs might only be done by robots and people will look back in amazement that humans were ever allowed to perform the task by hand, as they do now.

Here are five highly dangerous jobs that robots can do safely. Giving these jobs to a robot will immediately improve the safety of human workers.

Lifting Very Heavy and Medium-Heavy Objects

Robots are capable of lifting very heavy objects. The FANUC 2000iA robotic manipulator, for example, was presented lifting a car during the Automatica 2018 trade fair.[2] These lifting tasks wouldn't just be dangerous for human workers, they would be impossible.

Despite this, the most dangerous heavy lifting tasks are those that involve lighter objects that humans are capable of lifting but shouldn't lift for health reasons. Repeated lifting of objects between about 15–50 kg can be extremely dangerous for workers. A strong worker will be able to lift a 30 kg weight. However, if they lift this weight many times a day, they are likely to overexert themselves and there is a chance of permanent injury.

Manual handling health and safety guidelines recommend that people not lift over 25 kg,[3] but there is no hard-and-fast limit. Even lighter weights than this can be dangerous when lifted repeatedly over a long work shift.

Robots are perfect for repetitively lifting heavy loads. There are now even collaborative robots that can handle very heavy lifting, such as the FANUC CR-35iA (payload: 35 kg) and the Comau Aura (payload: 170 kg). As collaborative robots, they are capable of working safely alongside human workers without extra security measures in place as long as an adequate risk assessment has been performed.

Stirring 2,000°C Molten Metal

Some jobs are still performed by humans, but probably shouldn't be, as they are detrimental to health. One such job is "furnace tapping," which can be extremely unsafe when the furnace is large. This job involves stirring molten metal to remove a waste by-product called slag.

Furnace tapping requires workers to approach the molten metal (which can be at temperatures of 2,000°C) and agitate it with a long oxygen lance. The working environment can be extremely hot and filled with noxious fumes. As the worker "taps" the furnace, they are drenched in a cascade of burning sparks.

The Robotics Industries Association highlights that this repetitive and highly demanding job can now be achieved by robots.[4] Robot arms are wrapped in a heat-protective covering that allows them to withstand the high-temperature environment. They are equipped with custom-built oxygen lances to remove the slag.

Collecting and Packaging Radioactive Waste

The handling of radioactive materials is a job that has involved robots since before the 1980s.[5] It's easy to see why: it's impossible for humans to handle any radioactive materials without some risk to their health.

With small doses of radioactivity (e.g., in medical applications), workers' safety is managed by limiting the total radiation that they receive per year. However, with very high levels of radiation (e.g., when handling radioactive waste from nuclear power plants), the only option is robotics.

The most common robots used in the nuclear industry are telerobots, which are remotely controlled by human workers. The environment is often too unpredictable and the tasks too varied for autonomous robotics. The MARS robot,[6] for example, is a remotely operated arm that uses a high-powered jet of liquid to clean up radioactive waste at a facility in Hanford, Washington.

Robotic manipulators have also been proposed (Leggett 2015) as a way to package radioactive waste,[7] in an operation similar to packaging in a manufacturing environment.

Working in Contaminated, Dusty Environments

In some jobs, the cause of danger is not the task itself but an unsafe working environment. For example, jobs like industrial battery manufacture can produce toxic dust,[8] which would be highly hazardous if inhaled by human workers. Some drug manufacturing processes can also produce fumes and dust that are filled with dangerous chemicals.

Dust is a well-established danger to human health. Workers in very dusty environments are prone to respiratory problems and have an increased chance of contracting lung cancer.[9] This is true even when the dust contains no contaminants. However, the risk is increased even further when the dust contains chemicals, radiation, or harmful substances.

Robots can improve worker safety in contaminated, dusty environments. The main consideration that robot users need to make is ensuring that the robot itself can operate within dusty environments. This is ensured by using a robot with an appropriate Ingress Protection (IP) Rating.[10]

Repeating Physical Motions

Robots are perfectly suited to perform repetitive motion. Repetition could even be said to be a core function of industrial robotic manipulators.

Humans, by contrast, are very badly suited to repeating the same physical motions. Cognitively, such tasks are extremely dull for people. However, the physical repetition can also become highly dangerous over time. There is a real risk of conditions like repetitive strain injury (RSI) and other musculoskeletal disorders. If the safety of the task is not improved, these disorders can turn into long-term, or even permanent health problems.

For companies looking to justify incorporating robots into their facilities, this health danger presents a compelling business case. Musculoskeletal disorders account for around 35% of workdays lost to injury,[11] and so can also have a huge financial cost to businesses.

Robots are a highly effective tool to remove this danger. They excel at repetitive tasks and don't suffer from musculoskeletal disorders.

The five tasks presented here demonstrate the range of danger levels that can indicate whether a task is well-suited to robotic automation. In all cases, using a robot instead of a human worker will be better for the life and health of the human workers and will likely be more profitable.

Notes

1. Shea, Charlotte. (August 17, 2016). Robots tackling the three D's of industry. *Robotiq* (blog). https://blog.robotiq.com/robots-tackling-the-three-ds-of-industry.
2. Owen-Hill, Alex. (June 19, 2018). Automatica 2018 begins with the future of cobots. *Robotiq* (blog). https://blog.robotiq.com/automatica-2018-begins-with-the-future-of-cobots.
3. Health and Safety Executive UK (HSE). (2020.) Manual handling at work: A brief guide. https://www.hse.gov.uk/pubns/indg143.htm.
4. Anandan, T. M. (January 25, 2019). Robots to extremes. Robotics Industries Association, now the Association for Advancing Automation. https://www.robotics.org/content-detail.cfm/Industrial-Robotics-Industry-Insights/Robots-to-Extremes/content_id/7701.
5. Husseini, Talal. (December 4, 2018, last updated June 18, 2020). From Cherno-bots to Iron Man suits: The development of nuclear waste robotics. Power Technology. https://www.power-technology.com/features/cleaning-up-nuclear-waste-robotics/.
6. Newcomb, Tim. (November 7, 2011). MARS, the robot arm cleaning radioactive waste at Hanford. *Popular Mechanics.* https://www.popularmechanics.com/science/environment/a7291/mars-the-robot-arm-cleaning-up-radioactive-waste-at-hanford/.
7. Leggett, Theo. (August 11, 2015). How Sellafield's radiation-proof robots do our dirty work. BBC News. https://www.bbc.co.uk/news/business-33849026.
8. Anandan, Robots to extremes.
9. Moshammer, H. (2004). Lung cancer and dust exposure: Results of a prospective cohort study following 3260 workers for 50 years. *Occupational and Environmental Medicine* 61, 157–162.
10. Bernier, Catherine. (August 13, 2014, last updated May 5, 2016). IP terminology and ratings demystified – Robotics. *Robotiq* (blog). https://blog.robotiq.com/bid/66529/IP-Terminology-and-Ratings-Demystified-Robotics.
11. Owen-Hill, Alex. (May 26, 2016). Robots can help reduce 35% of work days lost to injury. *Robotiq* (blog). https://blog.robotiq.com/robots-can-help-reduce-35-of-work-days-lost-to-injury.

Answers to Sample Questions

Chapter 1

1. d
2. c
3. a
4. d
5. a
6. e
7. c
8. a
9. e
10. d

Chapter 2

1. d
2. d
3. b
4. d
5. e
6. e
7. d
8. e
9. a
10. a

Chapter 3

1. b
2. e
3. b
4. c
5. b
6. e
7. b

8. b
9. d
10. d

Chapter 4

1. b
2. a
3. c
4. d
5. d
6. a
7. c
8. c
9. d
10. b

Chapter 5

1. d
2. c
3. b
4. c
5. c
6. a
7. c
8. c
9. d
10. d

Chapter 6

1. b
2. b
3. d

4. b
5. a
6. a
7. d
8. c
9. c
10. d

Chapter 7

1. b
2. d
3. d
4. c
5. b
6. b
7. d
8. d
9. c
10. a

Chapter 8

1. c
2. b
3. d
4. d
5. d
6. a
7. c
8. c
9. b
10. d

Chapter 9

1. d
2. c
3. d
4. c
5. b
6. a
7. b
8. a
9. a
10. b

Chapter 10

1. b
2. c
3. a
4. b
5. d
6. c
7. b
8. a
9. b
10. c
11. d
12. d

Chapter 11

1. a
2. d
3. a
4. d

5. True
6. c
7. b
8. a
9. d
10. c

Chapter 12

1. a
2. b
3. b
4. The build material is jetted from the printing head in material jetting, while the binder material is jetted from the head onto the powdered material.
5. c
6. a
7. d
8. 4D printing refers to 3D printed parts that deform when external stimuli are applied.
9. d
10. b

Chapter 13

1. c
2. c
3. a

4. c
5. a
6. b
7. b
8. a
9. a, b, c, d, e

Chapter 14

1. c
2. e
3. a
4. c
5. c
6. b
7. d
8. b
9. b
10. b

Chapter 15

1. b
2. a
3. b
4. b
5. b
6. a, c, d
7. d
8. a, c
9. a, c, d
10. b, c, d

Links to Continuous Improvement Templates

The link below is to the master Excel file with links to over a dozen continuous improvement templates and samples. It can be helpful to use templates and then modify them to your application.

https://docsend.com/view/av6hyddji89igpb5

Index